BIOTECHNOLOGY IN THE FEED INDUSTRY

In fond acknowledgement of all that we owe to our parents, this book is dedicated to Peggy and Tommy Lyons. They were my guiding stars.

Pearse Lyons

Margaret Lyons, 1916–1997
Thomas K. Lyons, 1914–1996

WITHDRAWN

Biotechnology in the Feed Industry

Proceedings of
Alltech's Thirteenth Annual Symposium

Edited by TP Lyons and KA Jacques

Nottingham University Press
Manor Farm, Main Street, Thrumpton,
Nottingham, NG11 0AX, United Kingdom

NOTTINGHAM

First published 1997

ISBN 1–897676–646

Typeset by The Midlands Book Typesetting Company, Loughborough, Leciestershire, England
Printed and bound by Redwood Books, Trowbridge, Wiltshire, England

TABLE OF CONTENTS

INTRODUCTION

SECTION 1:

THE ROLE OF AGRICULTURE IN THE WORLD OF THE 21ST CENTURY

SECTION 3:

ORGANIC SELENIUM

SECTION 4:

MANIPULATING IMMUNE FUNCTION TO IMPROVE HEALTH AND FEED EFFICIENCY

A NEW ERA IN ANIMAL PRODUCTION: THE ARRIVAL OF THE SCIENTIFICALLY PROVEN NATURAL ALTERNATIVES

PEARSE LYONS

Alltech Inc., Nicholasville, Kentucky, USA

On February 19, 1997 Jacques Santer, president of the European Commission promised 'little short of a revolution' in the way we raise animals and a return within nine months to a program which would be friendly to the consumer, environment and animals. Demanding that we no longer view BSE as an accident of nature but rather the possible consequence of a policy of maximizing productivity without concern for negative effects, these powerful words from a powerful man were emphasized by statements from consumer groups around the world. A condemnation by Jim Murray of the European Consumer's Association that the 'routine use of antibiotics in feed was simply a substitute for poor hygiene and bad veterinary practices'gives a feeling of the consumer's attitude to these materials. Given this perspective it is no wonder that the three billion dollars spent on feed antibiotics is being scrutinized.

By any standards events of the last twelve months must rank as the worst public relations nightmare the animal production industry has ever faced (Table 1). Already facing rising feed costs and diminished raw material supplies, the fallout from the Mad Cow disaster just does not seem to disappear. Shock waves have been felt around the world, even in countries where BSE is unknown. For example, witness the defensive approach taken by the US FDA in banning meat and bone meal use in cattle diets.

Media coverage of agriculture problems and scandals has been relentless. In all of its forms, the media has endorsed and strengthened a single message: that the consumer, our ultimate boss and employer, has lost faith in how we rear and feed animals. Once the mainstay of an old and trusted way of

Table 1. Recent events in European agriculture

European Union bans use of avoparcin
BSE costs Ireland 300 Million pounds
1 Million beef cattle slaughtered in the UK
US FDA bans use of meat and bone meal
Russia refuses Swiss and Irish beef
Swiss supermarkets say no to antibiotics

life, the farmer is now often vilified as a polluter of both the environment and the food chain. His real or imagined abuse of both animals and antibiotics is in danger of encouraging a whole generation of vegetarians.

The marketplace is responding to the situation in both commercial and political terms. Russia has done significant damage to the beef markets of Ireland and Switzerland by refusing entry of beef from those countries. Great Britain meanwhile overflows with beef animals awaiting slaughter while Germany, France and the rest of Europe are feeling the pinch through reduced beef demand. Against this background, a cavalier approach to animal production continues with some farmers considering themselves above the law. Farmers in the EU are still being apprehended for possession and use of banned growth promoters and unapproved repartitioning agents. Though the backlash from the BSE scare increased pig and poultry consumption, some producers lost this opportunity by ignoring withdrawal times and allowing drug residues in these meats. In at least one member country 10 to 20 farmers are now serving jail sentences for violations.

The financial magazines are meanwhile constantly reminding us that agriculture is one of the few industries continuing to receive EU funding. While governments can in some ways be blamed for our plight, it was however not the government that introduced BSE. It was the feeders of meat and bone meal who believed that science is understood by the customer. The feed industry, the farmer and the animal nutritionist now face an uphill battle. The reality is that competition is both free and hard. The consumer demands to know what goes into the food chain and is well aware of his freedom to express his preferences in the marketplace. Supermarket buyers can and will be influenced by competition; and regardless of existing scientific data, they will respond to the consumer opinion and demand. Today's consumer no longer trusts science, and who can blame him based upon our performance this year?

Antibiotics are of course critical tools in animal production. Ionophores work well by manipulating gut and rumen microflora. Inorganic minerals are essential for animals. Smells are a normal consequence of raising animals in large (or small) numbers. None of these statements are in dispute. However what is in dispute is whether we have over-relied on them instead of adopting proven alternatives. The message from natural biotechnology is also not in dispute: natural alternatives exist and they have been proven both scientifically and practically. An explanation of what these alternatives are and how some of these alternatives work is the purpose of this paper.

Bioscience Centers: Maximizing the energies of universities and industry in a common cause

Necessity, the mother of invention, is the driving force behind Alltech's three fully operational Bioscience Centers in North America, Ireland and China (Figures 1a, b and c). Focused on finding solutions to both technical and consumer-related problems in agriculture, the Bioscience Centers act to focus the creative and scientific efforts of our company and shorten the timeline between an idea and a useful product reaching the marketplace. A few highlights in the following pages attest to the success of the Bioscience Center programs. They will also illustrate some of the natural alternatives now available.

Beijing Bioscience Center
Help for natural feeding programs: learning from China's 2000 years of experience with herbs and spices.

North American Bioscience Center
The study of ruminant and gastrointestinal function.

European Bioscience Center
Elucidating the biochemical nature of products, processes.

Figure 1a. North American Bioscience Center, Nicholasville, Kentucky

EGGSHELL QUALITY: A LESSON THAT BIOACTIVITY, NOT BIOAVAILABILITY, IS THE KEY TO OPTIMIZING MINERAL USE

The layer industry typically loses as much as 8–10% of eggs due to poor shell quality. This includes breakage due to hairline cracks or breakage during processing. On average it is estimated to account for one dollar per bird per year. Our European Bioscience Center, in cooperation with the Mendel Institute in the Czech Republic, focused on two enzymes involved in egg shell quality. Carbonic anhydrase (Figure 2) and mucopolysaccharide synthetase. The first enzyme is responsible for depositing calcium in the shell, while the latter is responsible for the overall shell matrix. By combining two enzyme

Figure 1b. European Bioscience Center, National University, Galway, Ireland

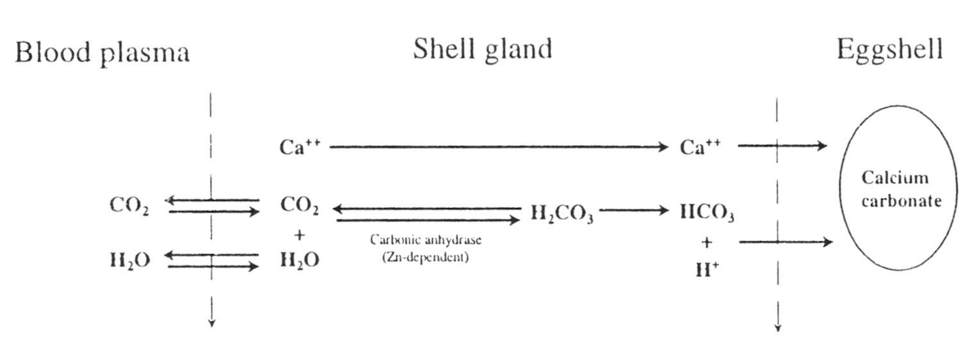

Figure 2. Carbonate production for eggshell formation.

activators in the Eggshell 49 product, the result is an improvement in not only shell quality, but in the case of brown eggs, shell color (Table 2). The enzyme activators are based on bioplexed manganese and zinc. This natural process protects the minerals through the harsh conditions of the gastro intestinal tract in order that they arrive intact at the site of absorption. This all-natural product went from concept to market in less than nine months, ironically about the time it takes in the layer production cycle for shell quality to begin to decline with greatest economic impact. The natural approach of Bioplexes is simply one of less total mineral but more active materials and hence less environmental burden. Similar examples can be given for Bioplex Zn where relatively small quantities (compared to the inorganic form) lead to improvement in hair coat, foot condition or tissue

Figure 1c. Beijing Biosciences Center, Beijing, China

Table 2. Effect of Eggshell 49 supplementation on egg grade.

	Control*	Eggshell 49†	Difference (%)
Number of birds	120,000	72,000	
Grade A eggs, %	81.90	84.60	+3.30
Checks	14.00	12.03	–14.00
Loss	4.10	3.37	–17.80

* Birds were 61 weeks of age at the beginning of the experiment.
† Eggshell 49, Alltech Inc., 1 kg/tonne.

repair. Practical consequences are not just a reduced environmental burden but reduced incidence of mastitis and lower somatic cell counts. By using copper in a similar approach, interference with iron and molybdenum can be overcome. At the cellular level it is now known that Biopolex minerals have a profound influence on immune status leaving the animal better able to withstand the rigors of intensive farming.

NATURAL ENZYMES TO MAXIMISE PROTEIN AND CARBOHYDRATE UTILIZATION AND MINIMIZE POLLUTION

A reviewer of the animal feed literature looking for enzyme information might be forgiven for believing all useful enzymes were either glucanases or pentosanases. These enzymes, while important, were first promoted as long ago as the mid 1970s. Acceptance in the industry took 20 years. While on one hand this represents a remarkable turnaround on the part of industry, on the other it reflects an incredible reluctance to take advantage of a new developement. Nonetheless, it has opened an opportunity for a second generation of enzymes, one which can release more protein from feed grains and oilseed meals and another to improve fat digestion.

The first enzyme, called Vegpro, was designed in conjunction with Scotland's Roslin Institute and the British Agricultural and Development and Advisory Service (ADAS) to improve protein and energy utilization in poultry. It focused on α-galactosidase, an enzyme which degrades oligosaccharides in soy and other sources thereby releasing up to 15% more energy and an endopeptidase allowing utilization of additional protein for improved amino acid digestibility (Table 3).

Key to successful use of this enzyme is to assign soy protein plus Vegpro as a separate ingredient with the higher energy and amino acid digestibility values. The additional amino acid specification can be reduced by up to 7% since digestibility is improved by 7%.

Though originally Allzyme Vegpro was designed for poultry diets, recent

Table 3. Effect of two addition levels of Allzyme Vegpro on true amino acid digestibility of broiler diets.

Amino acid	Control	1 kg/tonne	2 kg/tonne
Alanine	72.4	72.8	76.2
Asparagine	67.0	69.6	75.5
Cysteine	41.6	57.9	51.3
Glutamine	82.7	84.6	86.8
Histidine	54.7	68.5	68.9
Isoleucine	81.0	82.8	85.4
Leucine	80.6	82.0	84.3
Lysine	82.1	86.0	85.6
Methionine	65.4	69.0	67.1
Phenylalanine	86.3	88.9	90.6
Proline	70.0	77.4	79.7
Serine	78.6	83.9	84.7
Threonine	72.7	78.5	79.5
Tyrosine	74.3	77.1	78.4
Valine	93.7	89.6	86.4

data from the University of Kentucky have shown the potential for improving utilization of corn/soy diets fed pigs. Pigs supplemented with Vegpro during the growing and finishing phases improved feed conversion by once again approximately 7% (Figure 3).

Another enzyme on the horizon is Allzyme Lipase. In metabolism trials at Massey University Monogastric Research Centre in New Zealand addition of Allzyme Lipase improved apparent metabolizable energy (AME) of rice bran. This enzyme opens up the possibility of using larger quantities of rice bran and other less expensive raw materials with a significant impact on production costs (Table 4). Both Allzyme Vegpro and Allzyme Lipase are now in widespread use across the world. While these enzyme supplements are stable to pelleting, liquid forms are available for spray application post-extrusion.

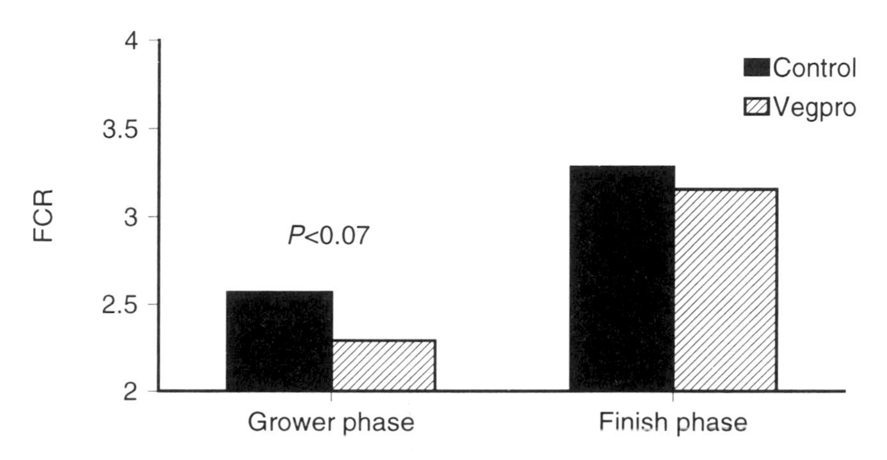

Figure 3. Effect of Allzyme Vegpro on feed:gain ratio of pigs during the growing and finishing phases

Table 4. Effect of Allzyme Lipase on AME of rice bran and excreta energy content of adult cockerels.

	Control	Allzyme Lipase
AME, kJ/g	14.31 ± 0.193	14.60 ± 0.223
Excreta dry weight, g	20.24	20.68
Excreta energy content, kJ/g	16.57	15.78

CHINA: 2000 YEARS OF EXPERIENCE WITH HERBS AND SPICES, ANOTHER NATURAL ALTERNATIVE

For years the western world has heard tales of the benefits of Chinese herbal medicine. The same herbs have found favor in the animal production sector with both performance and health improvements. Alltech, as part of its brief to the Chinese Bioscience Center, has identified some of the active ingredients in useful herb and spice preparations. These products will undoubtedly form part of natural feeding programs in the future (see Newman, Section 4).

PLANT DERIVED ALTERNATIVES TO PLASMA PROTEIN

Plasma protein has become a standard ingredient in many pig starter diets worldwide. Improved feed intake, reduced disease and fewer days to market have been consistently reported. However importing blood-derived products is surely not a viable option given the spectre of BSE and the possibility of indirect disease transfer from one area to another. An alternative solution was to use enzyme-hydrolyzed cereals and yeast protein along with immunoglobulins from egg. By using selected protein sources all the essential amino acids can be presented in the diet. In effect we are returning to Cole's concept of the ideal protein. Comparisons of performance and costs have shown that this new protein source is extremely useful in modern pig-rearing diets (Figure 4).

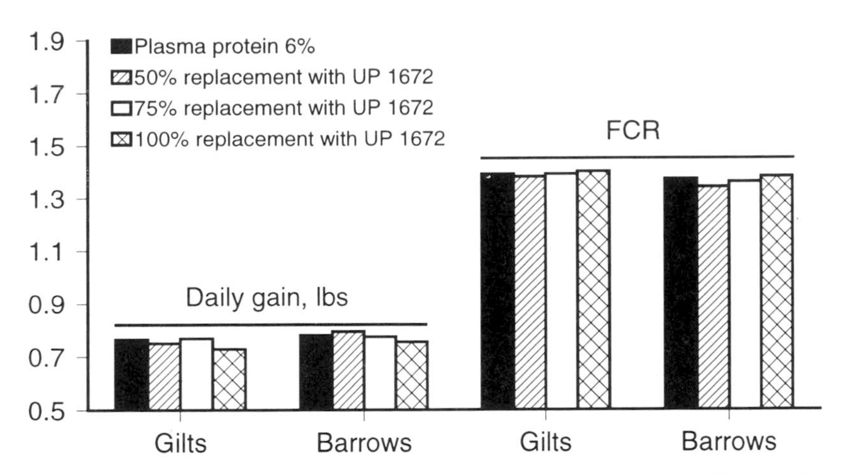

Figure 4. Comparison of plasma protein with 50–100% replacement with Ultimate Protein 1672

NATURAL ORGANIC CHROMIUM FROM YEAST

Chromium (Cr) has been recognized as an essential nutrient for animals since the late 1950s, although only recently have we begun adding Cr to livestock feeds. Just recently approved in the US, Cr has already been approved for use in many countries around the world, including Australia, India, Malaysia, New Zealand, the Philippines, South Korea, Taiwan and Vietnam. Alltech's approach has been to use yeast as the means of producing the so-called glucose tolerance factor, the bioactive form of chromium.

Needed as a cofactor for insulin function, Cr is critical for efficient use of modern high energy diets if genetic potentials for either reproduction or carcass lean are to be reached. In Korea earlier this year, research demonstrated that organic Cr supplementation increased average daily gain, feed intake and liveweight gain when included in the diets of finisher pigs (Table 5). The importance of this natural ingredient on sow reproduction (increased piglet production) has been documented by Lindemann and others.

Table 5. Effects of Cr yeast on performance of finisher pigs.

	Liveweight (kg)		Gain (g/day)	Intake (kg)	F:G
	Initial	Final			
Control	66.8	91.8	676	2.41	3.59
200 ppb Cr	66.8	94.6	750	2.64	3.56
400 ppb Cr	66.9	92.5	692	2.50	3.62
800 ppb Cr	66.9	95.9	785	2.69	3.43
Average	66.9	93.7	726	2.56	3.55

ORGANIC SELENIUM: A NON-POLLUTING/NON-TOXIC AND MORE BIOAVAILABLE SOURCE OF SELENIUM

Sodium selenite has long been accepted as the primary source of selenium for animal feeds. However, research in recent years has shown that organic selenium is more bioavailable. Dr Don Mahan of Ohio State University has found that feeding organic selenium to pigs resulted in higher selenium levels in the milk of sows and in piglets both at birth and at weaning, resulting in lower piglet mortality. Work from Dr Edens at North Carolina State University has demonstrated that supplying selenium to broilers from selenium yeast reduced drip loss from broiler meat (Figure 5). Muñoz and co-workers in Spain at the University of Murcia showed a similar result with an antioxidant supplement including selenium in the form of selenium yeast from Sel-Plex 50 (Figure 6).

Another interesting response to selenium yeast replacement of selenite has

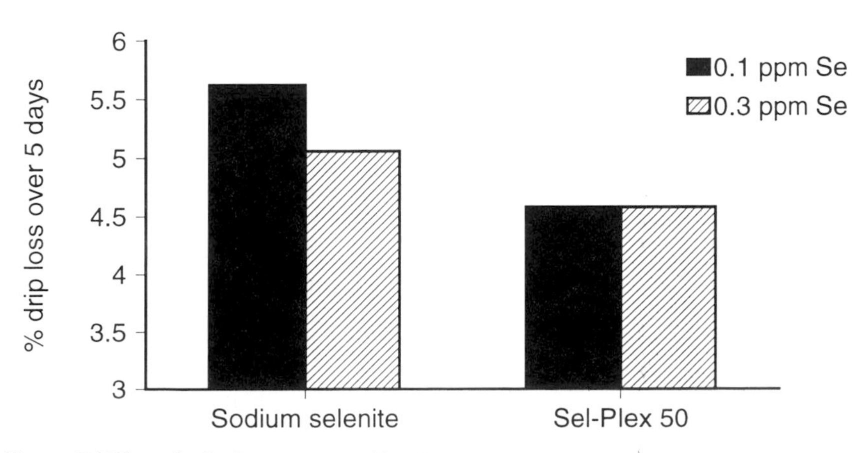

Figure 5. Effect of selenium source and level on drip loss from broiler meat

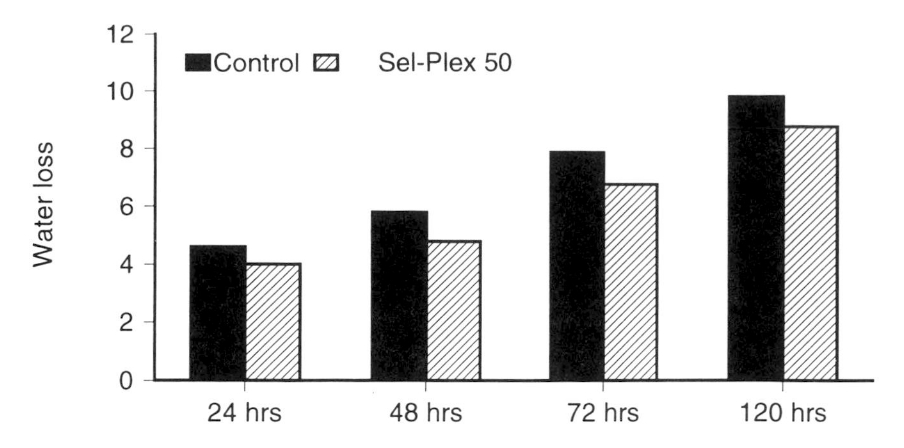

Figure 6. Effect of antioxidant supplementation on water loss from *L. dorsii* steaks at 24, 48, 72 and 120 hours post-mortem

been improvements of 2–4% in broiler breeder performance. A secondary effect of organic selenium, improved feathering (+ 1–2 feather score units), has economic implications due to the relationship between feather cover and feed efficiency. The change in feathering can mean as much as 5% less energy required by the bird. Recent work with catfish examining comparative effects of selenium sources has shown that supplying selenium in organic form improves immune status (Table 6, See Lovell and Wang, Section 3). Again, the animal given Sel-Plex 50 was better able to withstand the stress of disease challenge.

Table 6. Effect of selenium source on antibody titers of catfish.

Dietary Se level, mg/kg	Selenite	Sel-Plex 50
0	64	64
0.03	80	75
0.06	149	320*
0.20	192	554*
0.40	365	717*

* Means differ, P< 0.05.

Unlike many other trace mineral complexes that are organically bound, selenomethionine must be formed biologically during yeast growth. This product has already proven itself on the farm where reduced somatic cell counts in dairy cows, improved sows fertility and improved piglet survival have been observed.

YEAST CULTURES: UNIVERSITY PROVEN ALTERNATIVES/ADDITIONS TO IONOPHORES

When the yeast strain 1026 was introduced in 1982 to the feed industry it signaled the arrival of scientific scrutiny to a natural additive which had existed for many years, namely the addition of living yeast to feed. The product has withstood the test of time with reports of efficacy from such prestigious institutes as the Rowett and an ever-lengthening list of universities.

Yea-Sacc ^{1026} is now known to stimulate certain microbial populations in the rumen and gastrointestinal tract thereby promoting digestion of fiber and removal of lactic acid from the fermentation. The activity survives pelleting and can be used in conjunction with ionophores in order to stimulate feed intake. Performance improvements are on the same order of magnitude as ionophores at 5 to 8% increases in milk yield or weight gain. Application, however, is not limited to ruminant diets as indicated by positive effects on fertility in broiler breeders and results in India demonstrating ability to alleviate negative effects of mycotoxins.

Based on a thorough understanding of the mode of action of strain 1026, other strains have now been selected based on specific effects for specific diets. A strain which predominantly improves cellulose digestion and a second which increases numbers of, and lactate utilization by, Selenomonas are now available. Should ionophore usage come under fire, the industry has available a cost-effective solution in Yea-Sacc^{1026} and the newly available strains.

Yea-Sacc^{1026}: Proven at Universities and Institutes:
- Kentucky
- Florida
- New Hampshire
- Rowett Research Institute
- Irish Agricultural Institute

Proven mode of action
- Stimulates cellulolytic, lactic-utilizing species
- Increased mineral digestibility

Broad applications
- Dairy
- Beef
- Poultry and ratites
- Pigs
- Pets

MANNANOLIGOSACCHARIDES: A NATURAL IMMUNOMODULATOR WITH EFFECTS ON GASTROINTESTINAL INTEGRITY

The gastrointestinal tract is the only organ in the body which must work in non-sterile conditions, cope with an ever-changing diet and defend the body from an array of chemical attacks. Integrity of this key defense system is vital. It is here that the latest natural alternative comes into it's own. Unlike antibiotics, which work by changing the microflora spectrum in favor of benevolent bacteria by killing bacteria or inhibiting their growth, mannanoligosaccharides in Bio-Mos alter microbial populations by selective nutrition and by binding certain pathogens to prevent colonization. Additionally, Bio-Mos enhances immune response by stimulating cell-mediated immunity. Mycotoxins have been shown to be bound by Bio-Mos mannanoligosaccharides including zearalenone and aflatoxin and more recently aflatoxin in studies at the University of Guelph and in the Agricultural Institute in India. Integrity of the gastrointestinal tract, as reflected in effects on crypt depth and (or) villi length, indicate less turnover and more energy for growth or production (Figure 7, see Spring, section 6 of this book). In performance terms, work with Bio-Mos in turkeys led the product to be described as having a "300 g/15 point" effect on growth and feed conversion. This pattern of response, along with its stability to pelleting and extrusion, has led Bio-Mos to become standard in many diets as an alternative to in-feed antibiotics.

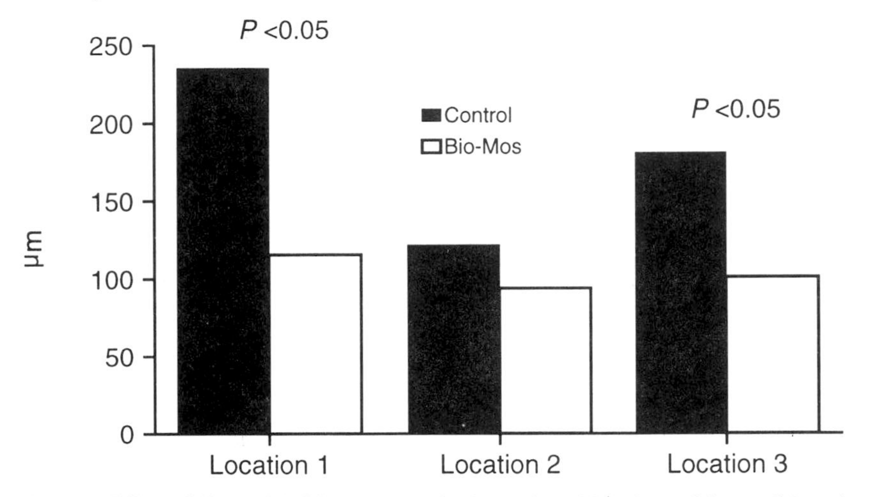

Figure 7. Effect of dietary Bio-Mos on crypt depth at selected locations of the small intestine in 8-week old turkeys.

ENVIRONMENTAL POLLUTION: A SOLUTION FROM YUCCA GLYCOCOMPONENTS TO THE NUMBER 1 COMPLAINT FROM FARM NEIGHBORS: SMELL

Yucca-derived products have long been used by native North Americans. Modern agriculture now knows the benefits with countless studies reporting reduced odors in pig and poultry units. Lower ammonia is associated with less respiratory disease incidence leading to improved performance. Again, research and science have enabled us to improve on a natural product not with genetic engineering but by harvesting the plant at the point when maximum levels of ammonia-binding glycocomponents are present. Long familiar to garden enthusiasts and desert dwellers, the *Yucca schidigera* plant is now a powerful tool in animal husbandry.

Conclusion

As we set out to achieve Mr Santer's promise of a return to natural feeding programs, biotechnology offers many solutions. These approaches work through:

- Improved utilization of non-conventional raw materials
- Improving animal efficiency and production economics
- Reducing the pollution potential of animal agriculture
- Increasing meat quality and consumer perception

Natural biotechnology, which does not rely on genetic engineering but rather on utilization of natural raw materials, is the key to the future. Our message to Mr Santer is that natural alternatives are available and they have been proven both scientifically and practically. Our message to feed industry professionals is that we must use these alternatives before consumer and governments mandate them. Change and challenge are really only opportunity viewed from another side. The successful feed company in the future will be the one who warmly accepts change. Since we now have natural tools there should be no hesitation on behalf of any forward-thinking feed or livestock producer. The future will always be for those most fit to survive; and we are reminded of the advice given over 100 years ago by Charles Darwin:

> ***'It is not the strongest of the species that survive, nor the most intelligent, but the one most responsive to change'.***

THE ROLE OF AGRICULTURE IN THE WORLD OF THE 21ST CENTURY

THE BIOSCIENCE CENTER CONCEPT: A GUIDELINE FOR COOPERATION BETWEEN UNIVERSITIES AND INDUSTRY

RONAN POWER

European Biosciences Center, National University of Ireland, Galway, Ireland

Introduction

Alltech opened the first of its three Bioscience Centers in April of 1991 at the company's headquarters in Kentucky. The primary aim of this Center was to establish research links between the company and academic institutions in the USA and abroad. Later that year the concept was extended through the establishment of a European Bioscience Center at the National University of Ireland's Galway campus. More recently, in 1995, a third Center was opened in Beijing, China. The fundamental goal of these three Centers is to promote a high level of cooperation between industry and academia by enabling university students to pursue Master's and Ph.D. programmes in research areas of direct relevance to the company. Student projects are closely monitored by both a company scientist and an academic supervisor, thereby ensuring that each project retains its original industrial focus yet benefits from the expertise and facilities available in a university setting. In addition to longer-term projects leading to the degree of M.Sc. or Ph.D., the Bioscience Centers are uniquely suited to address research topics or problems which require a rapid response. For such projects, laboratory placements of two to three months duration are provided for final-year undergraduate students in such disciplines as animal science, biochemistry, biotechnology, microbiology and environmental science. Having Centers in different global areas ensures that relevant local research topics are addressed directly but with the assistance of a network of technical staff and facilities worldwide. A selection of past and present research areas is presented in Table 1.

In 1996, four students graduated with the degree of Ph.D. and two with the degree of M.Sc. from the Alltech Bioscience programme. While this alone is proof of the success of the Bioscience Center concept, a review of some of these recent projects may provide the best indication of how industry and universities can interact in a mutually beneficial fashion.

Table 1. Key research areas at Alltech's Bioscience Centers.

- Yeast cultures as digestibility enhancers. Diet specific yeast cultures.
- Microbial enzymes for treatment of dietary constituents and for environmental improvement.
- Competitive exclusion products and toxin binders.
- Plant extracts as environmental control agents and as sources of novel biological compounds.
- Trace mineral proteinates to improve mineral availability and enhance animal performance.
- Selenium and chromium-enriched yeast.
- Silage inoculants.
- Bioremediation.
- Vegetable protein-based products as alternatives to plasma protein.
- Chinese herbs and spices. Production responses, modification of intestinal microflora, improvement of meat quality.

Mannanoligosaccharides

COMPETITIVE EXCLUSION PRODUCTS IN POULTRY

Methods to prevent pathogens from colonizing the alimentary tract of poultry and other species have been pursued for many years. A number of variables influence the population levels of microorganisms in the intestine including host factors, nutrition and of course the animal's environment. Many of the processes which enable a pathogen to become established in the gastrointestinal tract are complex and poorly understood. Such diversity makes it improbable that any one method will suffice for control of pathogens such as Salmonella. Several preventative measures are available ranging from the use of vaccines to the controversial though widely practised prophylactic administration of antibiotics in animal feed. One useful tool in the control of Salmonella infections in poultry is competitive exclusion (CE). The principle of CE has been long known and well reviewed (Impey *et al.*, 1982; Stavric and D'Aoust, 1993). It aims to create an environment in the gastrointestinal tract which prohibits colonization by enteric pathogens. Live bacterial cultures ranging from single strains of lactic acid bacteria to complex mixtures of poorly defined cultures derived from the caecal or faecal contents of adult animals have been used successfully as CE agents. Exclusion of enteric pathogens may also be effected through the use of simple sugars such as mannose which have been shown to inhibit adherence of certain pathogenic bacteria in the gastrointestinal tract by blocking bacterial lectin-epithelial receptor interaction (Ofek *et al.*, 1977; Oyofo *et al.*, 1989). Unfortunately, relatively high concentrations of mannose are required to control colonization by pathogenic bacteria and the inclusion of such levels of purified D-mannose in animal diets, even for short periods, is cost prohibitive. However, mannose-based carbohydrates are abundant in natural sources, such as yeast cell wall, and are relatively easy to extract and purify. Early results with such yeast cell wall preparations (Jacques and Newman, 1994; Sisak, 1994) showed promise in terms of reducing colonization by enteric pathogens, particularly those that express type-1 fimbriae. To extend the work, one of the past North American Bioscience Center Ph.D. candidates, Dr Peter Spring, working in conjunction with the

University of Kentucky and the Swiss Federal Institute of Technology, undertook detailed studies on Alltech's mannoligosaccharide (MOS) preparation from *Saccharomyces cerevisiae*. Some of the objectives of these studies were to screen a range of bacterial pathogens for their ability to agglutinate MOS and to determine the effect of MOS on caecal fermentation parameters, caecal microflora and enteric pathogen and coliform colonization in chicks under controlled conditions.

In a series of challenge trials using chicks, the effects of MOS on the prevalence of three strains of enteric bacteria that expressed type-1 fimbriae (*Salmonella typhimurium* 29E, *S. dublin* and *E. coli* 15R) were studied. The gastrointestinal microflora of each bird was standardized at the beginning of the experiment and birds were housed in bacterial isolation chambers. Birds received 10^4 CFU/bird of the challenge organism on day 3 and caecal parameters were analysed on day 10 (Spring, 1996). Among the results obtained was the finding that MOS-treated chicks had significantly lower caecal concentrations of *S. typhimurium* 29E (Figure 1) and the number of birds testing positive for *S. dublin* in the caeca was also significantly lower when MOS was part of the diet. Dietary inclusion of MOS (4000 ppm) also reduced the number of birds from which *E. coli* 15R could be recovered (75% versus 15%). Importantly, MOS had no effect on caecal concentrations of lactobacilli, enterococci and anaerobic bacteria.

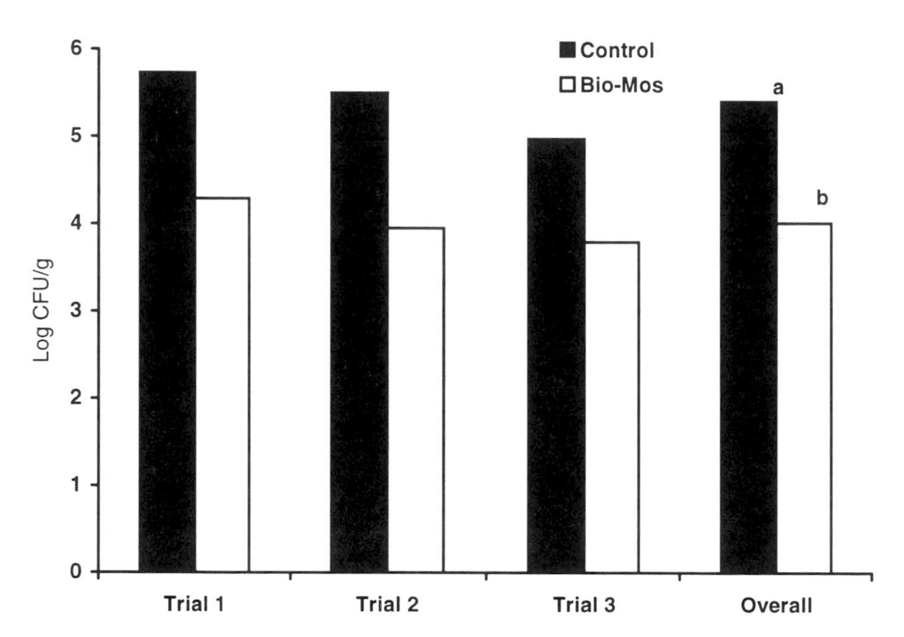

Figure 1. Effect of MOS on concentrations of *Salmonella typhimurium* in the ceca of chicks maintained in microbiological isolators (Spring, 1996ab). Values with different superscripts differ significantly. $P<0.05$.

MANNANOLIGOSACCHARIDES AS IMMUNOSTIMULANTS

In addition to its clear role as a CE agent, MOS has the interesting property of stimulating components of the immune system. It has long been established that glucans and mannans from the yeast cell wall have the ability to activate the complement cascade via the properdin pathway (Brade *et al.*, 1973; Nicholson *et al.*, 1974). Several observations have been made on the stimulation by mannans and glucans of macrophages and monocytes in treated animals (Ohya *et al.*, 1994; Jouault *et al.*, 1995; Karaca *et al.*, 1995). Recently it has become evident that MOS has the ability to stimulate the specific immune system in a variety of species. Savage and Zakrzewska (1996) reported increased concentrations of plasma IgG and bile IgA in turkeys fed diets containing MOS. Stimulation of such humoral immune responses by orally administered MOS probably occurs via interaction with gut associated lymphoid tissues (GALTs) such as Peyer's patches. M cells in these regions possess short microvilli and can endocytose antigens, inert particles and even microbes into the GALT. Distinct follicles are found under the dome region of the Peyer's patch which contain germinal centres where significant B cell division occurs. B cell conversion to IgA production and the process of affinity maturation occurs in these germinal centres.

Immune system stimulation by MOS is an active area of research in the Bioscience Centers. For example, a current Master's project is examining the dose-response relationship of dietary MOS inclusion on immune stimulation in laboratory animals. As anticipated, significant increases in the production of intestinal IgA were noted over time in laboratory rats fed MOS at dietary inclusions ranging from 2000 ppm to 10,000 ppm (Figure 2). Interestingly, unlike the aforementioned results of Savage and Zakrzewska (1996) in turkeys, no significant differences were noted in plasma IgG levels (Figure 3). This may indicate different species reactivity to MOS at an immunological level whereby the reaction in some species may be mainly restricted to secretion of mucosal IgA, whereas in others an additional systemic humoral response may be elicited. It is tempting to speculate that the latter may be brought about by microbe-mannan aggregates presenting a sufficiently immunogenic stimulus to trigger an IgG response in some species. Additional studies are in progress to evaluate the effect of MOS on immune stimulation in dogs and in gnotobiotic piglets.

Elucidating the mode of action of yeast culture

Yeast cultures have been used as animal feed supplements for decades and there is a wide body of published literature to support the efficacy of these preparations across many species. In cattle the beneficial effects of yeast cultures are associated with their ability to alter rumen function and the concentration of total anaerobic and cellulolytic bacteria in the rumen. Based on extensive *in vivo* and *in vitro* studies, Dawson (1992) proposed a now widely accepted model for the action of yeast culture in the rumen (Figure 4). The key step in this model is the initial stimulation of the growth and

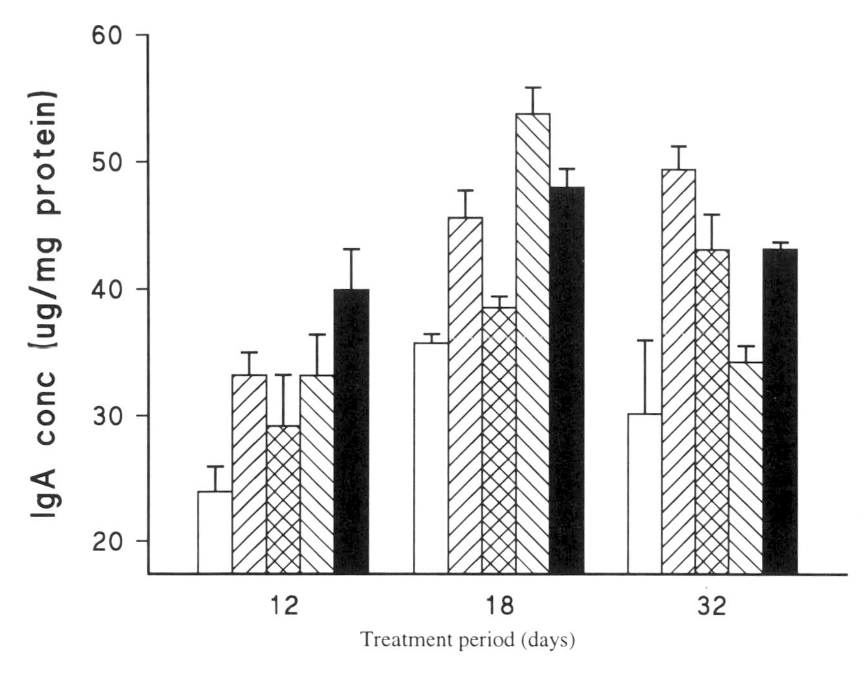

Figure 2. Effect of Bio-Mos on the intestinal IgA levels (mean ± S.E.) in Wistar rats fed diets containing graded levels of the oligosaccharide preparation. For all three sampling dates, pooled treatment means were significantly different to controls ($P<0.05$)

activities of ruminal bacteria by *S. cerevisiae*. However, the missing piece in this scheme has been the identification of the actual stimulus produced or brought about by the presence of the yeast culture. Nisbet and Martin (1991) proposed that malate was the stimulatory factor based upon results which showed that L-malic acid from a culture filtrate of Yea-Sacc[1026] stimulated the ruminal bacterium *Selenomonas ruminantium* by increasing lactate uptake. However, the levels of yeast culture used in these studies were very much greater than those which would be used under practical circumstances of yeast culture supplementation. Newbold *et al.* (1993) put forward the theory that it was the oxygen-scavenging ability of yeast which provided the stimulus for enhanced growth of ruminal bacteria. This seems unlikely, however, as previous work had shown that oxygen concentrations 20 to 50-fold higher than those found in the Newbold *et al.* study are need to inhibit the growth of several strains of ruminal bacteria such as *S. ruminantum*, *Butyrivibrio fibrisolvens* and *Megasphera elsdenii* (Loesche, 1969). Chaucheyras *et al.* (1993) showed that *S. cerevisiae*

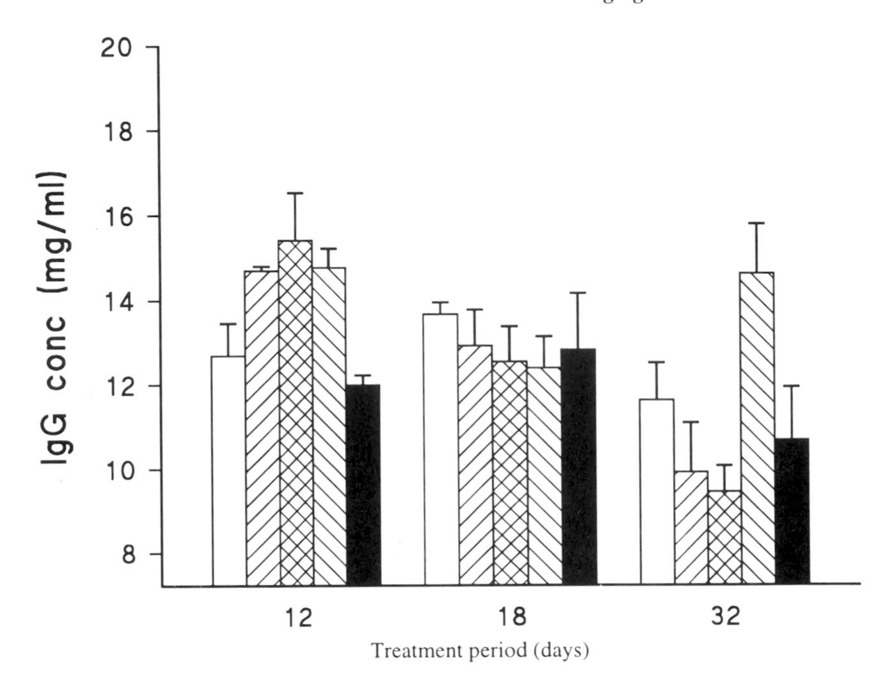

Figure 3. Effect of Bio-Mos on plasma IgG levels (mean ± S.E.) in Wistar rats fed diets containing graded levels of the oligosaccharide preparation.

stimulated the growth of the rumen fungus *Neocallimastix frontalis* in a vitamin-deficient medium, an effect which could be mimicked by culturing the fungus in the presence of mixed B vitamins. Again, however, the levels of yeast used to obtain the stimulation of fungal growth were far in excess of those which would be found under practical conditions.

With this background another recent graduate of the Bioscience Ph.D. programme, Dr Ivan Girard, conducted his studies on the characterization of the stimulatory activities from *S. cerevisiae* strain 1026 (Yea-Sacc^{1026}) on the growth and activities of ruminal bacteria (Girard, 1996). Part of his work involved the sequential ultrafiltration of 24-hour culture supernatants from Yea-Sacc^{1026} followed by the testing of fractions for stimulatory activity using the criterion of lag time reduction in cultures of *Ruminococcus albus.* Stimulatory fractions were further separated and characterized by gel filtration chromatography and HPLC. These fractionation experiments showed that the culture supernatants contained heat-stable stimulatory factors having molecular weights ranging from 400 to 650 Da (Table 2). These factors contained at least two ninhydrin-reactive residues per molecule and displayed

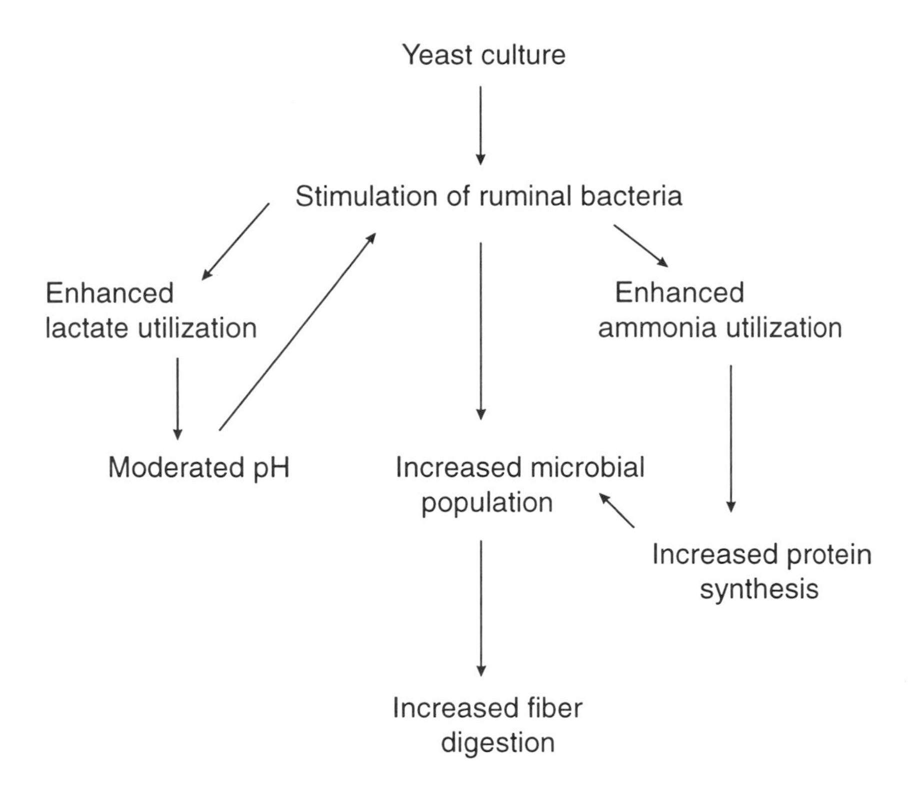

Figure 4. Model for yeast culture activities in the rumen (Dawson, 1992).

absorption maxima of 210 and 280 nm. HPLC analysis revealed that the stimulatory factors were indeed a mixture of peptides. This was confirmed by the fact that synthetic peptides of similar amino acid composition and molecular weight also reduced the lag time of cultures of *R. albus*. Such work represents an important advance in understanding the mode of action of yeast cultures. Furthermore, it provides several possibilities for enhancing the efficacy of yeast cultures in ruminants. Moreover, the identification of the amino acid sequence of these peptides makes possible the use of recombinant DNA technology to produce yeast strains which express bacterial growth stimulants at high levels. Further characterization of the mechanisms by which these peptides stimulate ruminal bacteria may well allow strain-specific stimulation to suit particular diets, thereby optimizing rumen function and animal productivity.

Yucca schidigera and ammonia control

Extracts of the desert plant *Yucca schidigera* such as De-Odorase have a variety of beneficial effects when included in the diets of farm animals such as pigs

Table 2. Characterization of low-molecular weight stimulatory factors from culture supernatants of *S. cerevisiae* 1026 (Girard, 1996).

Item	Characteristics
Source	Supernatant from cultures of *S. cerevisiae*^{1026}
Typical effect on lag time of *R. albus*	30% reduction
Heat stability	Stable at 121°C for 20 minutes
Cold stability	Stable at -20°C for at least 6 months
Acid reactivity	Destroyed or structurally altered following 24 hour hydrolysis in 12 M HCl
Passage through 10,000 Da filter	Yes (presence of stimulatory components in filtrate)
Passage through 1000 Da filter	Yes (presence of stimulatory components in both filtrate and retentate)
Molecular weight range by gel filtration	400-650 Da
Number of amino groups per molecule	At least 2
Absorbance peaks	210 nm and 280 nm

and poultry at levels as low as 100 ppm (Rowland *et al.*, 1976; Goodall *et al.*, 1979; Cromwell *et al.*, 1984; Headon *et al.*, 1991). These benefits include improved growth rate, feed efficiency and health. Many of the effects of dietary *Y. schidigera* supplementation are attributed to the ability of the extract to reduce ammonia levels in the gastrointestinal tract and manure of supplemented animals.

Given the great number of publications on the effects of such preparations, it is somewhat surprising that many hypotheses still exist as to the mechanisms by which these supplements reduce ammonia levels under practical conditions. Two of the most popular models postulate that these preparations work by direct inhibition of microfloral urease in the gut and excreta or that they exert their effect via steroidal saponins, which are relatively abundant in Yucca extracts. The large number of theories surrounding the mode of action of Yucca preparations has led to confusion and indeed scepticism within the animal feed industry about the quality, efficacy and scientific knowledge behind these products. Again, this provides a good example of how collaborative research between industry and universities has helped to improve the scientific profile of widely used but often poorly understood products.

Another of Alltech's past Ph.D. students, Dr Gerard Killeen, undertook his project work on the effects of De-Odorase and its constituent saponins on nitrogen metabolism. In the course of his studies, he investigated several of the aforementioned theories on the mode of action of Yucca. In the case of urease inhibition, he demonstrated that many of the published experiments on this topic had utilized phosphate-based buffers for the actual urease inhibition assays. In addition, these studies had utilized weakly buffered or non-buffered preparations of *Y. schidigera*. He demonstrated that phosphate itself was a potent inhibitor of bacterial urease and that the slightly acidic nature of *Y. schidigera* preparations can cause misleading decreases in urease activity simply by reducing the pH of the assay to below the pH optimum (pH 6.8 to 8.0) of bacterial ureases (Killeen *et al.*, 1994). When urease assays were performed in non-inhibiting buffers using pH adjusted *Y. schidigera* preparations, only a weak and non-specific inhibition was noted (Figure 5). This non-specificity was confirmed by the fact that α-galactosidase, an unrelated fungal hydrolase, was also weakly inhibited by the preparation. Similarly, no significant inhibition of urease by 4 ml per litre *Y. schidigera* preparation (40-fold typical feed inclusion levels) was observed over the pH range 3.0–8.5. Equating the weak inhibition of the small urease concentration used in this study to reported *in vivo* urea degradation rates in animals such as pigs (Mosenthin *et al.*, 1992), it was estimated that to achieve 10% inhibition of typical gut urease levels such animals would have to consume daily between 80 and 400% of their own body weight in feed containing the maximum recommended level of De-Odorase (Killeen *et al.*, 1994; Killeen, 1996). Likewise a role for steroidal saponins in the mode of action of these extracts seems unlikely. These may be completely removed from the preparation by extraction with *n*-butanol. *In vivo* experiments comparing the effects of the saponin-containing butanol fraction of De-Odorase to the residual aqueous or non-butanol fraction have shown that the non-butanol fraction contains the bulk of the biological activity of the preparation (Figure 6).

Ongoing work in the area of De-Odorase research includes a collaborative study between the European Bioscience Center and the University of Plymouth in the UK where one of our current Ph.D. students is examining the effects of De-Odorase on nitrogen metabolism in pigs. In addition to metabolism studies, the effects of De-Odorase on the emanation of volatile constituents from effluent is being assessed by gas chromatography/mass spectrometry techniques.

Enhancing trace mineral availability through the use of mineral proteinates

For several decades continuous efforts have been made to optimize the mineral requirements of animals through the scientific formulation of feed rations. Traditionally, trace mineral supplementation has been in the form of simple inorganic salts such as copper sulphate. Research has shown, however, that the biological availability and hence the performance enhancement achieved by trace mineral supplementation can be significantly improved if the mineral is complexed or chelated with organic molecules such as amino acids and peptides. Proteinates or Bioplexes are the products formed when a metal salt is reacted with a protein hydrolysate containing a mixture of amino acids and peptides. A great deal of university research has been conducted over the last five years to establish the superior efficacy of proteinated minerals over inorganic and other sources (Vandergrift, 1992; Hemken *et al.*, 1993; Jackson, 1993; O'Donoghue *et al.*, 1995; Boland *et al.*, 1996). In addition to these important studies, we continue to actively research the area of 'organic' minerals in an attempt to better define the mechanisms behind the observed increases in biological availability. To this end several students at the various Bioscience Centers have conducted research projects on the uptake and distribution of minerals from Bioplexes. Examples include radiotracer studies in laboratory animals using radiolabelled ^{65}Zn-Gly-Met chelates versus $^{65}ZnSO_4$ in

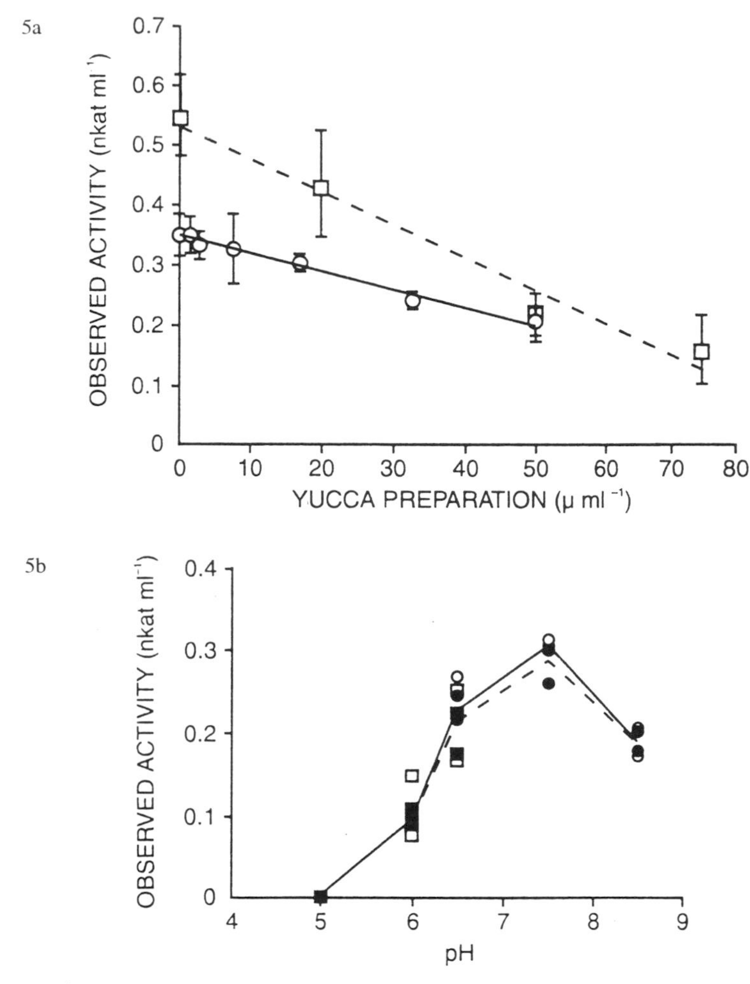

Figure 5a. The effects of varying the concentration of De-Odorase on activity of urease from *B. pasteurii* and on ß-galactosidase from *A. oryzae* at pH 6.5 and 4.5. Error bars represent ± 95% confidence limits. o: urease; □: ß-galactosidase.

5b. Effect of 4 ml/l De-Odorase on activity of urease from *B. pasteurii* in the pH range 5.0 – 8.5 in 150mM K-HEPES and K-citrate. □:K-citrate; ■:K-citrate plus *Y. schidigera*; o: K-HEPES; ●:K-HEPES plus *Y. schidigera* — Control (buffers without *Y. schidigera*). - - - Analytical (buffers with *Y. schidigera*).

rats. As shown in Figure 7, the final tissue distribution of absorbed ^{65}Zn in these experiments was quite different between the respective zinc treatments. Tissue uptake of ^{65}Zn from the zinc chelate was significantly higher than from $^{65}ZnSO_4$ in hair ($P = 0.002$), kidney ($P = 0.012$) and muscle ($P = 0.006$). The numerous reports in the literature in favour of 'organic' minerals suggest that they are more effective in targeting specific tissues than inorganic minerals. Laboratory studies such as the radiotracer study in rats lend support to these conclusions. For example, the results obtained for zinc

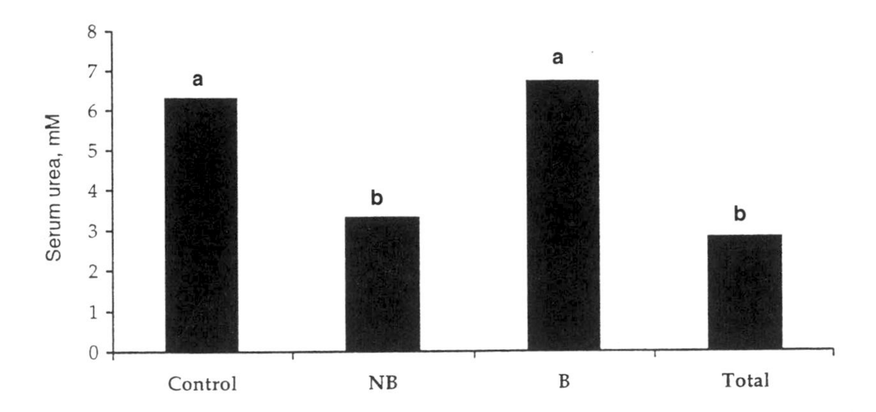

Figure 6. The influence of dietary supplementation with *Yucca schidigera* and fractions thereof on serum urea concentrations in rats (Killen, 1996). Control = untreated: B= treated with a dried butanol-extract of *Y. schidigera*; NB= treated with the non-butanol-extractable fraction of *Y. schidigera*; Total= treated with the extracted *Y. schidigera*. [ab] values with different superscripts differ significantly (P <0.05).

deposition in hair are quite interesting and represented almost a three-fold difference in uptake between the zinc treatments. Many of the positive reports on the use of organic zinc supplementation in farm animals note improvements in hoof quality and in coat condition. Preferential deposition of zinc derived from an organic zinc source in keratinaceous tissues such as hair is easily demonstrated by studies in model systems and provides scientific data to back-up field observations and testimonial literature.

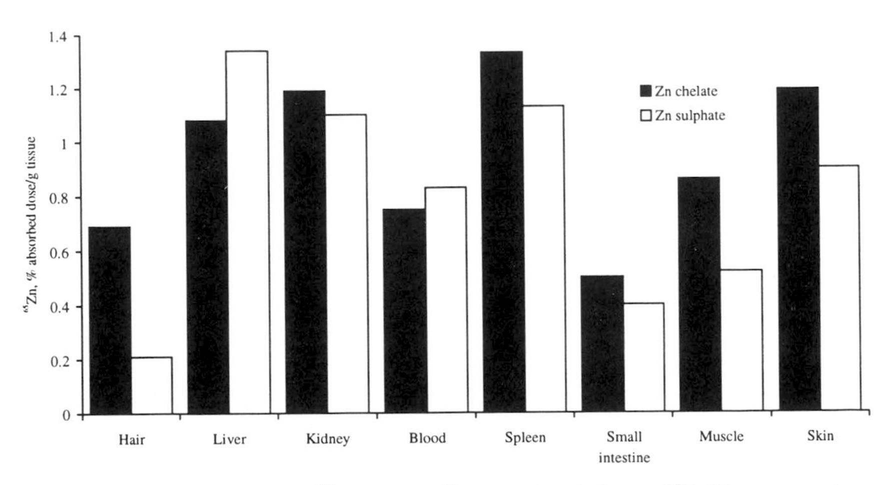

Figure 7. Rat tissue uptake of ^{65}Zn from a ^{65}Zn-labelled chelate or $^{65}ZnSO_4$ expressed as a percentage of absorbed dose.

Also in the area of enhanced mineral availability the company continues to actively research selenium-enriched yeast (Sel-Plex 50) through collaborative university-based studies and via its own pool of students based at the Bioscience Centers. The efficacy of this preparation as a source of selenium superior to sodium selenite has been well established and a discussion of the results obtained to date is beyond the scope of this text since ample evidence is presented elsewhere (Pehrson, 1993; Mahan, 1994; 1995). Of equal importance, perhaps, with preparations such as this which attract concern over potential toxicity and misuse, is the development of validated control methods and a complete chemical characterization of the product in order to allay regulatory, animal producer and consumer fears about the quality and safety of selenium products. One of our past Master's students completed his studies on process optimization and complete characterization of selenium-enriched yeast (Kelly, 1995). In the course of his studies, he identified and characterized the predominant forms and localization of selenium in Sel-Plex 50 and developed methods for the control of this preparation on an organic selenium basis. Much of this work has been incorporated into regulatory petitions in the USA and Europe where the company is seeking to have Sel-Plex 50 registered as a source of selenium for several target species. Again it represents a good example of how an applied focus can be brought to bear on a graduate research project in order to assist in the introduction of novel and effective products to the marketplace.

Other project areas and future goals

In addition to the topics described above, Alltech is actively involved in research and development work in many other project areas. Much effort is being directed towards the development of novel enzyme preparations with applications in the areas of enhanced digestibility of dietary constituents and in waste recycling. Examples of the contribution which the Bioscience Centers have made to new product development for the company include that of the Vegpro enzyme which is designed to enhance utilization of the non-cereal components of poultry and pig diets (Charlton, 1996). Additionally, many short-term laboratory projects focus on the development of effective enzyme cocktails for the treatment of a variety of feed ingredients such as rapeseed (Table 3). Enhanced production of enzymes from microbial sources is a continuing priority area for the company. Recombinant DNA technology has been employed to produce acid phosphatases with phytate-degrading activity in commercially useful quantities (Moore *et al*., 1995) and current projects using this technology include enhanced production of α-galactosidase and chitinase in heterologous expression systems.

Chromium research also represents a key area for future product development and current projects include an investigation of the biological activity of several novel chromium complexes by one of our Ph.D. students at the European Bioscience Center. It is anticipated that the recently established

Bioscience Center in Beijing will provide exciting possibilities in the utilization of traditional Chinese herbs and spices as animal production aids in western countries.

The contribution of the Bioscience Centers to company development in terms of generating a better understanding of existing products and developing new ones has to date been very significant. The concept represents an excellent model by which a company can expand its research database in a university setting while providing a resourceful pool of future technical support and scientific personnel already well versed in the company's products and capable of making a rapid transition from academia to industry.

References

Boland, M.P., G. O'Donnell and D. O'Callaghan. 1996. The contribution of mineral proteinates to production and reproduction in dairy cattle. In: Biotechnology in the Feed Industry. (T.P. Lyons and K.A. Jacques, Eds). Nottingham University Press, Loughborough, Leics. UK. 95–103.

Brade, V., G.D. Lee, A. Nicholson, H.S. Shin and M.M. Mayer. 1973. The reaction of zymosan with the properdin system in normal and C4-deficient guinea pig serum. J. Immunol. 111: 1389–1399.

Charlton, P. 1996. Expanding enzyme applications: higher amino acid and energy values for vegetable proteins. In: Biotechnology in the Feed Industry. (T.P. Lyons and K.A. Jacques, Eds). Nottingham University Press, Loughborough, Leics. UK. 317–326.

Chaucheyras, F., G. Fonty, G. Bertin and P. Gouet. 1993. Effects of live yeast cells on zoospore germination, growth and cellulolytic activity of *Neocallimastix frontalis* MCH3. Biennial Conference on Rumen Function, Chicago, IL. p. 29 (Abstr.).

Cromwell, G.L., T.S. Stahly and A.T. Monegue. 1984. Efficacy of sarsaponin for weanling and growing-finishing swine housed at two animal densities. J. Anim. Sci. 61 (Suppl.): 111.

Dawson, K.A. 1992. Current and future role of yeast culture in animal production: A review of research over the last six years. In: Supplement to the Proceedings of Alltech's Eighth Annual Symposium. T.P. Lyons (Ed.). Alltech Technical Publications, Nicholasville, KY.

Girard, I. 1996. Characterization of stimulatory activities from *Saccharomyces cerevisiae* on the growth and activities of ruminal bacteria. Ph.D. Dissertation, University of Kentucky.

Goodall, S.R., J.D. Echenblaum and J.K. Matsushima. 1979. Sarsaponin and monensin effects upon *in vitro* VFA concentration, gas production and feedlot performance. J. Anim. Sci. 49(Suppl.): 370–371.

Headon, D.R., K.A. Buggle, A.B. Nelson and G.F. Killeen. 1991. Glycofractions of the Yucca plant and their role in ammonia control. In: Biotechnology in the Feed Industry. Proc. Alltech's Seventh Ann. Symp. T.P. Lyons (Ed.). Alltech Technical Publications, Nicholasville, KY. 95–108.

Hemken, R.W., T.W. Clark and Z. Du. 1993. Copper: Its role in animal nutrition. In: Biotechnology in the Feed Industry. Proc. Alltech's Ninth Ann.

Symp. T.P. Lyons (Ed.). Alltech Technical Publications, Nicholasville, KY. 35–39.

Impey, C.S., G.C. Mead and S.M. George. 1982. Competitive exclusion of Salmonellas from chick caecum using a defined mixture of bacterial isolates from the caecal microflora of an adult bird. J. Hyg. Camb. 89: 479–490.

Jackson, S.G. 1993. Mineral proteinates: Applications in equine nutrition. In: Biotechnology in the Feed Industry. Proc. Alltech's Ninth Ann. Symp. T.P. Lyons (Ed.). Alltech Technical Publications, Nicholasville, KY. 91–98.

Jacques, K. and K.E. Newman. 1994. Effect of oligosaccharide supplementation on performance and health of Holstein calves pre- and post-weaning. J. Anim. Sci. 72 (Suppl. 1): 295.

Jouault, T., G. Lepage, A. Bernigaud, P.A. Trinel, C. Fradin, J.M. Wieruszeski, G. Strecker and D. Poulain. 1995. 1,-2,-linked oligomannosides from *Candida albicans* act as signals for tumour necrosis factor production. Infect. Immun. 63(6): 2378–2381.

Karaca, K., J.M. Sharma and R. Nordgren. 1995. Nitric oxide production by chicken macrophages activated by acemannan, a complex carbohydrate extracted from Aloe vera. Int. J. Immunopharmacol. 17(3): 183–188.

Kelly, M.P. 1995. The production and characterization of selenium-enriched *Saccharomyces cerevisiae*. M.Sc. Thesis, The National University of Ireland.

Killeen, G.F. 1996. The effects of dietary supplementation with extracts from *Yucca schidigera* and its constituent saponins on vertebrate nitrogen metabolism. Ph.D. dissertation, National University of Ireland.

Killeen, G.F., K.A. Buggle, M.J. Hynes, G.A. Walsh, R.F. Power and D.R. Headon. 1994. Influence of *Yucca schidigera* preparations on the activity of urease from *Bacillus pasteurii*. J. Sci. Food Agric. 65: 433–440.

Loesche, W.J. 1969. Oxygen sensitivity of various anaerobic bacteria. Appl. Microbiol. 18: 723.

Mahan, D.C. 1994. Organic selenium sources for swine, how do they compare with inorganic selenium sources? In: Biotechnology in the Feed Industry. (T.P. Lyons and K.A. Jacques, Eds). Nottingham University Press, Loughborough, Leics. UK. 323–333.

Mahan, D.C. 1995. Selenium metabolism in animals: What role does selenium yeast have? In: Biotechnology in the Feed Industry. Proc. Alltech's Eleventh Ann. Symp. T.P. Lyons and K.A. Jacques (Eds.). Nottingham University Press, Loughborough, Leics. UK. 257–267.

Moore, E., V.R. Helly, O.M. Conneely, P.P. Ward, R.F. Power and D.R. Headon. 1995. Molecular cloning, expression and evaluation of phosphohydrolases for phytate-degrading activity. J. Ind. Microbiol. 14: 396–402.

Mosenthin, R., W.C. Sauer, H. Henkel, F. Ahrens and C.F.M. Delange. 1992. Tracer studies of urea kinetics in growing pigs. The effect of starch infusion at the distal ileum on urea recycling and bacterial nitrogen excretion. J. Anim. Sci. 70: 3467–3472.

Newbold, C.J., R.J. Wallace and F.M. McIntosh. 1993. The stimulation of rumen bacteria by *Saccharomyces cerevisiae* is dependent on the respiratory activity of the yeast. J. Anim. Sci. 71 (Suppl. 1): 280 (Abstr.).

Nicholson, A., V. Brade, G.D. Lee, H.S. Shin and M.M. Mayer. 1974. Kinetic studies of the formation of the properdin system enzymes on zymosan: evidence that nascent C3b controls the rate of assembly. J. Immunol. 112: 1115–1123.

Nisbet, D.J. and S.A. Martin. 1991. Effect of a *Saccharomyces cerevisiae* culture on lactate utilization by the ruminal bacterium *Selenomonas ruminantium*. J. Anim. Sci. 69: 4628–4633.

O'Donoghue, D., P.O. Brophy, M. Rath and M.P. Boland. 1995. The effect of proteinated minerals added to the diet on the performance of post-partum dairy cows. In: Biotechnology in the Feed Industry. Proc. Alltech's Eleventh Ann. Symp. T.P. Lyons and K.A. Jacques (Eds). Nottingham University Press, Loughborough, Leics. UK. 293–297.

Ofek, I., D. Mirelmann and N. Sharon. 1977. Adherence of *E. coli* to human mucosal cells is mediated by mannose receptors. Nature 265: 623–625.

Ohya, Y., K. Ihara, J. Murata, J. Sugitou and T. Ouchi. 1994. Preparation and biological properties of dicarboxy glucomannon-enzymatic degradation and stimulating activity against cultured macrophages. Carbohydrate Polymers 25: 123–130.

Oyofo, B.A., J.R. De Loach, D.E. Corrier, J.O. Norman, R.L. Ziprin and H.H. Mollenhauer. 1989. Prevention of *Salmonella typhimurium* colonization of broilers with D-mannose. Poult. Sci. 68: 1357–1360.

Pehrson, B.G. 1993. Selenium in nutrition with special reference to the biopotency of organic and inorganic selenium compounds. In: Biotechnology in the Feed Industry. Proc. Alltech's Ninth Ann. Symp. T.P. Lyons (Ed.). Alltech Technical Publications, Nicholasville, KY. 71–89.

Rowland, L.O., J.E. Plyler and J.W. Bradley. 1976. *Yucca schidigera* extract effect on egg production and house ammonia levels. Poult. Sci. 55: 2086–2089.

Savage, T.F. and E.I. Zakrzewska. 1996. The performance of male turkeys fed a starter diet containing a mannanoligosaccharide (Bio-Mos) from day-old to eight weeks of age. In: Biotechnology in the Feed Industry. Proc. Alltech's Twelfth Ann. Symp. T.P. Lyons and K.A. Jacques (Eds). Nottingham University Press, Loughborough, Leics. UK. 47–54.

Sisak, F. 1994. Stimulation of phagocytosis as assessed by luminol-enhanced chemiluminescence and response to Salmonella challenge of poultry fed diets containing mannanoligosaccharides. Poster presented at Alltech's Tenth Ann. Symp. Alltech Technical Publications, Nicholasville, KY.

Spring, P. 1996. Effects of mannanoligosaccharides on different cecal parameters and on cecal concentrations of enteric pathogens in poultry. Ph.D. dissertation. Swiss Federal Institute of Technology, Zurich.

Stavric, S. and J.Y. D'Aoust. 1993. Undefined and defined bacterial preparations for competitive exclusion of Salmonella from poultry – a review. J. Food Protection. 56: 173–180.

Vandergrift, B. 1992. The theory and practice of mineral proteinates in the animal feed industry. In: Biotechnology in the Feed Industry. Proc. Alltech's Eighth Ann. Symp. T.P. Lyons (Ed.). Alltech Technical Publications, Nicholasville, KY. 179–192.

NATURAL FOODS: A TEN BILLION DOLLAR INDUSTRY AND GROWING. WHAT CAN WE LEARN FROM THE HUMAN FOOD SECTOR?

CASPAR WENK

Institute of Animal Sciences, Nutrition Biology, ETH Zurich, Switzerland

Introduction: enough food – high quality, natural food

The various ways in which human food is produced are increasingly discussed and questioned in a modern society (Figure 1). We expect food from plants, farm animals and microorganisms to be inexpensive, of good quality and healthy. Of increasing interest are environmental concerns to consumers, the food industry and politicians, because the costs of environmental care are becoming of great significance. Idealistic arguments come primarily from biological farmer organizations and organizations like Green Peace or World Wildlife Fund, but in general all of us expect our food to be as natural as possible and free of any undesired or toxic substances.

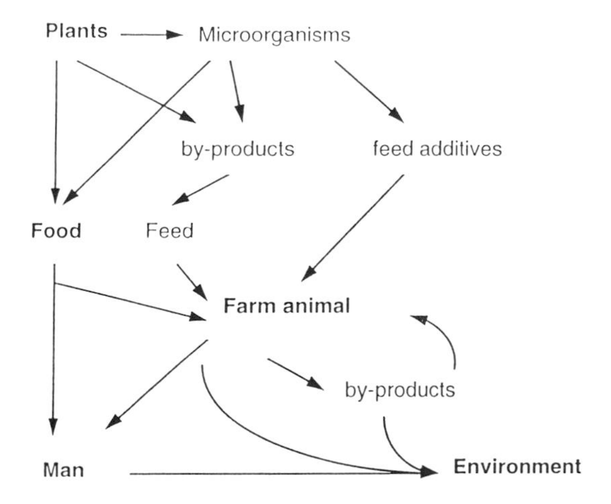

Figure 1. A schematic of human natural food production.

In highly developed countries (e.g. Western Europe and the USA) we do not always feel the impact of the steady growth in the world population. Twenty-five years from now there will be almost 9 billion inhabitants on earth who will expect enough food to meet nutritional needs. Today more than 800 million people suffer from hunger. The goal of sufficient food for everyone can only be achieved if world food production increases by about 2% every year. Furthermore, we must realize that the actual world cereal reserves supply current needs for less than 50 days. This is the lowest reserve in more than 20 years (FAO, 1996). World food production must grow without increasing the environmental waste load. In many parts of the world pollution, available water, soil structure and energy availability are the primary limiting parameters for agricultural production. The efficient use of all these resources must be achieved.

Today, world agricultural is at a point where production could be increased. On the other hand the consumers of highly developed countries have higher and higher expectations about quality and idealistic images of food that focus attention on issues other than yield. In Switzerland, a small country in Central Europe with about 7 million inhabitants, the debate over natural food is especially pronounced. After the occurrence of BSE (bovine spongiform encephalopathy) (242 cases as of January, 1997) in 1989 the use of all protein sources from animal origin was banned in ruminant diets. The two biggest supermarket chains with a market volume in Switzerland of 65% took it a step further and banned meat meal and other feedstuffs of animal origin also for pigs and poultry.

Consumers further expect that their food and animal feed is not genetically modified. Even synthetic amino acids, vitamins and other feed additives are banned in certain production programs. Since the risk of resistance against antibiotics is of concern in connection with the use of antibiotic feed additives, a general ban on these substances in animal nutrition is discussed. The use of avoparcin will end in April 1997.

We are increasingly aware that nutrition policy decisions are strongly influenced by idealistic and emotional arguments conveyed by media and politicians who lack sufficient scientific knowledge in animal production, hygiene and environmental considerations. Scientists in nutrition, hygiene or toxicology try to explain phenomena or calculate the benefits and risks of nutritional variables by first understanding fundamental physiological facts involved. They see nutrition clearly from the 'bottom-up' (Figure 2). The consumer (whoever that is) with an idealistic and emotional view of nutrition selects the food he believes to be the best for him. In addition to experience and expectation he builds his opinion mainly on the basis of advertisements or product appearance. During recent years media and food market leaders have stimulated wide-ranging debates on food and human nutrition. They see issues from the 'top-down' perspective. Problems arise due to these different points of view because frequently the consumer with the top-down view and the scientist with the bottom-up view do not speak the same language and therefore do not respect the considerations and arguments of the other.

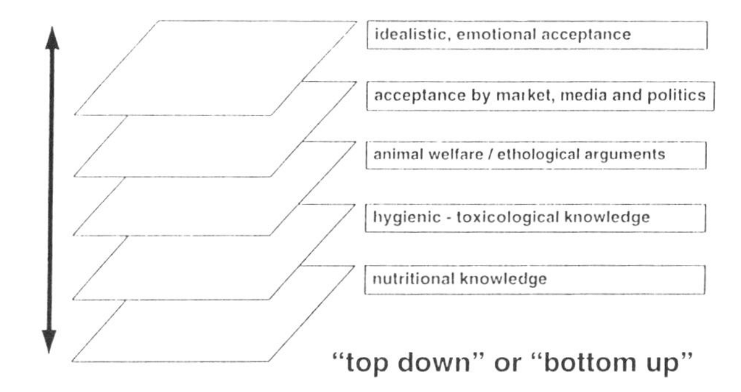

Figure 2. Decision making perspectives when evaluating food issues.

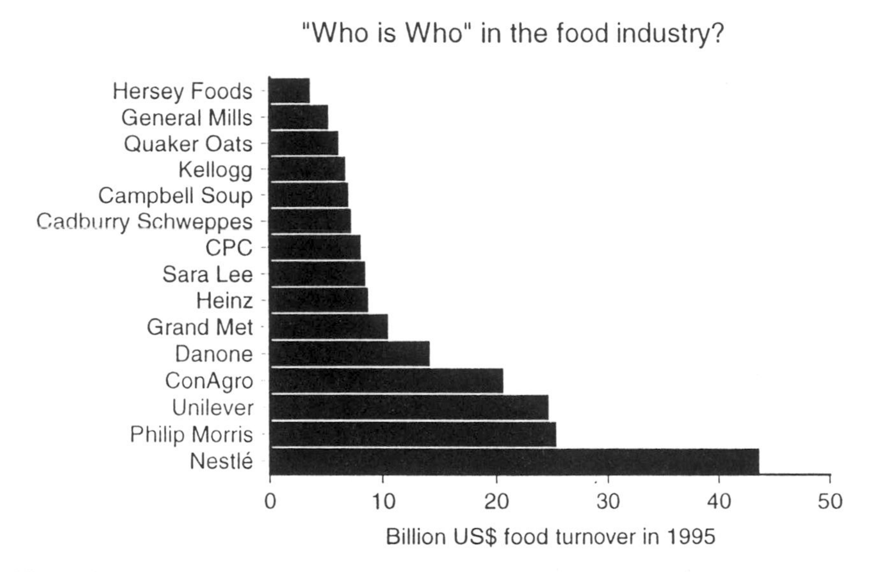

Figure 3. Financial turnover in 1995 for the top companies in the food sector.

Internationally the food industry is very powerful (Figure 3). The 15 biggest food industries had a combined turnover of 200 billion dollars in 1995. The smaller industries, distributors and local supermarket chains are not included in this figure; however, the power of local and international supermarket chains should not be underestimated. They play an important role in public opinion, especially in the field of nutrition.

What are the nutritional needs of man and animals?

From the view of the nutrition scientist many similarities between man and animals can be described. Food or feed contains nutrients and other desired or undesired substances. These ingredients are of nutritional relevance if the amount of food ingested contributes a significant amount of a certain substance. To meet the nutrient requirements of man or animals the nutrients must be available for digestion, absorption and intermediate metabolism.

While strict linearity between intake and energy utilization can be described for energy (energy balance), single essential nutrients are controlled by homeostasis, a regulation process including intake, absorption, intermediate turnover, storage or mobilization as well as excretion. Homeorhesis, another important regulation process, is responsible for prioritizing nutrient distribution among the various tissues in the body. Tissues of high priority are the brain and maternal tissues such as the uterus or mammary gland. Muscles or the adipose tissues have a lower priority.

Unlike human nutrition, the nutrition of farm animals aims to optimize health, fertility and performance (e.g. growth, lactation, reproduction) (Table 1). Furthermore, nutrition must be economic and should respect animal welfare as well as environmental concerns. Under these limitations and with respect to feed intake we optimize composition of animal diets in a way to minimize costs and nutrient excretion. The optimal combination of available feedstuffs with supplementation of nutrient and non-nutrient additives helps in achieving this goal.

In contrast, in human nutrition essentially everyone is his own dietition. Individuals decide for themselves what to eat. Food choices are based not necessarily on nutritional knowledge, but on experience, preference and pleasure. From a metabolic perspective, primarily hunger and satiety as well as cognitive factors limit or define nutrition. Only nutrition during growth, gestation and lactation or intensive physical exercise are considered in relation to performance.

The general recommendations in human nutrition can be summarized by meeting energy requirements with nutrients in an optimal combination (50–60% carbohydrates, 12–15% protein and 25–30% fat). Furthermore, food must contain enough essential nutrients in relation to energy content (nutrient density). From animal nutrition we learn that man has no steak or tofu needs

Table1. Key differences between human and animal nutrition.

Human nutrition considerations	Animal nutrition considerations
Hunger/satiety	Performance
Longevity, health	Reproduction
Knowledge, experience	Health
Preference, pleasure	Economy
Nutritional choices made by individuals	Ecology
Performance considered only in relation to exercise, growth, gestation	

but instead requirements for energy and nutrients. The same holds true for our farm animals.

Good and safe food from animal origin

As explained earlier, the quality of food from animal origin is determined by many different criteria. In modern human nutrition we are defining new categories of food and pharmaceuticals. Special properties of food in the context of health have led to terms such as functional foods, health foods or nutraceuticals. In Table 2 the most important arguments for food quality are summarized. From the 'bottom-up' perspective nutrient content, hygiene and health along with taste are the main determinants of meat quality or quality of other animal products. The 'top-down' perspective views image and origin of the products to be of major importance. But in the end, the final decision to buy a product is based on price in most cases. The ecological value is often mentioned in the context of origin, but instead should be based on a scientifically determined measurement of energy input and pollution. In most cases an objective determination of the ecological costs is missing. In this field we have to formulate new criteria which are generally relevant and can be accepted by both scientific and popular perspectives.

Based on food consumption figures we can calculate the contribution of food from animal origin used to meet the nutrient needs of the Swiss population (Figure 4). During recent years milk and meat consumption in Switzerland has decreased steadily. Only consumption of fresh fish has increased in the last 20 years. Questionnaires about the food habits of the Swiss population have shown that reduced meat consumption is mainly due to reduced number of meals with meat and not the amount of meat eaten per meal. Despite all the discussion surrounding BSE, the number of vegetarians has not increased during the last 10 years (GSF, 1996). Vegetarians constitute 2–3% of the whole population and up to 8–10% of the young female population in the German-speaking part of Switzerland.

On the basis of the Swiss Nutrition Report (1991) we have calculated proportions of the nutrient requirements supplied by the primary foods consumed in the Swiss adult population. Meat and milk are the major suppliers of some B vitamins as well as the trace elements iron and zinc. Considering the high availability of these trace elements (e.g. heme iron in meat), the foodstuffs from animal origin contribute even a higher proportion of those nutrients.

Table 2. The determinants of food quality.

Image
Price
Taste
Origin
Ecological value
Nutrient content
Health aspects
Hygiene

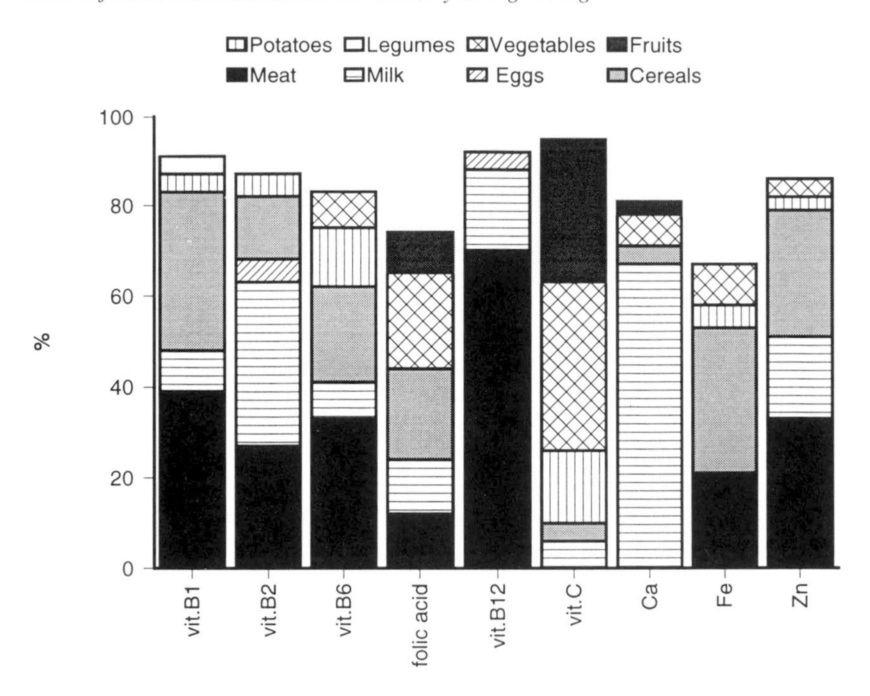

Figure 4. Contributions of various foodstuffs to nutrient supplies (Swiss Nutrition Report, 1991).

Foodstuffs of animal origin are the only suppliers of vitamin B12. For other vitamins like folic acid or vitamin C, foodstuffs from animal origin have either zero or only a modest contribution.

In a series of experiments we have studied the variability in nutrient content of different cuts of meat. We have learned that the mineral and vitamin content can vary over a wide range. Figure 5 shows the iron and zinc content of different meat cuts. Only red meat from beef (shoulder and back) and pork shoulder have a high iron content. Pork cutlet as well as cuts from veal calves and poultry are much lower in iron. Iron in heme form followed the same trend in the different cuts. Similar observations were made with zinc. Cuts with a high iron content were also high in zinc. Therefore, recommendations for meat consumption in the context of nutrient supply should take into consideration the different origins of meat. Vitamin content has also been shown to vary among different meat cuts (Leonhardt, 1996).

Effects of animal nutrition on essential nutrient content of animal products are well documented only for fat-soluble nutrients. It is therefore doubtful that the content of other nutrients can be altered in amounts relevant for human nutrition by special nutrition strategies.

In addition to nutrient content, other factors must be considered in describing the quality of meat and other products of animal origin. Some of the major failures of pork are PSE (pale, soft, exudative) meat and DFD (dark, firm and dry) meat (Table 3). Both failures are caused by disturbed carbohydrate

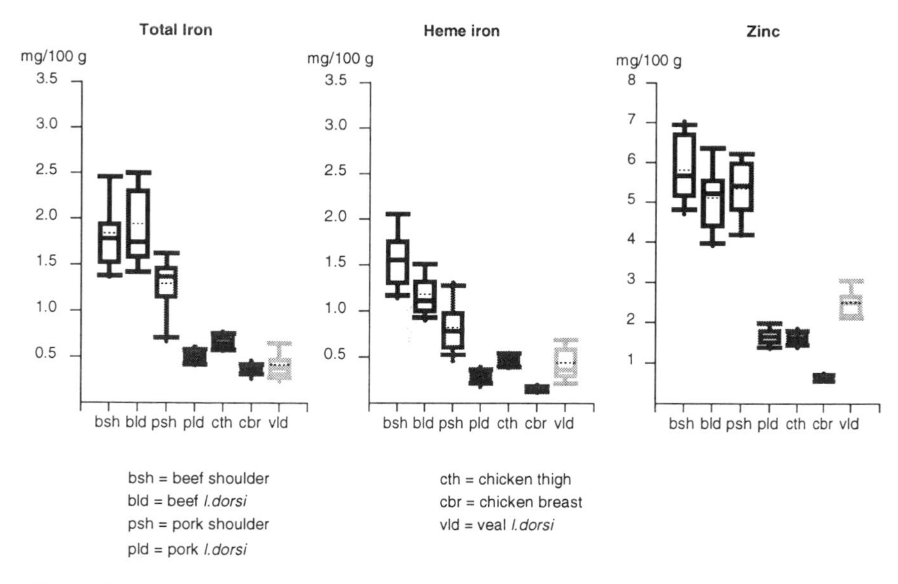

Figure 5. Iron and zinc content in various cuts of meat (Leonhardt, 1996).

metabolism before or immediately after death. The main cause is genetic, although transportation and slaughter management play a role. In pigs and other farm animals (beef and poultry) intramuscular fat content is often far too low. Fat content must be above 2% in lean meat to guarantee good taste (Bejerholm and Barton-Gade, 1986). Inadequate fat content is primarily due to genetics and age at slaughter. The younger an animal is at slaughter the lower its intramuscular fat content. Often we also observe low adipose tissue fat content. This phenomenon can cause severe problems in production of high quality meat products, especially sausages. The causes of that failure are not yet known and have only been studied to a very limited extent. Low intramuscular and adipose tissue fat content is difficult to alter through changes in feeding regimes.

Table 3. Frequent quality failures of meat and fat tissues in pigs.

Problem	Cause	Dietary influence
PSE and DFD*	Genetics, slaughter management	Selenomethionine?
Intramuscular fat content too low	Genetics, age at slaughter	Feeding?
Low fat content of adipose tissue	Genetics, age at slaughter	Feeding?
Poor oxidation stability	Genetics, age at slaughter, total fat deposition	Dietary polyunsaturates, vitamins E and C, selenium, other trace elements, phytase
Soft fat	Age at slaughter, total fat deposition	Dietary unsaturated fats

* PSE, pale soft exudative meat: DFD, dark firm and dry meat.

In human nutrition we are interested in increasing the amount of polyunsaturated fatty acid in the diet. Especially undesirable are saturated fatty acids with a chain length of 16 or less carbon atoms. In pork and poultry production we could easily change the fatty acid composition of the deposited fat; but a high content of polyunsaturated fatty acids increases oxidation potential and gives fat a soft consistency, two quality criteria which are of great importance in storage of meat and the production of sausages and other meat products. We therefore recommended that the pig industry in Switzerland reduce the content of unsaturated fats, especially polyunsaturated fatty acids, in the animal diet. Following our recommendations pig feed should contain less than 0.9 g polyunsaturated fatty acid per MJ digestible energy. In the major slaughterhouses in Switzerland fat consistency is a quality criteria affecting pork pricing. Farmers who produce pigs with soft fat therefore risk a reduced price. In Figure 6 the trends in pork quality in the slaughterhouse in St Gallen are plotted from 1988 through 1996

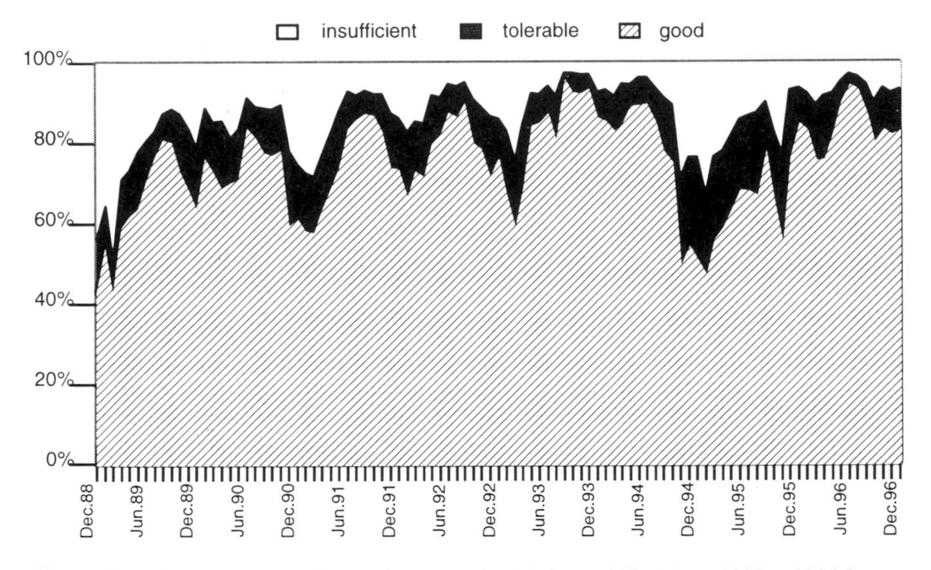

Figure 6. Pork quality scores (fat consistency and oxidation stability) from 1988 to 1996 from the slaughterhouse at St Gallen.

The quality score developed by our Institute has been used since December of 1988. At that time only about 55% of the pigs slaughtered at St Gallen satisfied the criteria for high quality meat products and to guarantee normal storage. Due to our recommendations and their applications in the feed industry the amount of meat with insufficient fat composition was reduced to about 10% in 1996. The variation in quality scores over time was mainly due to changes in the composition of diet in response to changing feed ingredient prices in the different years.

Food production and environmental aspects

In the discussion of the environmental impact of animal production the principal aspects of the nutrient cycles are often not properly considered. We are mainly interested in making animal production highly efficient. That typically means more units of output per unit of input. In intensive agriculture we try to maximize the yield per area, which in most cases minimizes the nutrient losses during production. One main argument for increasing intensity of production is therefore the reduction of nutrient losses. But the benefits of increased intensity must be balanced against an environmental burden when intensity of animal production is considered. The major arguments for a reduced environmental load in animal production are listed in Table 4.

Healthy animals are the first requirement for successful, environmentally friendly animal production. We can reduce environmental load by increasing fertility and performance. The feed conversion ratio can be increased by using more available nutrients or by a change in product composition (e.g. milk or meat). For instance protein deposition requires about six times less energy input than fat deposition. Feed additives are used to increase the health status, fertility and performance of farm animals. A further argument for the use of feed additives is the improvement in feed conversion ratio mainly due to increased digestibility of nutrients and energy.

Better nutrient availability can be achieved by supplying more available forms of nutrients (e.g. mineral proteinates) or with the use of special feed additives that increase nutrient digestibility. Enzymes can have a very positive effect on nutrient utilization when selected to appropriately match feed ingredients. Antimicrobials or probiotics can also increase nutrient availability.

A major contributor toward reducing environmental load is the use of home-grown feedstuffs and by-products from the food industry. With these groups of feedstuffs we help to establish close nutrient circles at the farm level or in a region. Since home-grown feedstuffs often contain a high fiber content, a combination with enzymes helps to increase feeding value, particularly for monogastric animals.

For 1993 we attempted to quantify the different by-products available from the food industry used in Switzerland (Table 5). The amounts used are separated into the direct use on the farms or by the feed industry after processing such as sterilization, drying or milling. In Switzerland by-products from the milk

Table 4. Animal nutrition and the environment: how can we reduce environmental load per kg of product?

Healthy animals
Higher performance
Increased feed conversion ratio
Better nutrient availability
More home-grown feedstuffs
More food industry by-products

industry and slaughterhouses contributed about 30% of total by-product use. By-products from vegetable foods amounted to more than 16%. The rest were derived from the production of beverages or specialties.

The pig is the primary farm animal for the utilization of by-products from the food industry. In Switzerland in 1993 the estimated contribution of by-products to diets for pigs amounted to about 60% and for poultry about 30%.

In 1993 most of the by-products from the food industry were fed to farm animals in the vicinities of the factories. Only negligible amounts of by-products were used as fertilizers or were exported. When a possible link between BSE and the Creutzfeldt-Jakob disease appeared in the media in spring of 1996, a ban on most feedstuffs of animal origin even for monogastric animals eventually resulted. Today only about 5% of meat meal is used in Switzerland. Around 70% of meat meal is exported and 25% is burnt in the cement industry.

With the expected restricted use or even banning of dietary antimicrobial agents, at least in special production programs, we must find new ways to improve and protect health status of farm animals, to improve animal performance and to increase nutrient availability. This goal can be attained with the best possible combination of the so-called pronutrients available including organic acids, digestive aids, herbs, highly available nutrient sources, alternative carbohydrate and protein sources.

Highly available nutrients (e.g. trace minerals as chelates or proteinates) can replace those sources currently used to meet the nutrient requirements of farm animals. This change allows reduction in total nutrient content in the diet. A reduction in the environmental load is the consequence. Trace elements like selenium when provided in the form of selenomethionine or chromium in the form of the biologically available glucose tolerance factor can have further specific effects on metabolism in addition to improved mineral retention and therefore help to improve health status.

Other kinds of additives work in different ways to affect health status or nutrient availability. Organic acids help to improve digestive processes especially for monogastric animals. With improved pH regulation, colonization of undesired microorganisms in the upper digestive tract can be prevented.

Table 5. By-products from the food industry in Switzerland (Chaubert, 1994).

	Direct use	Feed mill	%
Milk industry	85,000	15,000	16
Oil industry	5,000	85,000	15
Saughterhouses	20,000	70,000	15
Mills	5,000	150,000	25
Bakeries and biscuit industry	7,000	6,000	2
Restaurants and hotels (refuse)	20,000	–	3
Sugar and chocolate industry	45,000	25,000	11
Coffee industry	–	2,000	<1
Breweries	20,000	–	3
Fruit and vegetable industry	20,000	40,000	10
Tofu, mustard industries, others	1,000	–	<1
Total	228,000	393,000	100

Enzymes are mainly used in monogastric species to increase nutrient availability. Carbohydrases help to alleviate negative effects of dietary fiber. Enzymes allow the use of food industry by-products or of home-grown feeds with reduced risk of digestive problems. Proteases in optimal combination with carbohydrates increase the digestion processes especially of legumes like soybeans or lupines. Yeast cultures stimulate microbial activity in ruminants and horses and therefore help to optimize the digestion processes. Improved nutrient utilization and nutrient supply is the consequence. Enterococci and lactobacilli are mainly used in pigs and poultry. They stimulate and stabilize the digestion processes and help to increase competitive exclusion of undesired microorganisms in the digestive tract.

A wide range of other substances are increasingly used to optimize digestion processes and to improve animal health. Oligosaccharides can influence the microflora by enhancing competitive exclusion or by the supply of specific nutrients. Herbs may influence feed intake through flavor, influence the digestion processes due to certain antimicrobial effects or help to reduce oxidation of unsaturated fats in the digestive tract or in intermediate metabolism.

Conclusions

Food production must increase in order to supply the growing needs of the world population, but without increasing the environmental load. Farm animals will also play an important role in the future in the production of highly valuable foodstuffs and in the optimization of the nutrient cycles.

The consumers of a modern society expect food to be free of undesired ingredients and to be produced under environmentally friendly conditions.

Today many different supplements are discussed as alternatives for substances such as antimicrobial agents or hormones. Such feed additives must be standardized in composition and well documented as to their reputed effects of better health, increased performance or better efficiency of nutrient utilization. Often the claim of 'no undesired side effects' is difficult to validate or regulate. Only well-documented feed additives manufactured to high standards can be accepted as safe for man, animal and environment.

Finally, a well-informed society is essential if the goals of optimum yield and quality of food are to be met. Therefore, an open discussion among farmers and the feed industry with the food industry and consumers is necessary.

References

Bejerholm C. and P.A. Barton-Gade. 1986. Effect of Intramuscular Fat Level on Eating Quality of Pig Meat. 32nd Europ. Meat Res. Workers, Ghent.

Chaubert C. 1994. Sous-Produits de l'Agro-Alimentaire. In Proc. Symp. on By-products of Food Industry, ETH Zürich, 13–36.

FAO. 1996. Report of the FAO – World Food Summit Conference 11. 1996, Rome.

GSF. 1996. Annual Market Report of the Swiss Meat Board 1995, Bern, p. 136.
Leonhardt M. 1996. Lean Meat (Beef, Pork, Veal and Chicken) as a Source of Trace Elements and Vitamins in Switzerland, and Efficiency of Feeding Supplemental Vitamin E and C. Diss ETH Zürich No. 11.704.
Swiss Nutrition Report. 1991. Federal Office of Health, Switzerland.

THE ROLE OF CONSUMERISM: WHY WE MUST RESPOND TO MARKET CONSIDERATIONS

WESLEY V. JAMISON

Assistant Professor and Extension Policy Specialist, University of Arkansas, Fayetteville, Arkansas, USA

Introduction

Agriculture is facing a future that is at once exhilarating and threatening. Whereas in the twentieth century agricultural production focused on the singular goals of efficiency and cost, today agriculturalists face a myriad of complex and crosscutting goals. On the one hand, world population growth insures that industrialized, high-intensity agricultural production will continue in some form in the future. On the other hand, the parameters of production seem increasingly uncertain and obtuse. Citizens in affluent regions, e.g. western Europe, North America, etc., are driven by moral, ethical and political concerns which are uncoupled from the demand for increasing agricultural productivity. Should agricultural production continue to stress productivity and efficiency, or should it become increasingly predicated upon goals such as sustainability and aesthetic virtue? Likewise, scientists charged with answering that question also face uncertainty. On the one hand, scientific/technological advances such as genetic engineering and biotechnology indicate that agriculture will be able to meet the demand for a 'politically-correct' food supply which feeds many while offending few. On the other hand consumers in affluent countries are increasingly motivated to express their political views in their buying decisions, and those political views often oppose biotechnology. A case in point is the controversy surrounding the introduction of synthetic growth hormone (bovine somatotropin) to dairy feed in the United States. Should biotechnology be utilized to meet production goals, or should it be regulated or banned in response to fear? A telling example of this convoluted and vexing dilemma is illustrated by recent protests in Europe over genetically engineered foods and feedstuffs. The environmental group Greenpeace protested against companies like Nestlé, demanding 'natural' and 'unadulterated' products.

Hence, agricultural production must coexist within a political environment where consumers are willing to express their values in both their buying choices and their votes. Therefore, the importance of understanding the social forces and motivations behind political consumption and the necessity of understanding the role of the consumer in driving production parameters is paramount and cannot be overstated. This paper will briefly explore significant demographic and socio-political trends which influence consumer decisions. In so doing, its intent is not to predict the future, but rather to allow understanding of the decision making process that modern consumers in affluent cultures utilize in making buying decisions.

Current trends

INCREASED AFFLUENCE

It has been said that environmentalism is a disease of success. To put it another way, only those cultures that reach a certain level of material affluence are able to turn their attention from subsistence to quality of life issues (Westra and Wenz, 1995). First articulated in Maslow's *Hierarchy of Need*, this concept indicates that as affluence increases, interest in things like politics and policies also increases. Although the validity of such deterministic theories has been attacked, the emergence of opposition mass movements like environmentalism and the women's movement can certainly be linked to increasing levels of affluence (Lunch, 1987). This is to say that agriculturalists face a populace that has the financial and time resources to become interested in setting the parameters of production. Likewise, the increasing affluence of many countries frees consumers from the rigors of day-to-day subsistence existence and allows them the luxury of expressing abstract political beliefs like environmentalism in their buying decisions.

BETTER INFORMED

Social science data indicate that the general public in modern countries is increasingly well informed about issues which both indirectly and directly affect them (Lunch, 1987). Traditionally, policies concerning agricultural production, biotechnology and feed additives were captured within policy sub-governments whereby experts from affected industries, legislators from the industry's district, state, or province, and regulators from the industry's associated governmental bureaucracies formed policies in a consensual fashion (Berry, 1989). For example, western ranchers in the United States, congressmen from western states, and officials from the United States Bureau of Land Management would form cattle grazing policy in a consensual fashion. This style of consensual regulatory policy setting was partly a function of the required levels of information necessary to both understand, and be interested in, esoteric policies. In other words, very few people have traditionally had adequate information to make decisions about the legalization of feed additives, thus that realm of decision making was vested to experts, scientists, and the affected industries. Nonetheless, with the advent of mass media, and television in particular, highly complex and esoteric policies have gained wide exposure to public scrutiny. Whereas historically only westerners in the United States would have been exposed to cattle grazing policy, with the advent of mass communications people around the United States now know that beef cattle range on western public lands at greatly subsidized rates. To use an example from the feed industry, the use of animal by-products as a protein source in animal feeds has been traditionally overlooked by an ignorant and uninterested public. And yet, with the advent of mass communications and mass education, 'Mad Cow Disease' and its rumored causes were instantly spread throughout the world to a consuming public who is affluent enough to forgo the consumption of beef. Thus, in response to fears over the human side effects of eating contaminated beef, meat consumption in Europe plummeted in response to information which traditionally would have only interested physicians and scientists. This information, although arguably highly simplistic and generalized, nevertheless has provided a modicum of exposure to interested consumers where none had existed before. While it can be argued that the public knows more and more about less and less, it is undeniable that policy discussions like those relating to the feed industry have been directly influenced by the ability of mass media to rapidly disseminate information about the benefits and effects. This information dissemination has served to widen the circle of parties interested in policy formulation.

HIGHLY EDUCATED

Social science data also demonstrate that consumers in affluent cultures are increasingly educated. This has accomplished at least two results: increasing comfort with complex issues, and more important, increasing confidence in the individual's ability to make rational choices without outside expertise (Lunch, 1987; Wildavsky, 1991). With the advent of mandatory education and the widespread availability of college, consumers are exposed to topics such as chemistry and environmentalism which increase their affinity for topics implicit to agricultural production. If a little knowledge is dangerous, then Americans and Europeans may be among the most dangerous people in the world. This is only partly facetious; as consumers become better educated, their ability to sift, analyze and interpret various policy options tends to increase their opposition to policies which would have traditionally been implemented without general review. Similarly, as education levels increase, so does the confidence of consumers in their own ability to choose what is right and wrong for them.

Perhaps most important, demographers indicate that as educational levels increase, citizen interest in politics tends to similarly increase, particularly in situation-specific instances where policies directly affect highly educated consumers. In other words, education is a process whereby consumers are inculcated with a rational method of determining choices. This is not to say that people make rational decisions. Instead, increased education diminishes the population's dependence upon Authority (Divine or otherwise) and Expertise to make decisions. Hence, in the case of the feed industry, consumers become less and less dependent upon 'experts,' e.g. scientists and government authorities, and more and more dependent upon their own decision making process. Thus, the increasing educational level of the public has caused an increased interest in agricultural practices within affected communities.

TECHNOLOGICALLY COMPETENT

A fourth related phenomenon which directly influences agricultural policy is the increasing technological competence of the public. Once again, this is not to say that the public is competent about specific technologies, e.g. feed additives, yucca extracts, or rendered protein sources. Rather, the public is increasingly comfortable with the idea of seeking out esoteric information with which to feed their particular biases (Florman, 1981; Lunch, 1987; Berry, 1989). Indeed, the advent of rapid-dissemination technologies such as the Internet has democratized information about areas which were previously the exclusive purview of policy wonks (Taylor, 1995). In other words, information about feed additives and agricultural production parameters has become readily available to anyone able to access the World Wide Web. Likewise, the advent of mail sent directly to consumers as an opposition group political mobilization tool has served to bring a level of technological competence within reach of the general public (Taylor, 1995; Berry, 1989). In effect, opposition groups now have the technology to take their particular version of the facts before a wider audience, and thus attempt to gain widespread support for their policy.

ATOMIZED AND INDIVIDUALISTIC

Perhaps the most profound trend is the public's increasing level of individualism and atomization (Wildavsky, 1991; Fukuyama, 1992). Traditionally, consumers vested 'experts' such as scientists with extensive implicit authority to make decisions. Based upon the concept of the social contract, consumers understood that 'authorities' and 'experts' would act in their best interest or would be punished for failing to do so (Winner, 1986). This understanding of the authority and accountability of experts has largely eroded (Wildavsky, 1991). In its place the concept of the individual as the sole source of legitimate authority has ascended to preeminence in the political ethos in westernized, industrialized countries (Fukuyama, 1992). Endless social critics have commented on the rising individualism of the public, and even more have noted the seemingly atomized nature of the polity. For example, from *The Lonely Crowd* to *Habits of the Heart*, social scientists have noted a disturbing trend among the American public away from reliance upon community-based decision making processes and toward an individualistic decision process. Indeed, evidence indicates that people are trusting less in traditional community decisions and trusting more in themselves (Wildavsky, 1991; Sacks, 1996).

To put it another way, the tradition of public consensus arising from a commonly accepted 'fair and legitimate' process has largely been displaced by a culture of complaint where any individual who feels spited by the outcome of a community decision feels empowered and entitled to seek redress of his or her grievances. Indeed, no person can be expected to be an expert on many issues. Hence, highly specialized experts such as extension specialists were traditionally vested with extraordinary trust and authority to make decisions. This included such areas as poultry production. However, consumers who are highly individualistic are less likely to place their trust in 'experts' and are less likely to trust the outcome of a communitarian, democratic policy process which negatively affects them. This means that consumers are more and more likely to express their opinions in their buying choices and in their votes.

Conclusion and recommendations

For reasons of increasing affluence, information, education, and technological competence, consumers who once were complacent in accepting centralized, authoritarian policies are placing decreasing legitimacy in decisions derived from such processes. Likewise, the single ascendant political ethos of modern culture is the individual and his perceived entitlement to participate in, and thus alter, even the most obtuse and esoteric policies. This includes agricultural production. Consumers feel entitled and empowered to express their beliefs in their buying choices. Companies and industries which overlook the newly emergent power of political consumers are setting themselves up for rapid and disastrous shifts in governmentally and consumer driven changes in production parameters. Hence, managers and scientists should anticipate greatly increased frustration as individual consumers, through use of various technologies, not only become interested in agricultural production but attempt to alter or stop specific policies. Finally, scientists and industry managers can expect two general responses from the local community in which agricultural production takes place: apathy and antipathy (Lober, 1995). Attitudinal and behavioral responses of people living in proximity to agricultural production generally differ along these two lines, and individuals should pay careful attention to the source and outlet of each set of responses.

There is much uncertainty and haze surrounding the implications of these changes for scientists and managers. How should agriculture discern the attitude of various publics who consume their product? How will scientists and managers make decisions based upon public opinion, which is often fickle and transient? The questions are vexing. What is clear, however, is that scientists and agriculturalists must pay attention to public opinion concerning production parameters. This includes the realm of agricultural production. The regulatory and marketing landscape is littered with the corpses of companies, products, and production practices which failed to account for the attitudes of an increasingly affluent and politically motivated consumer.

In some ways, agriculture has become the victim of its own success. Because of the overwhelming successes of agricultural science in increasing efficiency and lowering costs, consumers have been liberated from the toil and drudgery of eking out a living, of being held hostage to famine and drought. And yet, with this economic liberation has come political and moral liberation, whereby consumers not only find themselves discontent with the source of their success, but increasingly choose to have a direct voice in changing that source. From bovine somatotropin to 'Mad Cow Disease,' from the odor associated with animal production to the ethics of consuming meat, consumers are forcing agriculture to pay attention to their demands. An affluent, highly educated and well-informed public asks this question: 'Why should agriculture be allowed to do whatever it wants? Why shouldn't we tell agriculture what *we* want it to do?' The answers to these questions are difficult. Indeed, during the next century, responding to the market requirements that are mandated by consumers will be the central issue facing agriculture. It is not a question of what seems rational, reasonable, and scientific to the agriculturalist. Instead, consumers will decide what is rational and reasonable, often in the face of contrary scientific evidence, and it therefore becomes the task of agriculture to respond. What is clear is that the perils of changing to conform to consumer expectations will be many. Nonetheless, the rewards of doing so will be great; for that same affluent society which demands environmentally and aesthetically pleasing production also has the money to pay a premium for those changes.

References

Berry, J. 1989. The Interest Group Society. Tufts University Press, Boston, MA.

Florman, S. 1981. Blaming Technology: The Irrational Search for Scapegoats. St. Martin's Press. New York.

Fukuyama, F. 1992. The End of History and the Last Man. Free Press, New York.

Lober, D. 1995. Why protest? Public behavioral and attitudinal responses to siting a waste disposal facility. Policy Studies Journal. V34, 3:499–518.

Lunch, W. 1987. The Nationalization of American Politics. The University of California Press, Berkeley, CA.

Sacks, S. 1996. Generation X Goes to College. The University of California Press, Berkeley, CA.

Taylor, B. 1995. Ecological Resistance Movements: The Global Emergence of Radical and Popular Environmentalism. State University of New York, Albany, NY.

Westra, L. and P. Wenz. 1995. Faces of Environmental Racism: Confronting Issues of Global Justice. Rowman & Littlefield, Boston, MA.

Wildavsky, A. 1991. The Rise of Radical Egalitarianism. Torch Books, San Francisco, CA.

Winner, L. 1986. The Whale and the Reactor: A Search for Limits in an age of High Technology. University of Chicago Press, Chicago, IL.

BOVINE SPONGIFORM ENCEPHALOPATHY: IMPLICATIONS FOR ANIMAL AGRICULTURE

D.W. HARLAN[1] and S.L. WOODGATE[2]

[1]*Taylor By-Products, Wyalusing, Pennsylvania, USA*
[2]*Beacon Research Limited, Clipston, Leics, UK*

Introduction

Bovine spongiform encephalopathy (BSE) was first identified in 1986 in Great Britain (Wells *et al.*, 1987). Ten years later, the discovery of a new variant form of Creutzfeldt-Jakob disease (nv-CJD), a related disease in humans, was postulated to be causally linked to BSE exposure in the UK (Will *et al.*, 1996). Further research (Collinge *et al.*, 1996) on the molecular similarities between nv-CJD and BSE strongly supports the link. The results of more definitive strain typing experiments are expected to be announced in 1997 (Taylor, personal communication). The serious concerns regarding this complex issue are frequently complicated by political, industry, trade and consumer issues. While providing a brief history of the BSE outbreak and global regulatory actions, the specific focus of this paper will be to discuss methods by which animal agriculture may deal with potential feed-borne transmissible spongiform encephalopathy risks to animal health in the future.

Overview of bovine spongiform encephalopathy

BSE belongs to a group of complex diseases known as transmissible spongiform encephalopathies (TSE). Known to occur in various animal species and man, TSEs are chronic degenerative diseases of the central nervous system. The incubation periods for TSEs are generally long with a median of five years for BSE (Wilesmith *et al.*, 1992a). The development of a reliable preclinical diagnostic test is hindered due to the lack of an immune response. Currently, definitive clinical diagnosis can only be determined by histopathological examination of brain tissue. While several theories exist, the infectious agent has not been identified (Prusiner, 1982; Gajdusek, 1996; Lasmezas *et al.*, 1997; Bastian, 1993). In addition, TSE agents are resistant to complete sterilization by normal procedures. The fact that many aspects of these perplexing diseases are still unclear impedes the development of clear solutions by animal health authorities and by the animal agricultural community as a whole.

OCCURRENCE IN THE UK

In November of 1986 the Central Veterinary Laboratory in Weybridge, England identified BSE as a new disease in cattle (Wells *et al.*, 1987). By late 1987 epidemiological studies (Wilesmith *et al.*, 1988) concluded that meat and bone meal (MBM) contaminated with a scrapie-like agent was the only viable hypothesis explaining the emergence of BSE. In the UK, MBM had been used extensively in ruminant rations since the early 1940s. The relatively abrupt cessation, around 1980, of the solvent extraction process used in the production of MBM was suggested as the principal factor that allowed for ample survival of the causative agent (Wilesmith *et al.*, 1991). The initial cases of BSE are thought to be due to sheep scrapie exposure. However, existence of a previously undetected TSE in cattle was not ruled out as the source. Once established, the recycling of BSE-contaminated bovine material clearly drove the outbreak (Wilesmith *et al.*, 1992b).

Ruminant MBM was banned for use in ruminant feed as a control measure in 1988 (HMSO, 1988). However, 'leakage' of this product into cattle rations occurred well into the early 1990s until further controls were implemented. By late 1989, specified bovine offals (SBO) were banned for human consumption primarily due to concerns with baby food. The SBO tissues were selected based upon distribution of scrapie infection in sheep. Less than a year latter SBOs were banned in all animal feeds due to experimental transmission of BSE to swine (Dawson *et al.*, 1990). By July of 1993 over 100,000 cases of BSE had been confirmed. The effectiveness of the 1988 feed ban became apparent in 1994 when actual BSE cases declined (Figure 1). However, a significant number of cattle born after the 1988 feed ban continued to succumb to BSE (Figure 2) . These 'born after ban' animals can be attributed to incomplete compliance with feed regulations and possibly maternal association.

Currently, BSE has been diagnosed in over 165,000 head of UK cattle from over 31,000 herds. About 60% of UK dairy herds and 15% of beef herds experienced at least one clinical case. The eradication of BSE from the UK cattle herd is projected to occur soon after the year 2000.

BSE OUTSIDE THE UK

The potential for BSE to spread outside the UK due to exports of live cattle and contaminated MBM was extensive. During the period between 1985 and 1990 over a half million breeding cattle were exported from the UK. An estimated 1700 cases of BSE would have been confirmed from these exported animals if they had remained in the UK (MAFF, 1996a). In addition, exportation of MBM by the UK during the 1980s has been attributed to some cases outside the UK. Curiously, very limited numbers of BSE cases have been reported relative to expectations. Most reported cases are located in Europe where BSE exposure was highest.

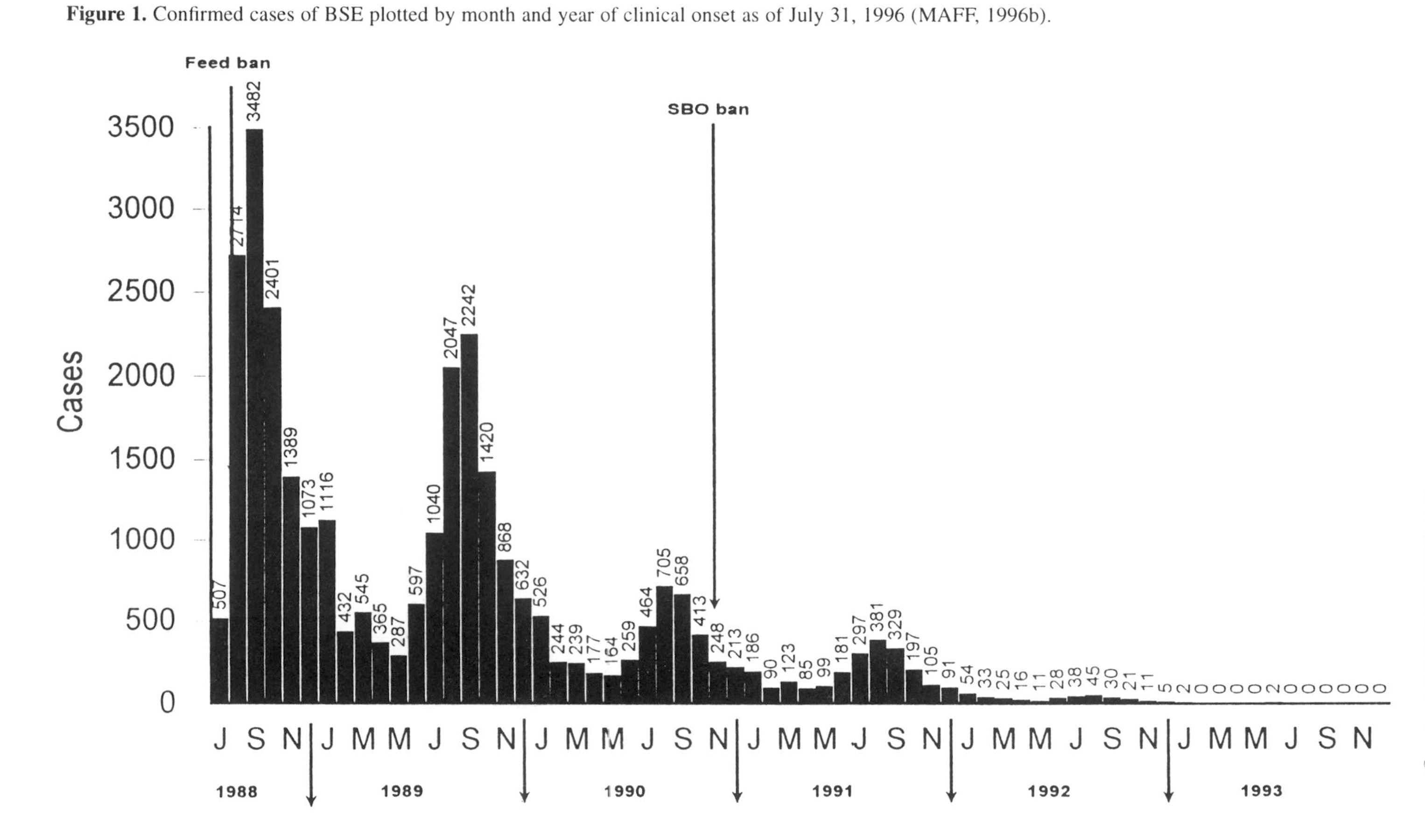

Figure 1. Confirmed cases of BSE plotted by month and year of clinical onset as of July 31, 1996 (MAFF, 1996b).

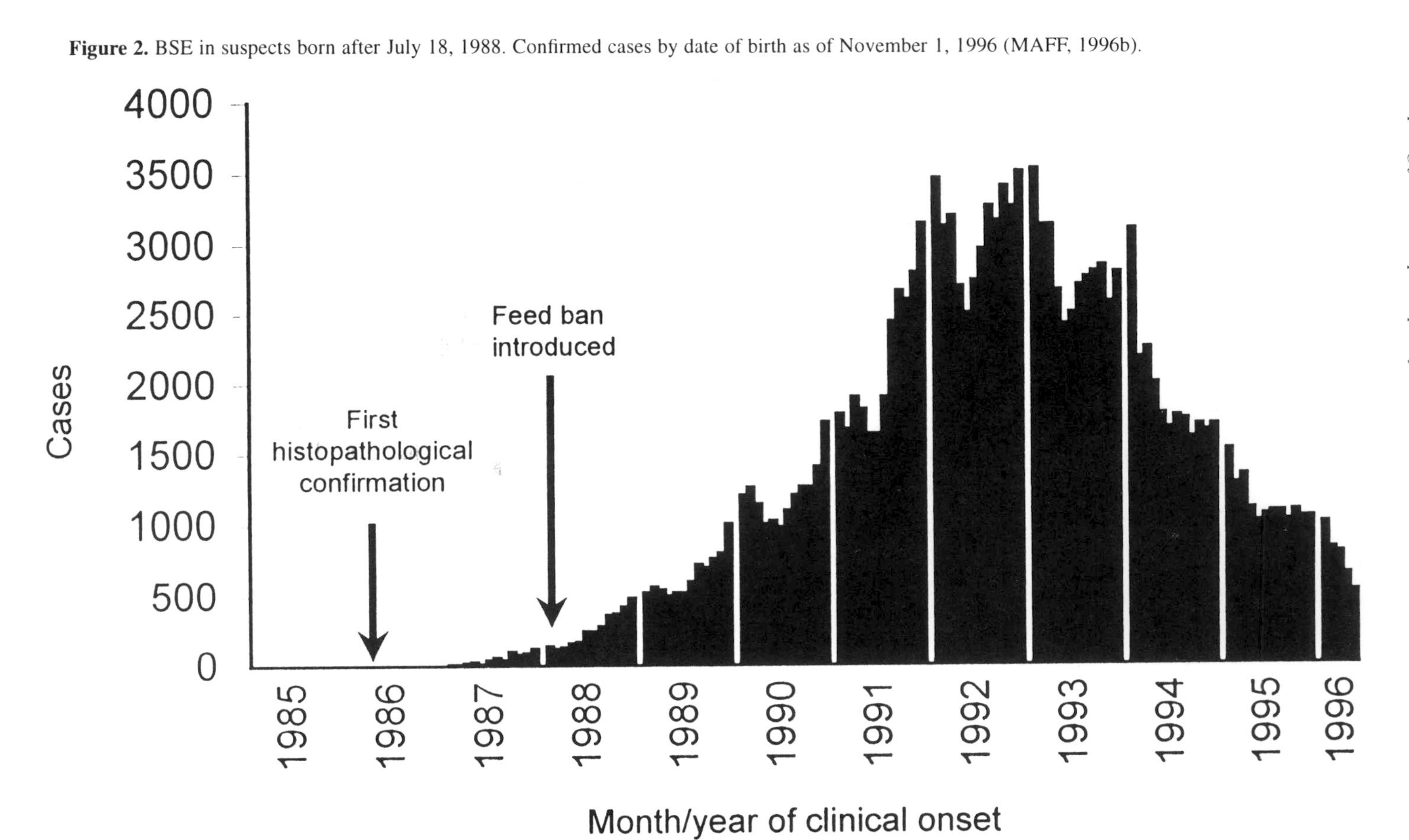

Figure 2. BSE in suspects born after July 18, 1988. Confirmed cases by date of birth as of November 1, 1996 (MAFF, 1996b).

EUROPEAN UNION DIRECTIVES ON RENDERING

With the completion of the Common Market drawing near, the European Economic Commission (EEC) started to establish veterinary checks for intra-Community trade of animals and animal products in 1989. At first the directives were developed with routine animal health issues in mind. However, by the early 1990s directives relative to TSEs were passed.

In 1990, rules for the disposal and processing of animal waste were set in directive 90/667/EEC. This directive established the classification of animal waste tissues into two categories; high and low risk. Low risk material must be fresh and from animals which pass veterinary ante-mortem inspection and are subsequently presented for post-mortem inspection at slaughter. Since low risk material is basically from animals 'fit for human consumption', the EU did not consider these materials to present serious risks of spreading communicable diseases to animals or humans. High risk material contains any other tissues, all or in part, such as those from dead and (or) diseased animals. High risk material was to be processed at 133°C for 20 minutes at 3 bar pressure with a maximum particle size of 50 mm. Alternative systems for processing high risk materials were permitted, but specific processing conditions were not defined. Processing conditions for low risk material were not specified. However, microbiological and facility hygiene standards were instituted for material from both risk groups.

In November of 1992, the Commission described the details of specific alternative processing systems for high risk materials (92/562/EEC). Hazard analysis and critical control point (HACCP) programs were established for seven alternative systems. Microbiological and hygiene standards from the previous rule were maintained. This allowed for the continued use of various systems which were operating throughout the EU. Only Germany and the Netherlands routinely operated pressure cooking systems described in the 1990 directive.

When a BSE pilot scale rendering deactivation study was completed in 1994, two of the seven alternative processing systems approved in 1992 were found to be either totally or partially ineffective against the BSE agent (Taylor *et al.*, 1995). The EU promptly prohibited (94/382/EEC) one system and mandated that both high and low risk ruminant material be processed by any of the five remaining alternative systems (including the pressure cooking process) to validated processing conditions. Low risk material for pet food use, gelatin, blood, milk and rendered fats were exempt. In light of limitations on the sensitivity of the deactivation study and the spread of BSE within the EU, the feeding of mammalian proteins to ruminant animals was prohibited (94/381/EEC). However, member states such as Denmark which could produce pure non-ruminant mammalian proteins (porcine) were permitted to take a ruminant protein-to-ruminant animal prohibition approach.

In July of 1996, soon after the results of the scrapie deactivation study were reported to the EU, pressure cooking systems were mandated for all mammalian waste (96/449/EC) (Taylor *et al.*, 1996). Exemptions similar to the previous directive were allowed. Full implementation of this directive occurred on April 1, 1997. Additional risk controls such as the mandated removal of specified risk tissues of bovine and ovine origin from the food chain were being discussed within the EU Scientific Veterinary Committee at publication time.

WORLD HEALTH ORGANIZATION

In April of 1996 the World Health Organization (WHO) met to review the possible human health implications of TSEs, with special emphasis on BSE. In light of the probable link between BSE and the 10 reported cases of nv-CJD, the WHO provided the following recommendations (WHO, 1996):

- No part of any animal which has shown signs of TSE should enter any food chain, human or animal.
- All countries establish continuous surveillance and compulsory notification for BSE as established by Office International Des Epizooties (OIE).
- Countries where BSE exists in native cattle should not permit tissues that are likely to contain the BSE agent to enter any food chain, human and animal.
- All countries should ban the use of ruminant tissues in ruminant feed.
- Milk, gelatin and tallow are considered safe.
- Previous measures to minimize the risk of transmitting the BSE agent in pharmaceuticals are still applicable.
- Research of TSE should be promoted.

WORLD ORGANIZATION FOR ANIMAL HEALTH

The OIE functions as the leading international animal health authority which advises the veterinary services of 124 member nations. In May of 1996 a chapter of the *International Animal Health Code* was revised concerning BSE (OIE, 1996). Changes allowed for the unrestricted trading of milk and milk products, hides and semen from healthy cattle and properly processed by-products such as gelatin, collagen and tallow from countries where BSE has occurred. Import restrictions for cattle from countries with BSE were also set.

The code defines the specific criteria for countries to be categorized as either 'free of BSE', 'low incidence of BSE' or 'high incidence of BSE'. In order to be considered free of BSE, countries must have an effective and continuous surveillance program in place and make BSE a notifiable disease. Cases of BSE detected must have originated from the importation of live cattle, otherwise the country is reclassified as low incidence. Low incidence countries must have additional control measures in place such as a ruminant meat and bone meal to ruminant feed ban in order to export cattle. More stringent requirements for the UK, the only high incidence country, were set for export of cattle and beef.

US PREVENTION MEASURES

USDA

The United States Department of Agriculture (USDA) has restricted the importation of live ruminants and ruminant products from countries where BSE is known to exist. Between 1981 and 1989, 496 cattle were imported from the UK. All but 33 have been accounted for with no reported signs of BSE. In addition, the USDA continues to conduct an ongoing surveillance program which as of January 23, 1997 had examined 5342 brains of suspect cattle with no BSE detected (Detwiler, personal communication). Surveillance figures are used to periodically update government risk assessments. The first series of USDA risk assessments, released in 1991 (USDA, 1991a, b), concluded that there was 'little evidence to support a broad risk of BSE among a large portion of the dairy population'. The most current USDA risk assessment update (USDA, 1996a) concluded that 'the overall risk of BSE in the U.S. has decreased'. However, these assessments were based on the now-questioned assumption that domestic sheep scrapie is the only risk factor for BSE in the US.

The USDA has developed a BSE response plan in the event a case is ever detected in the US (USDA, 1996b). This detailed plan outlines the exact steps to be taken in order to effectively handle the situation. The three hypothetical situations that have been prepared for are:

- BSE in an imported British cow.
- BSE in a domestic cow.
- BSE in progeny of an imported British cow.

FDA

In 1994 the Food and Drug Administration (FDA) published a notice of proposed rule-making (NPR) which would have declared specific tissues of adult sheep and goats to be 'not generally recognized as safe' for use in ruminant feed. The proposal, opposed by industry groups, was not finalized. After the announced link between BSE and nv-CJD in 1996, the FDA published an advance NPR to solicit comments on the issue of using ruminant proteins in ruminant feed. In early 1997 an NPR was published which proposed to prohibit the use of ruminant and mink tissues in ruminant feed. Exemptions for milk, blood and gelatin products were proposed. The FDA stated that the intent of the proposed rule was 'to prevent the establishment and amplification of BSE in cattle in the United States, and thereby minimize any risk which might be faced by animals and humans'. Mandatory labeling and retention of paperwork from renderer to animal producer was proposed as the enforcement method. Procedures were proposed in order to prevent cross contamination of ruminant proteins into ruminant feeds.

The 1997 NPR contains an extensive review of the literature justifying FDA's action. Three potential TSE sources to cattle were identified as:

- BSE from cattle or sheep imported from the UK.
- TSEs found in US ruminants and mink.
- Spontaneous TSE of cattle.

As previously mentioned, 33 cattle imported from the UK have not been accounted for. In addition, prior to 1996 imported cattle were not restricted from entering the food cycle. While no clinical symptoms were observed in these imported cattle, the occurrence of pre-clinical cases capable of transmitting the disease cannot be ruled out. Also, recent work demonstrated the potential to transmit BSE to sheep by the oral route (Foster *et al.*, 1996). The pathogenesis of the disease and tissue distribution of the infection were not distinguishable from natural scrapie, but strain typing clearly identified the agent as BSE. This has led to concerns that BSE may have infected the UK sheep population and is now a potential threat to animal and human health. Thus, sheep imported from the UK pose a theoretical risk.

Several animal species which have been diagnosed with TSEs in the US are a potential source for BSE. Sheep scrapie is known to occur at a low level and chronic wasting disease of mule deer and elk is endemic in a small area of Colorado and Wyoming (Williams and Young, 1993). Transmissible mink encephalopathy has not been reported since 1985, although the FDA has included mink tissues in the proposed prohibition. The occurrence of BSE, among other TSEs, has been postulated to occur spontaneously in about one in a million adult cattle (Gajdusek, 1996; Gibbs, 1996). The FDA points to the occurrence of BSE in some domestic cattle found in Switzerland and Northern Ireland in which the animals apparently had no direct or indirect exposure to the BSE agent. The fact that sporadic CJD occurs at a very constant annual rate of one in a million worldwide is supportive of this hypothesis.

The FDA did not rule out the possibility of taking other actions and included six alternatives to the proposed rule in the NPR:

1. Partial ruminant to ruminant feed prohibition.
2. Mammalian to ruminant feed prohibition.
3. US species with known TSE to ruminant feed prohibition.
4. Sheep and goat tissue rule as proposed in 1994.
5. Other alternative approaches to achieve acceptable risk reduction.
6. No action.

It is clear from the NPR that the FDA does not consider the third, fourth and sixth options as viable. The first option deals with the feasibility of removing only the specific tissues which have been found to present a risk of transmitting TSE. This option is not viable for sheep and goats due to the broad tissue distribution of the scrapie agent (Pattison and Millson, 1962; Hadlow *et al.*, 1982). However, this approach is potentially feasible for bovine material due to the limited distribution of the tissues which are known to contain the BSE agent (Fraser and Foster, 1994). Another approach would be to take a mammalian to ruminant feed prohibition approach which is suggested to be the simplest from an enforcement standpoint. However, there are no scientific data which suggest any risk from swine and the FDA is aware that some facilities currently process only porcine material. In addition, the NPR specifically requested comments on other alternative approaches which would achieve the agency's regulatory objectives.

Prevention of feed-borne TSE

Extensive steps to protect human and animal health are well advised in regions where BSE exposure has occurred. In the UK complete disposal and incineration of at-risk animals has occurred in order to provide near absolute assurances to the public. The UK renderer is no longer a recycler of valuable nutrients, they are part of a disposal industry. Clearly the current UK situation is neither sustainable nor nutrient efficient, but necessary until BSE is eradicated. After eradication it is unlikely that MBM will be allowed as a feed ingredient due to consumer concerns. Future uses for this product will be highly dependent upon the consequent outcome of nv-CJD.

Preventative approaches need not mimic disease eradication measures. However, the dilemma presented by TSEs is in part due to the very long incubation period measured in years, frequently decades for the related human disorders. Even with the best surveillance programs, the course of the disease can be well established prior to the first case being confirmed. Clearly, waiting for the first case to take preventative action is not appropriate. The development, validation and adoption of systems which prevent the amplification of potential feed borne TSEs are crucial.

MATERIAL CONTROLS

Risk of obtaining TSE infection in materials processed by a rendering facility can be reduced through several methods. Divergent raw material sources have the potential to contain different levels of TSE risk. Material from species which have been diagnosed with a TSE within a region are of a higher risk than material from species with no detected TSE. While purely qualitative, assurances that these known species are avoided or otherwise managed are important.

Sources of animal material may be categorized into several classes with different potentials for TSE risk. These classes relate to the status of the animals which make up the pool of material to be processed. On-farm fallen or in-transit fallen animals passing ante-mortem veterinary inspection (EU low risk) and age are some category examples. Material from calves would obviously have a very low TSE risk while mature cattle showing clinical signs of an unknown neurological disorder are much higher. Pre-clinical TSE animals will not be avoided using these strategies, although the quantity of infection they contain is lower. By managing the source of material received in this manner, individual facilities can qualitatively lower, but not eliminate, potential TSE risk.

The removal and diversion of specific tissues which are known to carry TSE infection is a useful risk reduction method which is easily quantified. This approach is mandated by OIE in countries which have BSE. Unlike sheep scrapie, detectable BSE infection is restricted to central nervous system (CNS) tissues and the retina. However, the distal ileum has been demonstrated to be capable of harboring the BSE agent under relatively high oral dose conditions (MAFF, 1995). The relative titer of infection found under these experimental conditions was significantly lower than that of CNS tissue. Due to the sensitivity limits of the mouse bioassay used to quantify infection, trace infection of non-CNS tissues on the order of 10^{-5} relative to CNS levels cannot be ruled out. Trace infection in the lymphatic system is assumed to occur as this system is suspected to play a role in establishing the initial infection.

PROCESS CONTROLS

The reduction of TSE infection by commercial processing methods is an integral part of any risk reduction strategy. However, there are several experimental limitations in full reliance on processing steps alone. It is difficult to prove absolute deactivation due to inherent limitations in existing TSE detection methodologies. Currently, only a live animal bioassay can detect the presence of TSE infection. Infection level is reported not in numbers of TSE agents, but in ID_{50} units; number of 'infective doses' in which 50% of animals succumb. Lack of detection does not necessarily imply no agent is present, only that the level and or strain of dose administered is insufficient to demonstrate the disease in the animal model used. The real world model (demonstrated in the UK epidemic) with no species barrier, realistic infection levels and a near infinite test population size is not a practical approach to future research.

Both BSE and scrapie deactivation studies have been completed in the UK (Taylor *et al.*, 1995, 1996). Both studies tested conditions which approximated various processes found in the EU. Pilot scale rendering equipment was used to mimic operating parameters found both within and among the different systems. Temperature and time profiles, maximum particle size and fat recycle parameters were measured. The processing systems can be categorized into six general methods:

- Batch under atmospheric conditions.
- Batch under pressure.
- Continuous flow under atmospheric conditions with no added fat.
- Continuous flow under atmospheric conditions with added fat.
- Continuous flow under vacuum conditions with added fat.
- Continuous flow wet rendering under atmospheric conditions.

Measurement of the retention time in a batch process is relatively straightforward. However, not unlike feed passage in the rumen, material entering a continuous rendering system will flow at different rates based on many factors such as particle size, density, mixture of material processed and quantity of added fat. Rare earth marker studies can be used to quantify the retention curve and then calculate minimum and average time and temperature conditions (Woodgate, 1991). These studies independently tested the minimum and average retention times for the continuous systems.

In the studies, pooled brain homogenate was used to spike normal raw materials at a 10% brain inclusion level. Final raw material infectivities of $10^{1.7}$ and $10^{3.5}$ ID_{50} were achieved for the BSE and scrapie studies, respectively. The unexpectedly low infection of the BSE material was partially due to the removal of specific brain sections for histopathology which are proportionally higher in infection. The detection capabilities of the mouse bioassay used to quantify infection were about 10^{-2} for BSE and 10^{-3} for scrapie compared to the input levels.

As previously mentioned in the EU directives section, two of the seven systems tested in the BSE study produced material with detectable infection. The conditions found in a continuous flow system under vacuum with added fat were found to be completely ineffective in reducing BSE dose. In addition, the continuous flow system at atmospheric pressure with no added fat failed under minimum operating conditions but passed under average processing conditions. This latter system was upgraded while the former was prohibited. The scrapie study found infection in all systems except for those operating under pressure cooking conditions. The systems which had previously passed in the BSE study, but failed for scrapie were found to reduce the scrapie dose by two orders of magnitude. A resulting scrapie positive MBM was used to evaluate the various processing conditions encountered in the now-discontinued solvent extraction process. Infection in the extracted MBM was detected leading the researchers to conclude that these processes, either individually or in combination, appeared to provide no more than a small level of deactivation (Taylor, 1995). These results do not necessarily dispute the earlier epidemiological findings, rather they suggest that only a relatively small 'barrier' needed to be removed in order for clinical BSE to emerge in the UK.

Research is underway in the Netherlands to specifically assess the efficacy of Dutch pressure systems (Schreuder *et al.*, unpublished). The researchers documented the actual process conditions achieved at the two domestic rendering facilities. A laboratory processing protocol which can be statistically applied to these commercial systems was developed. In order to obtain a higher 'spike' of infection, brainstems in place of whole brain are being used at 10% of the raw material mix. This step should enhance the detection capability of the study by almost one order of magnitude over the UK scrapie deactivation study.

TRANSMISSION CONTROLS

Restrictions on the use of a TSE-sensitive animal protein can take the form of a 'species barrier' or age constraint. Ruminant and mammalian feed prohibitions reduce TSE risk by diverting these sensitive products to species which are not known to be at risk. One concern, especially for potentially non-emerged TSEs, is that the species barrier is difficult to predict in advance. This point was made clear in the US FDA's 1997 NPR which stated 'transmission between any two species is difficult to predict based on available data because of variability in species barriers'. In retrospect, it is now evident that species barriers should not be relied upon as the only control measure especially during TSE epidemics.

One high risk practice, associated with BSE in the UK, was the feeding of MBM to young calves. A study by Wilesmith and colleagues (1992c) suggested that a high proportion of all BSE-affected cattle were infected as calves during the first six months of life. Avoidance of this apparent age susceptibility would substantially reduce risk. In addition, feeding rates of a TSE-sensitive ingredient would be directly related to potential level of exposure.

Evaluating preventative approaches

Hueston (1996) suggested a systems approach to reducing overall TSE risk from the recycling of animal derived nutrients. The concept being that overall risk is a product of risk from inputs, processes and transmission. By reducing risk in one or more of these segments, overall risk is reduced. Overlapping safety is achieved when very low risk is concurrently obtained in each segment. This concept has been further advanced with the development of a basic TSE risk reduction model (Harlan, 1996).

MODEL CONCEPT

In modeling risk reduction one needs to first establish the relative probabilities of receiving material from a TSE-affected animal into a rendering stream. The maximum number of TSE-affected animals entering the rendering stream can be calculated from national TSE surveillance figures, animal population and cull statistics and preventative practices employed. The number of affected animals can be further divided into the various raw material sources entering the rendering stream. Assumptions on the effectiveness of controls, such as ante-mortem veterinary inspection, can be used to evaluate the associated risk reduction or lack thereof.

While appropriate, procedures to prevent the entrance of TSE-affected animal material into the feed cycle are not absolute. A TSE dose calculation model is a useful tool to evaluate risk reduction strategies. A dynamic approach takes into account possible tissue removal practices, deactivation potentials, feed mixing and distribution patterns and animal feeding rates.

Total amount of TSE infection can be quantified by accounting for individual tissue weights and associated level of infection. Trace infection titers can be assigned for tissues with no detectable infection. Removal practices can be independently applied to the various tissues while accounting for incomplete removal. In addition, a factor can be applied to demonstrate resulting dose if multiple TSE animals were to enter a facility at the same time. The entrance of multiple animals would only be expected in regions with known TSEs. Deactivation factors, obtained from research, are applied based upon various process conditions. As research into TSE agent deactivation advances, a more dynamic deactivation sub-model can be incorporated.

During and after processing, a TSE dose is dispersed within a larger volume of MBM based upon individual equipment distribution patterns which can be quantified using rare earth markers. If a risk reduction approach were to be taken, finished product mixing is suggested to be an important factor in producing homogeneous non-infective doses (Bradley and Taylor, personal communication). Daily and cumulative residual TSE dose consumed by an individual animal with the highest intake of MBM in a herd is calculated based upon animal and average MBM feeding rates, herd size and quantity of feed or ingredient purchased.

MODEL RESULTS

Evaluation of the historical changes in the UK which led to the emergence of BSE is important in order to evaluate future risk. Prior to the 1970s high temperature rendering with solvent extraction was the primary process. During the 1970s continuous systems, predominately low temperature, replaced many older facilities as the rendering industry consolidated. Then, rather abruptly, around 1980 the solvent extraction process was largely discontinued. For simplicity, daily oral doses are discussed in terms of 'equivalent brain infection' (EBI) which is defined as mg of brain at 10^5 ID_{50} units. Maximum mature dairy cattle doses for the time periods of 1940 to 1970, 1970 to 1980 and 1980 to 1988 are calculated to be 2, 168, 532 mg EBI, respectively. Doses for calves consuming MBM are approximately one order of magnitude lower than mature cattle. Oral dose studies, in process, have demonstrated that a dose as low as 1 g of raw brain to calves can cause BSE, although only few of the animals have succumbed at this time (Bradley, personal communication).

When evaluating existing US rendering systems, potential doses of 3 and 266 mg EBI are demonstrated by the model for high and low temperature systems, respectively. Several strategies to reduce potential TSE dose to very low levels are proposed. The adoption of pressure cooking systems would reduce the dose to a maximum of 0.3 mg EBI. Another strategy would be to remove the brain, spinal cord and eyes and process the remaining material in a validated high temperature rendering process. This approach would reduce potential TSE dose to around 0.2 mg EBI. Absolute removal of these tissues would not be required, allowable tissue residues could remain making this a practical approach for packer-based renderers. Proof of safety cannot be stated and the authors do not imply the sole use of these strategies for eradication purposes. However, these alternative strategies appear to provide a high margin of safety for preventative TSE purposes when incorporated into a total risk reduction program.

Conclusions

The emergence of BSE and subsequent strong linkage to nv-CJD has had global implications to animal agriculture. BSE may very well have been just a UK phenomenon with multiple high risk factors merging at the same time. However, with this new knowledge in hand, animal agriculture must proceed by providing very high assurances that our animals and their products are safe. Much research is needed to enhance our understanding of these complex and intriguing diseases. While science can never provide absolute assurances, significant steps can be taken to prevent future occurrences of feed-borne TSEs while maintaining an efficient agricultural system.

References

Bastian, F.O. 1993. Bovine spongiform encephalopathy: relationship to human disease and nature of the agent. ASM News 59:235.

Collinge, J., K.C.L. Sidle, J. Meads, J. Ironside and A. Hill. 1996. Molecular analysis of prion strain variation and the aetiology of 'new variant' CJD. Nature 383:685.

Dawson, M., G.A.H. Wells, B.N.J. Parker and A.C. Scott. 1990. Primary parenteral transmission of bovine spongiform encephalopathy to the pig. Vet. Rec. 127:338.

Foster, J.D., M. Bruce, I. McConnell, A. Chree and H. Fraser. 1996. Detection of BSE infectivity in brain and spleen of experimentally infected sheep. Vet. Rec. 138:546.

Fraser, H. and J.D. Foster. 1994. Transmission to mice, sheep and goats and bioassay of bovine tissues. In: Transmissible Spongiform Encephalopathy. (R. Bradley and B. Marchant, eds.) Commission of the European Communities, Brussels, p.145.

Gajdusek, D.C. 1996. Infectious amyloids: subacute spongiform encephalopathies as transmissible cerebral amyloidoses. In: Fields Virology. 3rd ed. Lippencott-Raven, Philadelphia, p. 2851.

Gibbs, C.J. 1996. Debate: Is BSE endemic? Presentation at: International Symposium on Spongiform Encephalopathies. Georgetown University, 12–13 Dec.

Hadlow, W.J., R.C. Kennedy and R.E. Race. 1982. Natural infection of suffolk sheep with scrapie virus. J. Infectious Dis. 146:657.

Harlan, D.W. 1996. A preventative approach instead of a ban? Presentation at: International Symposium on Spongiform Encephalopathies. Georgetown University, 12–13 Dec.

HMSO. 1988. The bovine spongiform encephalopathy order. Statutory Instrument 1988, No. 1039. London.

Hueston, W. 1996. Overview of risk management applied to TSE. Presentation at: Tissue Distribution, Inactivation and Transmission of Transmissible Spongiform Encephalopathies of Animals. USDA:APHIS and FDA-CVM, Riverdale, Maryland USA, 13–14 May.

Lasmezas, C.L., J.P. Deslys, O. Robain, A. Jaegly, V. Beringue, J.M. Peyrin, J.G. Fournier, J.J. Hauw, J. Rossier and D. Dormont. 1997. Transmission of the BSE agent to mice in the absence of detectable abnormal prion protein. Science 275:402.

MAFF. 1995. Bovine spongiform encephalopathy: a progress report, November 1995.

MAFF. 1996a. Bovine spongiform encephalopathy: a progress report, May 1996.

MAFF. 1996b. Bovine spongiform encephalomalacia: a progress report, November, 1996.

OIE. 1996. New recommendations on bovine spongiform encephalopathy. Press release, 28 May 1996.

Pattison, I.H. and G.C. Millson. 1962. Distribution of the scrapie agent in the tissues of experimentally inoculated goats. J. Comp. Path. 72:233.

Prusiner, S.B. 1982. Novel proteinaceous infectious particles cause scrapie. Science 216:136.
Schreuder, B.E.C., J.A.A.M. van Asten, L.J.M. van Keulen, P. Enthoven and A.D.M.E. Osterhaus. Assessment of the efficacy of the rendering procedures in use in the Netherlands with regard to the inactivation of scrapie and BSE agents. Unpublished.
Taylor, D.M. 1995. Research reports: decontamination and disinfection; transmissible spongiform encephalopathies: rendering methods. Institute for Animal Health, Edinburgh, Scotland, UK, p.140.
Taylor, D.M., S.L. Woodgate and M.J. Atkinson. 1995. Inactivation of the bovine spongiform encephalopathy agent by rendering procedures. Vet. Rec. 137:605.
Taylor, D.M., S.L. Woodgate and A.J. Fleetwood. 1996. Scrapie agent survives exposure to rendering procedures. Assoc. of Veterinary Teachers and Research Workers, July Scarborough Meeting. (Abstract).
USDA. 1991a. Qualitative analysis of BSE risk factors in the United States.
USDA. 1991b. Quantitative risk assessment of BSE in the United States.
USDA. 1996a. Bovine spongiform encephalopathy: Implications for the United States, A follow up.
USDA. 1996b. USDA bovine spongiform encephalopathy response plan.
Wells, G.A.H., A.C. Scott, C.T. Johnson, R.F. Gunning, R.D. Hancock, M. Jeffrey, M. Dawson and R. Bradley. 1987. A novel progressive spongiform encephalopathy in cattle. Vet. Rec. 121:419.
WHO. 1996. Report of a WHO consultation on public issues related to human and animal transmissible spongiform encephalopathies, with participation of FAO and OIE. Geneva, Switzerland.
Wilesmith, J.W., G.A.H. Wells, M.P. Cranwell and J.M.B. Ryan. 1988. Bovine spongiform encephalopathy: epidemiological studies. Vet. Rec. 123:638.
Wilesmith, J.W., J.M.B. Ryan and M.J. Atkinson. 1991. Bovine spongiform encephalopathy: epidemiological studies on the origin. Vet. Rec. 128:199.
Wilesmith, J.W. and J.M.B. Ryan and M.J. Atkinson. 1992a. Bovine spongiform encephalopathy: recent observations on the age-specific incidences. Vet. Rec. 130:491.
Wilesmith, J.W., J.M.B. Ryan, W.D. Hueston and L.J. Hoinville. 1992b. Bovine spongiform encephalopathy: epidemiological features 1985 to 1990. Vet. Rec. 130:90.
Wilesmith, J.W., J.M.B. Ryan and W.D. Hueston. 1992c. Bovine spongiform encephalopathy: case-control studies of calf feeding practices and meat and bone inclusion in proprietary concentrates. Res. Vet. Sci. 52:325.
Will, R.G., J.W. Ironside, M. Zeidler, S.N. Cousens, K. Alperovitch, S. Poser, M. Pocchiari, A. Hofman and P.G. Smith. 1996. A new varient of Creutzfeldt-Jakob disease in the UK. Lancet 347:921.
Williams, E.S. and S. Young. 1993. Neuropathology of chronic wasting disease of mule deer and elk. Vet. Path. 30:36.
Woodgate, S.L. 1991. Pilot plant studies in BSE/scrapie deactivation. In: Subacute Spongiform Encephalopathies. R.Bradley, M. Savey and B. Marchant (eds.). Dordrecht, Kluwer, p.169.

OVERVIEW OF POULTRY MEAT INDUSTRY GROWTH AND FEED INGREDIENT DEMAND BEYOND 2000 IN THE ASIA PACIFIC REGION

STEPHEN BOURNE

Alltech Inc., Wrexham Industrial Estate, Wrexham, UK

Introduction

The poultry industry in the Asia Pacific Region comprises both locally produced and imported poultry meat and based on consumption trends is growing at a rate of over 10% per year (Henry and Rothwell, 1995). This compares to a world average annual growth of 5% and in large part is due to increasing population and per capita disposable income.

With the Asia Pacific population due to increase over the next 15 years by 788 million (Watt Poultry Statistical Yearbook, 1996) and per capita consumption of poultry meat today averaging 4 kg (Henry and Rothwell, 1995), this equates to an increased demand of 3.1 million tonnes of poultry meat with total demand being around 16.7 million tonnes. Given that an 'average broiler chicken' in Asia may convert feed at 2.0 kg for each 1 kg of body weight and has a 70% dressing percentage, an increased supply of 8.8 million tonnes of feed ingredients are required to meet demand with a total feed demand of 47.7 million tonnes in 2010. Should the level of poultry consumption per capita reach the world average of 9 kg in 2010 (Watt Poultry Statistical Yearbook, 1996), then in order for production to meet demand a further 60 million tonnes of feed would be required. A total of 185 million tonnes would be required if per capita consumption reaches the forecast figure of 15.5 kg.

The questions to be addressed are, 'Will the Asia Pacific Region attain the forecasted per capita consumption figure? If so, how can demand for feed ingredients be met when the region today includes 60% of the world population, 43% of the world's livestock units on only 33% of the world's arable land (Poultry International, 1996)?'

Poultry beyond 2000

World compound feed production in 1995 was approximately 550 million tonnes (Figure 1) of which the Asia Pacific Region amounted to 130 million tonnes (Table 1).

Table 1. Percentage of world and Asia Pacific feed production by species*.

Species	World	Asia Pacific Region
Beef	11	4
Aqua	3	11
Poultry	32	56
Pig	31	18
Dairy	17	6
Other	4	5

*Rabobank Nederland (1995).

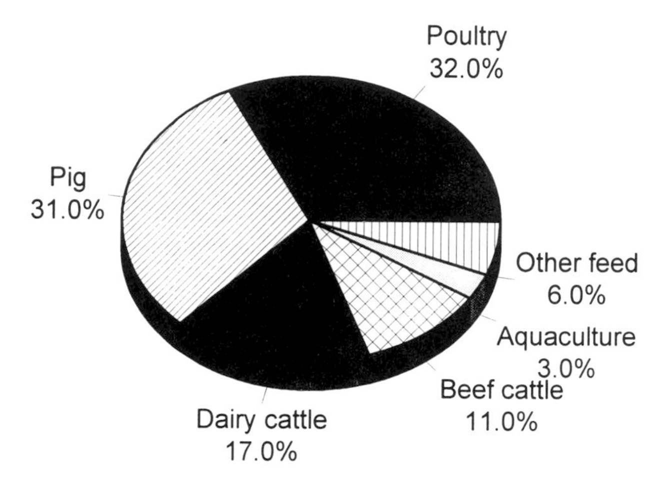

Figure 1. World compound feed production (Rabobank Nederland, 1995).

When feed production is broken down by species, poultry comprises 32% of feed production in the world (Table 1) whereas in the Asia Pacific Region it comprises 56% (Tables 1 and 2, Figure 1). These figures indicate that world poultry compound feed production is today 176 million tonnes of which the Asia Pacific Region contributes 73 million tonnes or 41% of the total. When looking 'Beyond the year 2000' through to 2010, the region's poultry industry is set for expansion which will have direct effects not only on production but on the supply and use of both conventional and non-conventional raw materials which could total as much as 185 million tonnes should per capita consumption reach 15.5 kg.

TRENDS IN POULTRY MEAT CONSUMPTION

Poultry meat consumption is determined by factors including population, economic growth and disposable income, urbanization, consumer preference, religion and legislation. Of the total world population (5.72 billion), the Asia Pacific Region represents 59% (3.39 billion), and is forecasted to

Table 2. Compound feed production in the top 10 countries of the Asia Pacific Region*.

Country	Feed produced (million tonnes)
Australia	3.2
Malaysia	3.3
Philippines	4.2
China	50.3
Japan	28
Korea	13.8
Thailand	7.5
India	7.3
Taiwan	7.8
Indonesia	4.5
Total	129.9

*Rabobank Nederland (1995).

remain at this percentage when in 2010 the total world population is predicted to reach 7.03 billion and the Asia Pacific Region will contain 4.17 billion people (Table 3). This equates to an increase of 788 million people in the Asia Pacific compared to increases of 10 million people in North America and 7 million people in Europe.

The population in the Asia Pacific Region is forecast to increase at 2% per annum between 1995 and 2000, decreasing to 1.5% per annum up to 2010. As noted earlier, per capita poultry meat consumption is increasing at 10% per annum and expected to reach 15.5 kg by 2010. These effects are indicated in Figure 2, illustrating a total poultry meat consumption increase from 13 million tonnes to 64 million tonnes over the period. In perspective, this increase represents the total amount of poultry meat consumed in the world today.

Vietnam, Indonesia and India are below the Asia Pacific regional average per capita consumption level (Table 4). It is in these countries and China, which though just above the average represents a large portion of the population, that a small per capita increase will have a large effect on total consumption in the Asia Pacific Region. Economic growth and per capita income have direct effects on total meat and poultry meat consumption (Figures 3 and 4). As the

Table 3. World population by region (million people)*.

	1995	2000	2010
North America	263	275	297
South America	293	320	346
Asia	3386	3659	4174
China	1221	1284	1388
India	935	1022	1189
Africa	728	831	1069
Europe	506	511	513
Former Soviet Union	292	295	305
Oceania†	28	31	35
Total world	5717	6158	7032

*Watt Poultry Statistical Yearbook (1996).
† Oceania: Polynesia, Guam, New Caledonia, Tonga, Samoa, Fiji, Australia,

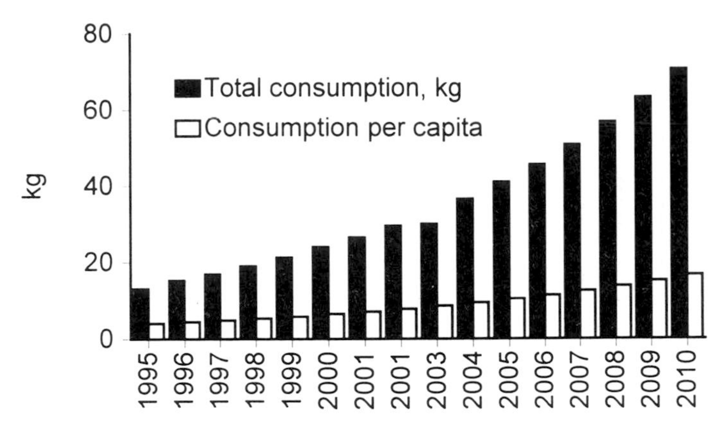

Figure 2. Poultry meat consumption in the Asia Pacific Region (Adapted from Henry and Rothwell, 1995 and Watt Poultry Statistical Yearbook, 1996).

Table 4. Per capita poultry meat consumption in 1992*.

	kg
World	7.96
Asia	3.82
Europe	15.33
USA	43.07
China	4.24
India	0.43
Indonesia	2.83
Japan	13.93
Korea	8.23
Malaysia	19.21
Philippines	4.77
Thailand	9.02
Vietnam	2.51
Australia	25.10
New Zealand	18.27

*Watt Poultry Statistical Yearbook (1996).

economic status of people living in the Asia Pacific countries continues to increase, so will poultry meat consumption. A similar picture emerges when comparing consumption and gross national product (GNP) on a per capita basis (Figure 4). Meat consumption is elastic in relation to income, especially in developing countries. When income rises, meat is often consumed in place of grain.

Of the other factors influencing poultry meat consumption, urbanization is occurring to a greater extent in many Asia Pacific countries. This stimulates livestock product consumption, creates a demand for convenience foods and increases the sale of poultry meat to domestic consumers and fast food retail outlet customers. The 'healthy image' of poultry meat, coupled with attractive pricing, is increasing its proportion of total meat consumption, especially in relation to red meat which is losing its market share in the wake of a

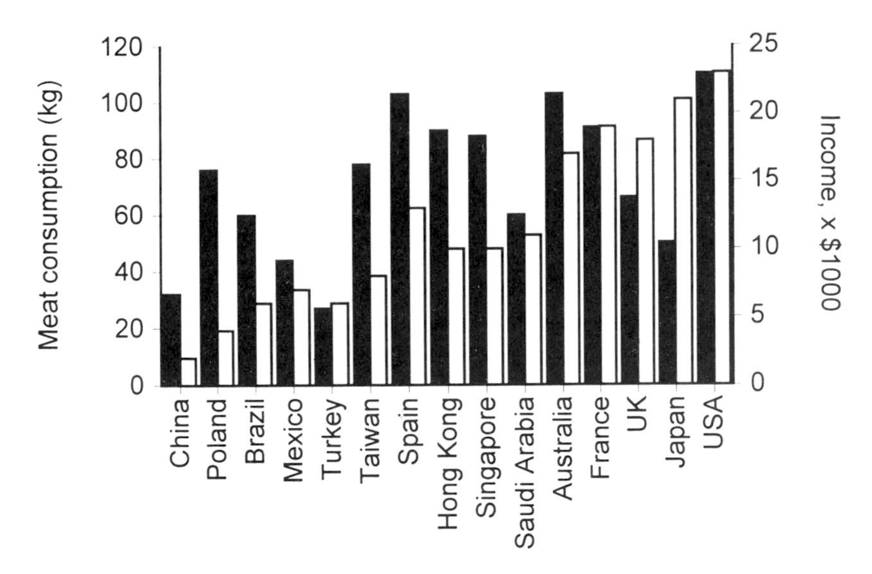

Figure 3. Meat consumption per capita versus real income per capita in 1993 (Adapted from Rabobank Nederland, 1993a, 1995).

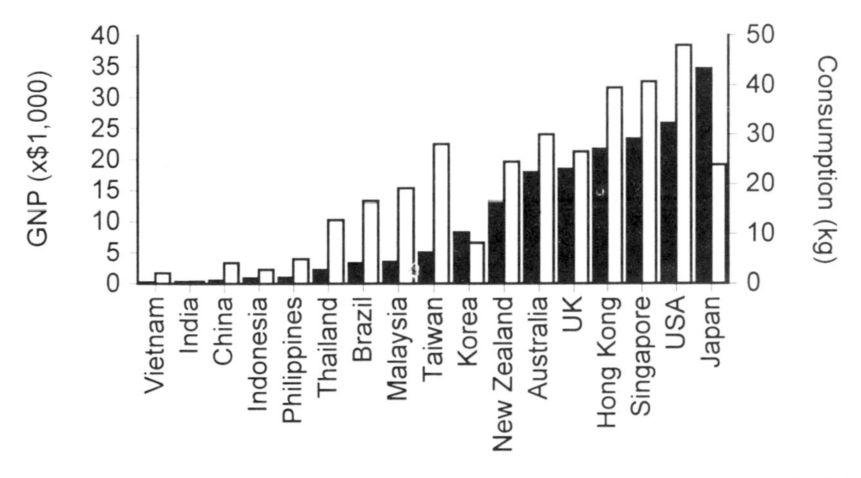

Figure 4. Poultry consumption per capita vs GNP per capita (Adapted from Watt Poultry Statistical Yearbook, 1996).

comparatively 'unhealthy' nutritional image and more recently the BSE scare. Poultry meat also benefits from its excellent organoleptic qualities and co-ordinated marketing programmes.

Religious beliefs and a dominant vegetarian culture in the Asia Pacific countries have a significant effect on animal product output, an example of which is the prohibition of pork in Muslim countries, most notably in Indonesia and India. These countries together constitute over 30% of the region's population. Legislation today is a minor issue, although the problems of environmental pollution are becoming more prominent and can only be expected to increase in the future.

When considering the region as a whole in terms of food markets, growth exists at 5–7% per year compared to 0–1.5% in developed markets. This figure is greater than the percentage rise in population and indicates increased per capita consumption. The Asia Pacific meat market has an overall deficit of production related to requirements of approximately 135,000 tonnes of meat per year (Rabobank Nederland, 1993b). It is generally Europe that acts as an 'exporter' of meat to the world market; although as the Asia Pacific demand increases and European supply remains static or even decline, greater emphasis will be placed on self-sufficiency for meat and also feed ingredient production.

TRENDS IN COMPOUND FEED PRODUCTION AND RAW MATERIAL DEMAND

Increasing per capita consumption in an increasing population will have a profound effect upon the livestock and feed industries (Table 5). When considering the increase in poultry meat consumption and the resultant need for higher production levels to meet demand, attention must also be paid to the supply of feed raw materials both locally produced and imported. Whilst considerable amounts of corn and soyabean are produced within the region, in many countries imports are now required. The total world use of coarse grains is greater than 1 billion tonnes per year and is expected to grow at 2.1% per year for the next decade, although 1995–1996 consumption is expected to exceed production with world stocks down 31% at the year end compared to 1994–1995.

While there is potential to increase grain production and crop yields in certain Asia Pacific countries, others must rely on imports to support a growing compound feed industry, ie. Japan, Korea and India. The tendency to date

Table 5. The effect of per capita consumption of poultry meat on compound feed requirements (million tonnes)*.

Per capita consumption of poultry meat	Total poultry meat consumption		Equivalent weight of broiler chicken produced		Compound feed requirement	
	1995†	2010‡	1995†	2010‡	1995†	2010‡
4.0 kg (1995 Asia Pacific average)	13.5	16.7	19.3	23.9	38.7	47.7
9.0 kg (1995 global average)	30.5	37.6	43.6	53.7	87.1	107.4
15.5 kg (2010 forecast)	52.5	64.7	92.5	92.4	150.0	184.9
45.0 kg (1995 USA average)	152.4	187.8	268.3	268.3	434.4	536.6

*Assumes FCR of 2.0, 70% dressing percentage.
† Asia Pacific population (1995) = 3386 million.
‡ Asia Pacific population (2010) = 4174 million.

has been to rely on imports, though in the future, supply problems may dictate an increase in local production to meet demand.

China was the world's second largest grain exporter in 1992, exporting 12 million tonnes (Simpson *et al.*, 1994). Today, China is the world's fourth largest importer and it is forecast that in 2004 China will import 18 million tonnes, a net demand on supply of 30 million tonnes. Thailand will switch from being an exporter to an importer, as will Indonesia and Vietnam. Korea's purchases of imported raw materials could increase by 26% by the year 2004 as well.

Much of imported raw material is produced in the USA, though decreasing production of corn and soya may mean that the current supply cannot meet the needs of the growing compound feed industry in the Asia Pacific Region (Table 6).

Unless supplies of conventional raw materials can be increased, more emphasis will be placed on the use of non-conventional raw materials and by-products which the region has in abundance. Rice bran, copra meal, peanut meal, palm kernel, canola meal and animal offal are examples (Table 7). Often these raw materials are of limited use in livestock diets due to anti-nutritional factors, although biotechnology may in certain instances overcome these problems. If suitable for use, the increasing price of conventional raw materials will make these alternative feedstuffs financially attractive.

To date, world production of conventional raw materials has kept pace with the demands of a growing world population by increasing crop land area and yields. Regions such as Asia and the former Soviet Union, as major 'importers' offer considerable scope to increase both land area and yield to improve domestic supply. The reality is that the world cereal producing industry should respond to increasing demand by increasing yields and (or) by bringing land back into production.

Factors influencing compound feed demand and hence raw material requirements are related to per capita meat consumption and the production efficiency of livestock (Table 8). Instead of breaking down livestock and compound feed production by country or by species (since the Asia Pacific region is so diverse) if one divides total compound feed produced by the population the resultant figure gives an indication of the factors influencing meat consumption and also indicates the production efficiency of livestock (Table 9). Per capita compound feed production, whilst providing an interesting

Table 6. Major crop market outlook in the USA for 1994–96*.

	Area planted (million acres)	Million bushels					Farm price $/bushel
		Output	Total supply	Domestic use	Export	End stock	
Maize							
1994–95	79.2	10,103	10,963	7205	2200	1558	2.26
1995–96	71.3	7541	9110	6425	2000	685	2.90
Soy							
1994–95	61.7	2517	2731	1551	845	335	5.45
1995–6	62	2191	2531	1511	800	220	6.75

*Gardner (1995).

Table 7. Production of traditional and non-traditional energy grains and protein sources (million tonnes) for livestock and poultry in developing Southeast and South Asian (DSESA) region in 1994*.

DSESA countries	Traditional cereals (maize)	Non-traditional cereals		Other energy-rich feeds		Traditional protein meal (soy)	Non-traditional protein meal (rape, sunflower, sesame, cotton)
		Sorghum	Millets	Cassava	Other roots and tubers		
Cambodia	0.1	–	–	0.1	146.6	–	–
China	103.6	4.9	3.0	3.5	0.1	16.3	31.8
Indonesia	6.6	–	–	15.0	2.9	1.6	1.1
Korea (RK)	2.2	–	–	–	3.0	0.6	–
Ioas	0.1	–	–	0.1	0.3	–	–
Malaysia	–	–	–	0.4	0.1	–	–
Myanmar	0.3	–	0.2	0.1	0.2	–	0.8
Philippines	5.4	–	–	1.8	0.9	–	–
Thailand	3.8	0.3	–	19.1	0.2	0.5	0.2
Vietnam	1.0	–	–	2.6	0.3	0.1	0.4
Bangladesh	–	–	0.1	–	1.9	–	0.3
Bhutan	–	–	–	–	0.1	–	–
India	10.5	12.5	10.3	5.3	16.2	3.3	23.3
Nepal	1.3	–	0.3	–	0.9	–	–
Pakistan	1.3	0.2	0.2	–	1.5	–	4.4
Sri Lanka	–	–	–	0.3	0.2	–	–
DSESA region[†]	136.0	17.9	14.1	48.3	175.4	22.4	62.3
	(24%)	(29%)	(54%)	(32%)	(41%)	(16%)	(46%)
Asia (total)[†]	139.1	18.6 x	14.2	48.5	192.3	22.7	62.6
	(24%)	(45%)	(54%)	(32%)	(45%)	(16%)	(46%)
World (total)	569.6	61.0	26.0	152.5	430.2	136.7	136.3

* Reddy (1996). Countries producing less than 50,000 tons are omitted.
[†] Figures in brackets indicate the percentages of world totals.

comparison among markets, is general and does not account for the fact that some populations consume higher levels of ruminant products which are produced less efficiently than either pig or poultry meat.

Per capita of compound feed production in different parts of the world reflects large differences between Europe and the USA on one hand and the Asia Pacific

Table 8. Factors affecting compound feed production required.

Per capita meat consumption	Income Age structure Religion Consumer patterns Population growth
Production efficiency of livestock	Species Nutrition Health Genetics Management Biotechnology

Region on the other. When one considers growth in the future, if per capita compound feed production remains at 40 kg (Asia Pacific), then based on population growth alone a further 38 million tonnes of compound feed will be required in the year 2010 (Table 9). However, this may be regarded as conservative in view of the impact of other factors which increase meat consumption. If in the year 2010 the Asia Pacific population is 4.2 billion, and other determinants of meat consumption are such that the region moves closer to per capita meat consumption levels in the USA and Europe or even South America, the demand for increased compound feed production will be enormous (Table 10). If consumption equal to that in Europe is achieved, an extra 1.1 billion tonnes of compound feed would be required. Based on consumption levels equal to South America, 216 million tonnes would be needed.

The increased demand for compound feed will obviously have a direct effect on feed ingredient needs. Typical poultry diets in the region include principally corn and soya (Table 11). Given the potential shortages of these raw materials in the future, increasing demand could mean that diets may include alternative raw materials excluding corn and soya or may be a balance based on supply and price (Table 12).

Using 1995 as a comparison with 113 million tonnes of compound feed produced (40 kg per person), requirements of raw materials are as stated in Table 13. If diet formulations remain similar to that of today, i.e. corn/soya, in the year 2010, an extra 26 million tonnes of corn and 11 million tonnes of soya would be required at the same per capita compound production (Table 14). Without corn and soya, high emphasis is placed on the use of by-products. Should compound feed production per person increase to 300 kg (i.e. European standard), there is a dramatic increase in demand for raw materials equalling 1260 million tonnes (Table 15). In such an instance, can supply meet demand?

A predicted increase in demand for soya of 11 million tonnes between 1995 and 2010 will have a large impact on supply and price given that the world's total production of soyabean meal was 113 million tonnes in 1993–1994 (Table 16).

In the future, should corn and soya no longer form the bulk of poultry diets, by-product and non-conventional raw material usage will increase as will that of synthetic amino acids. The energy component of the diet will increase in cost and greater reliance will be placed on the rendering industry to produce

Table 9. Compound feed production forecast for 2010.

	Population (millions)	Compound feed per person (tonnes)	Total compound feed, (million tonnes)*
Europe	513	0.3	154 (152)
North America	538	0.33	176 (138)
South America	397	0.1	40 (30)
Asia Pacific	4200	0.04	168 (130)
Total	5648		538 (423)

* 1995 figures in brackets.

Table 10. Compound feed production demands forecast (million tonnes) in Asia (year 2010) based on North American, European or South American per capita meat consumption.

Asia Pacific population (millions)	Per capita meat consumption (tonnes per person)			
	Europe	North America	South America	Feed production needed (million tonnes)
4200	0.3	–	–	1,260
4200	–	0.33	–	1,386
4200	–	–	0.1	420

Table 11. Typical broiler grower diet – 1996.

Ingredient (%)		Nutrient (%)	
Corn	67.0	Crude protein	20.50
Soya	25.4	Lysine	1.13
Meat and bone meal	5.0	Methionine	0.40
Limestone	1.6	Methionine + cysteine	0.70
Premix	0.5	Threonine	0.86
		ME, kcal	1362
		Calcium	1.05

Table 12. Broiler grower diet (no corn/soya).

Ingredient (%)		Nutrient (%)	
Peanut	10	Crude protein	18.2
Cassava	30	Lysine	1.0
Rice Bran	37.5	Methionine	0.40
Canola	5.2	Methionine + cysteine	0.7
Meat and bone Meal	7.5	Threonine	0.64
Feather meal	1.0	ME, kcal	1250
Fat	5.0	Calcium	0.95
Copra meal	2.2		
Limestone	0.6		
Premix	0.5		
Amino acids	0.3		

both energy and protein sources. Use of rendered materials will also efficiently utilize animal carcass waste.

In addition, the role of biotechnology will become ever more important. The use of transgenic animals and plants can increase product yields and assimilate nutrients more efficiently, though public perception will determine acceptance of this. The importance of biotechnology was recently emphasized by Mr Tony Petch (Dalgety, UK) who stated 'Forecasts for 2030 point to a world population of 8000 m, an annual rise of 4%. Food production growth for the past 20 years has been around 2% per year. To meet the needs of the world population, biotechnology will become more important as a means to improve yields and quality quickly.'

Table 13. Raw material usage (million tonnes) in Asia Pacific poultry diets in 1995 (130 million tonnes compound feed, 0.04 tonnes compound feed/person*.

	Typical broiler diet	No corn/soya
Corn	87	–
Soya	32	–
Meat and bone meal	6.5	10
Cassava	–	39
Rice bran	–	49
Fat	–	6
Peanut meal	–	13
Canola meal	–	6
Feather meal	–	1
Copra meal	–	2

* Rabobank Nederland (1995).

Table 14. Raw material usage (million tonnes) in Asia Pacific predicted for the year 2010 (168 million tonnes compound feed, 0.04 tonnes compound feed/person).

	Typical broiler diet	No corn/soya
Corn	113	–
Soya	43	–
Meat and bone meal	8	13
Cassava	–	50
Rice bran	–	63
Fat	–	8
Peanut meal	–	17
Canola meal	–	9
Feather meal	–	2
Copra meal	–	4

Table 15. Raw material usage in Asia Pacific predicted for the year 2010 if per capita feed production equals that in Europe (1260 million tonnes compound feed, 0.3 tonnes per person).

	Typical broiler diet	No corn/soya
Corn	844	–
Soya	320	–
Meat and bone meal	63	94
Cassava	–	378
Rice bran	–	472
Fat	–	63
Peanut meal	–	126
Canola meal	–	65
Feather meal	–	13
Copra meal	–	28

Today's biotechnological solutions come in the form of approved feed additives which, for the poultry industry, are primarily enzymes to increase utilization of conventional and non-conventional raw materials and to improve utilization of rendered products. Enzymes to enhance utilization of wheat and barley are widely accepted in poultry production; and the use of second generation enzymes to more efficiently utilize soyabean meal, sunflower, fat, rice bran and other raw materials are now commercially available.

Table 16. Top five countries in soyabean production (million tonnes) in the world*

Year	USA	Brazil	Argentina	China	India	World total
1989–90	52.4	20.3	10.8	10.2	1.8	107.4
1990–91	52.4	15.8	11.5	11.8	2.6	104.1
1991–92	54.1	19.3	11.2	9.7	2.5	107.0
1992–93	59.5	22.3	11.0	10.3	3.1	116.4
1993–94	49.2	23.8	12.2	13.0	4.5	113.1

* Gardner (1995).

Conclusions

In conclusion, meat consumption is increasing in Asia Pacific countries and poultry will remain a significant proportion of this consumption. Whether or not forecasted per capita poultry meat consumption can be reached remains to be seen, though trends and indicators are positive. Assuming such increases, greater local livestock production will be needed to meet demand. The compound feed industry will have to significantly increase output and may need to adapt current dietary formulations from conventional raw materials to other feed ingredients.

References

Henry, R. and G. Rothwell. 1995. In: World Poultry Industry Report. World Publications, 1818 H Street NW, Washington DC, USA.

Gardner, B. 1995. World Agricultural Supply and Demand Estimates: US Marketing Years for Exports. In: Agra Europe Special Report Number 83. Agra Europe (London) Ltd, 25 Frant Road, Tunbridge Wells, Kent, TN2 5JT, UK.

Rabobank Nederland. 1993a. The World Poultry Market. In: Rabobank Nederland. PO Box 17100, 3500 HG Utrecht, Nederland.

Rabobank Nederland. 1993b. FAO and USDA in: Competitiveness in the Pig Industry. In: Rabobank Nederland. PO Box 17100, 3500 HG Utrecht, Nederland.

Rabobank Nederland. 1995. The Compound Feed Industry. In: Rabobank Nederland. PO Box 17100, 3500 HG Utrecht, Nederland.

Reddy, C.V. 1996. Feed Resource Management in Asia. Poultry International, November. Watt Publishing Co, 122 S Wesley Avenue, Mount Morris, Illinois, 61054–1497, USA.

Simpson, J., X. Cheng, and A. Miyazati. 1994. China's Livestock and Related Agriculture – projections to 2025. CAB International 1994, Wallingford, Oxon, UK.

Watt Poultry Statistical Yearbook. 1996. Poultry International, Volume 35(8). Watt Publishing Co, 122 S Wesley Avenue, Mount Morris, Illinois, 61054–1497, USA.

NATURE'S BIOCATALYSTS: NEW APPLICATIONS FOR ENZYMES IN ANIMAL FEED

RELEASING ENERGY FROM RICE BRAN, COPRA MEAL AND CANOLA IN DIETS USING EXOGENOUS ENZYMES

J.R. PLUSKE[1], P.J. MOUGHAN[1], D.V. THOMAS[1],
A. KUMAR[2] and J. DINGLE[2]

[1] *Monogastric Research Centre, Massey University, Palmerston North, New Zealand*
[2] *Department of Animal Production, The University of Queensland, Gatton College, Lawes, Queensland, Australia*

Introduction

There is increasing concern and hence a growing awareness that, next century, there will be a shortage of conventional feedstuffs for use in animal production. This is because forecasts to the year 2030 suggest a world population of eight billion people, which is an annual rise of 4%, yet the production of food for mankind in the previous 20 years has been increasing at only 2% per annum.

To meet the food demands of the burgeoning world population, and in particular the demand for meat products, biotechnology and its implementation in animal industries will become increasingly widespread and, most likely, essential, if supply is to meet demand. Opportunities for biotechnology in animal production include:

- maximizing the efficiency of conventional raw material use,
- maximizing the use of non-conventional feedstuffs. It is advantageous that many of these products, such as rice bran and copra meal, are produced in the greater Asian region where the forecasts for an increased demand for human food commensurate with increased *per capita* income are greatest,
- reduced environmental pollution,
- increased consumer awareness of the value of biotechnology in areas such as meat quality and genetic manipulation of feedstuffs and animals.

The purpose of this paper is to review the use of several by-products for potential use in animal production, and examine ways where judicious intervention with exogenous enzymes increases the availability of energy, and particularly that of fat. This is because fat contains a high energetic value (35–39 MJ/kg), and hence only small increases in the efficiency of its use are required to increase inclusion levels in diets and (or) reduce the contribution of other ingredients in diets. We wish to focus specifically on the meat chicken because, world-wide, the meat chicken industry uses about one-third of the world's total production of compounded feed. Additionally, there is a move away from the consumption of red meat toward that of poultry (and pigs), and it is the meat of choice in the fast-growing Asian economy.

But first, what are by-products? By-products are generally considered residual feedstuffs obtained when a feed or food crop is processed. They can be used on a commercial basis but are generally inefficient in terms of animal utilization when compared to conventional feedstuffs. Amongst these products can be included rice bran and copra meal which are produced in the Asian and Pacific regions, and canola meal, which is produced predominately in Canada and has relevance to North America. When one considers the growing shortage of conventional feedstuffs for use in animal industries, as outlined previously, there is great attraction in the use of feed by-products as raw materials for animal diets. However, this may require some form of biotechnological intervention to increase their utilization.

Rice bran

INTRODUCTION

Rice (*Oryza sativa L.*) was first cultivated some 7000 years ago in eastern parts of China and India. It is the staple cereal grain for two-thirds of the world's population, with 90% of the world's production of more than 425 million tonnes grown in the Asian region (Saunders, 1986). White rice is milled from brown rice because very little brown rice is consumed by the human population. Milling removes the outer layers of the rice caryopsis producing white rice which is almost all endosperm (Farrell, 1994). The approximate product fractions from standard milling in Australia are depicted in Figure 1.

In Australia, the by-product of white rice milling is referred to as *rice pollard*, and includes the true bran and polishings. For the remainder of this paper, we will use the term *rice bran* which is the by-product remaining after the milling of brown rice for white rice (i.e. it is synonymous with rice pollard). Rice bran is about 10% of brown rice, and may contain 20–25% of the total protein, 80% of the total oil, more than 70% of the total minerals and vitamins, and up to 10% of the starch endosperm (Houston, 1972). Since little rice bran is consumed by humans, the 40–45 million tonnes of rice bran produced annually in the Asian countries (90,000 to 100,000 tonnes produced annually in Australia) represents an enormous waste of important nutrients which, if harnessed appropriately, would represent a valuable feed source for intensive animal industries.

CHEMICAL COMPOSITION AND THE USE OF RICE BRAN AS A FEEDSTUFF

Using Australian rice bran, Warren and Farrell (1990a) reported that the chemical composition of rice bran from several rice cultivars grown in Australia was reasonably uniform. Crude protein ranged from 134 to 173 g/kg and ether extract from 204 to 234 g/kg. Mean neutral detergent fiber (NDF) was 256

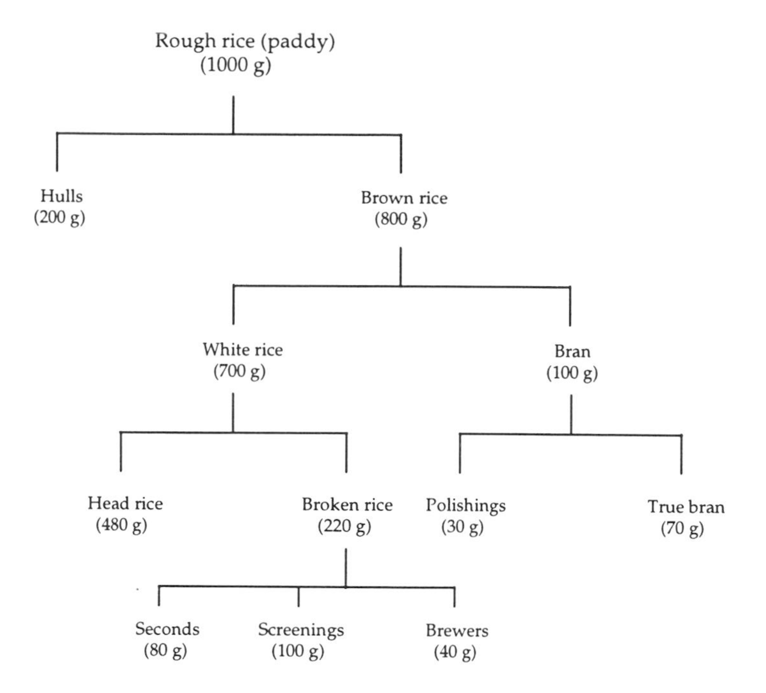

Figure 1. The various fractions of paddy rice caused by the milling of brown rice (adapted from Warren, 1985; cited by Farrell, 1994).

g/kg, acid detergent fiber (ADF) 122 g/kg, and ash 105 g/kg. Australian rice bran also contained > 17 g/kg phosphorus; however more than 50% of this is in the phytate form and hence has a reduced availability. The profiles for some essential amino acids show a good balance, but none are present in large amounts (Warren and Farrell, 1990d).

Indications suggest, however, that the nutritional value of rice brans may differ among regions and different harvest years despite the chemical composition of the rice brans being similar (Warren and Farrell, 1990a). In a comparison of two varieties of Australian rice bran with a variety of de-fatted rice bran, Warren and Farrell (1990c) reported significant reductions in the utilization of dry matter, energy, ether extract and nitrogen in young broiler chickens compared to adult cockerels. Consequently, the AME of rice bran was lower (by 4.6 MJ/kg on average) in broiler chickens than with adult cockerels, a result that reflects a reduced apparent digestibility of fat in rice bran. Coefficients of metabolizability of ether extract in broilers were, on average, 0.37 for the two varieties compared to 0.94 for adult cockerels (Table 1). Similar results were reported by Askbrant and Farrell (1987), who determined that the AME of oil in rice bran increased significantly with the age of the bird.

In general, inclusion of more than 200 g/kg of rice bran in diets for broiler chickens depresses performance (see Farrell, 1994 for a review). For example, Warren and Farrell (1990b) reported few ill effects when up to 200 g/kg of rice bran were included in diets for chickens aged between 21 and 49

Table 1. Coefficients of apparent metabolizability of dry matter and other nutrients, apparent retention of nitrogen, and the AME (MJ/kg DM) for three rice brans in adult cockerels and broiler chickens aged between 15 and 20 days*.

	Full-fat rice bran		De-fatted rice bran	
	'Calrose'	'Starbonnet'	'Calrose'	SEM
Dry matter				
Cockerels	0.58[a]	0.61[a]	0.39	0.065
Chickens	0.42[b]	0.47[b]	0.46	0.055
Energy				
Cockerels	0.69[a]	0.69[a]	0.50	0.049
Chickens	0.43[b]	0.50[b]	0.47	0.066
Ether extract				
Cockerels	0.94[a]	0.93[a]	–	0.035
Chickens	0.31[b]	0.43[b]	–	0.112
Nitrogen				
Cockerels	0.30[a]	0.51[a]	0.32[a]	0.201
Chickens	0.36[a]	0.38[b]	0.32[a]	0.071
Starch				
Cockerels	0.99[a]	0.94[a]	0.96[a]	0.020
Chickens	0.95[a]	0.98[a]	0.99[a]	0.036
NDF				
Cockerels	0.18[a]	0.22[a]	0.09[a]	0.067
Chickens	0.14[a]	0.13[a]	0.12[a]	0.029
ME				
Cockerels	14.7[a]	15.0[a]	9.4[a]	0.33
Chickens	9.6[b]	10.8[b]	7.3[b]	0.49

*From Warren and Farrell (1990c). Rice brans were included in diets at 400 g/kg.
[a,b] Within columns and within items, values not having the same superscript differ ($P < 0.05$).

days of age. However, a linear decline in growth of 0.5 g/day and feed intake of 0.8 g/day occurred for each 10 g/kg increase in dietary rice bran above this level.

Reasons for this decline are not known, although were apparently unrelated to the lipid fraction or a crude water-soluble extract (Warren and Farrell, 1990b). Annison *et al.* (1995) demonstrated that the depression in growth of broiler chickens on a rice bran diet was not associated with the non-starch polysaccharide (NSP) component, because addition of graded levels of isolated NSP (predominately arabinose and xylose) had no detrimental effects on performance. In fact, AME slightly increased in response to addition of purified NSP to the sorghum-based diets. Similar results were reported recently by Aboosadi *et al.* (1996) who found no increase in AME or performance when cell-wall degrading enzymes were added to the diets of broiler chickens. The fact remains, however, that the AME content of rice bran is lower than that anticipated on the basis of its gross energy concentration.

Given this low AME, there would appear to be enormous potential to incorporate an exogenous lipase in diets containing rice bran to increase the digestibility of fat, and hence raise the AME. The number of studies conducted with lipase, however, is surprisingly low, which may be a reflection of the current perception of rice bran as a feedstuff for the animal industries. In one

Table 2. Excreta energy, excreta weight, and the AME of Australian rice bran in the absence or presence of Allzyme Lipase (Alltech Inc., added at 100 mg/kg) for adult cockerels. Results are mean ± SEM*.

Treatment	Excreta energy (MJ/kg DM)	Excreta weight (g DM)	AME (MJ/kg, 'as is')
Rice bran *without* Allzyme Lipase (n = 12)	16.6[a] ± 0.18	20.2[a] ± 0.33	12.2[a] ± 0.18
Rice bran *with* Allzyme Lipase (n = 11)	15.8[b] ± 0.20	20.7[a] ± 0.34	12.5[a] ± 0.17

*From Thomas *et al.* (1997).
[a, b] Within columns, means not having the same superscript differ ($P<0.05$).

of the few studies conducted, Martin and Farrell (1993; cited by Farrell, 1994) showed no beneficial effects on the performance of broiler chickens growing between 4 and 23 days of age when a feed lipase was added to a sorghum-soyabean meal diet containing 400 g/kg rice bran.

In contrast, recent research from the Monogastric Research Centre at Massey University, New Zealand, has clearly demonstrated that the inclusion of a lipase developed by Alltech Inc. in diets for both broiler chickens and adult cockerels is beneficial to performance and AME. In an experiment where adult cockerels were intubated with 40 g of rice bran and total excreta collected over the ensuing 48 hours, Thomas *et al.* (1997) found that the addition of 100 mg/kg Allzyme Lipase caused a 5% reduction (0.8 MJ; $P < 0.05$) in fecal energy concentration concomitant with a 3% higher (0.3 MJ; $P = 0.203$) AME value (Table 2). It should be noted that only a limited number of birds were used per treatment. Experimentation with a larger number of birds may have detected a statistical difference in the AME of rice bran for adult cockerels.

Given that the apparent utilization of fat increases with the age of the bird (Krogdahl, 1985), a greater response to supplementary lipase in diets containing rice bran may be expected in rapidly growing broiler chickens. At the Monogastric Research Centre, the effect of lipase supplementation (Allzyme Lipase, 100 mg/kg) was investigated in diets containing either 200 g/kg or 400 g/kg of Australian rice bran. Fifteen replicates per dietary treatment were used with six birds per replicate, and diets were fed from 1 to 14 days of age.

In the presence of lipase, and between days 1 and 7:

- growth rate was improved ($P < 0.05$) by 11 and 15%, respectively, in chickens receiving diets containing 20 or 40% rice bran,
- (FCR) was improved ($P < 0.05$) by 9 and 15%, respectively, in chickens receiving diets containing 20 or 40% rice bran,
- AME was improved ($P < 0.05$) by 9 and 17%, respectively, in chickens receiving diets containing 20 or 40% rice bran.

In the presence of lipase and between 7 and 14 days of age:

- growth rate was improved ($P < 0.05$) by 6 and 9%, respectively, in chickens receiving diets containing 20 or 40% rice bran,
- FCR was improved ($P < 0.05$) by 5 and 6%, respectively, in chickens receiving diets containing 20 or 40% rice bran.

Table 3. The effect of Allzyme Lipase (100 mg/kg) added to diets containing 200 g/kg or 400 g/kg of Australian rice bran on the performance and AME of broiler chickens between 1 and 14 days of age. Results are interaction least-squares means ± SEM*.

Item	Full-fat rice bran (200 g/kg)			Full-fat rice bran (400 g/kg)		
	No lipase	Allzyme Lipase	SEM	No lipase	Allzyme Lipase	SEM
Days 1–7						
Growth rate, g/day	16.3[a]	18.1[b]	0.29	11.2[a]	12.9[b]	0.21
Food intake, g/day	21.0[a]	21.2[a]	0.31	18.6[a]	18.2[a]	0.29
FCR, g food per g gain	1.29[a]	1.17[b]	0.052	1.66[a]	1.41[b]	0.061
AME, MJ/kg[†]	10.0[a]	10.9[b]	0.31	8.7[a]	10.2[b]	0.29
Days 7–14						
Growth rate, g/day	21.9[a]	23.2[b]	0.44	12.6[a]	13.7[b]	0.32
Food intake, g/day	32.8[a]	33.3[a]	0.34	21.8[a]	22.2[a]	0.33
FCR, g food per g gain	1.50[a]	1.43[b]	0.038	1.73[a]	1.62[b]	0.041

*From the Monogastric Research Centre, Massey University, New Zealand (unpublished data).
[†] AME determined between days 3 and 7.
[a, b] Within main effects of rice bran inclusion level, means not having the same superscript differ ($P < 0.05$).

These data are presented in Table 3. No interactions were present between the level of rice bran included and the absence or presence of lipase. These data confirm our previous experiment using adult cockerels, and show quite clearly that the bioavailability of energy in rice bran may be increased by the use of an exogenous lipase. These data also show that the response to lipase was greatest when 400 g/kg of rice bran was included in the diet, although overall performance was depressed at this rate.

These data are exciting, and substantiate the use of lipase in diets containing Australian rice bran for broiler chickens. We are continuing our research with lipase by investigating the possible synergy between lipase and supplementary bile acids in the young growing chick. It is well recognized that pancreatic secretion of lipase is limiting in the young bird (see Ketels, 1994, for review) which, in turn, reduces apparent lipid digestibility (Smits, 1996). Long-chain saturated fats, particularly 16:0 and 18:0, are utilized poorly in young birds in comparison to unsaturated fats, and thus represent a considerable production penalty. Dietary addition of bile acids improves the digestibility of fat and increases AME (e.g. Gomez and Polin, 1974; Polin *et al.*, 1980). Rice bran oil contains 18–20% long-chain saturated fats (Gunstone, 1996), so addition of lipase in conjunction with bile acids may cause a response in excess of that achieved by addition of lipase only.

Copra meal

INTRODUCTION

The coconut palm (*Cocos nucifera*) is widely distributed throughout the tropics. The primary coconut product available for use in world trade is copra,

Table 4. Variation in nutrient content of copra meals (as a percent of air-dry meal)*

	Expeller meal from		
Nutrient	Philippines	Solomon Islands	UK
Dry matter	90.2	89.4	95.4
Crude protein	20.8	19.2	19.0
Ether extract	8.1	11.7	16.1
Crude fiber	11.6	8.8	8.8
Ash	6.2	5.1	4.1
Calcium	0.10	0.05	0.05
Phosphorus	0.59	0.49	0.45
Lysine	0.42	0.70	0.58
Methionine+cysteine	0.36	0.77	0.49

*From Thorne *et al.* (1990).

the sun- and then oven-dried kernel of the coconut, which is used as the raw material for the production of coconut oil. Around 30–40% by weight of the copra which is used for oil extraction remains as a residue, this being termed *copra meal* (Thorne *et al.*, 1990), and when dried contains between 5 and 10% moisture. In 1994, total world production of copra meal was estimated at 1.67 million tonnes, with 85% of this being produced in Asia (A. Kumar, personal communication).

Copra meal is produced predominately by mechanical means using expeller presses which extract the coconut oil from dried coconut kernels. The copra is heated to temperatures in excess of 120°C during the mechanical extrusion process. An alternative to expeller presses, solvent extraction, removes greater quantities of oil from the meal comparetl to mechanical extraction. Most copra mills in the South Pacific use mechanical expellers, with the efficiency of oil extraction depending considerably on the extent of wear of the expellers. This is reflected in the chemical composition of the copra meal (Table 4).

COPRA MEAL AS A FEEDSTUFF

By virtue of these different extraction techniques and the efficiency of mechanical extraction, copra meal is a variable product (Table 4). Typically, it contains 180–200 g/kg crude protein, 80–160 g/kg oil, and 80–120 g/kg crude fiber. The balance of amino acids tends to be poor, with low levels of lysine and the sulfur-containing amino acids, and this requires supplementation with synthetic amino acids. NDF content was 530 g/kg and ADF was 269 g/kg (Kumar and Dingle, personal communication). In a classical balance trial, Baidya *et al.* (1995) determined the AME value of expeller copra meal (containing 12.75% ether extract) included in diets for layers at 200 g/kg and 300 g/kg to be 8.74 MJ/kg and 9.90 MJ/kg, respectively. In 28-day-old broilers fed expeller copra meal at the same levels, AME values were 8.87 MJ/kg and 7.16 MJ/kg, respectively.

The feeding value of copra meal for broiler chickens (and also pigs; see

Thorne *et al.*, 1988) has been recognized for many years, especially in those countries producing the product (Fronda and Mallonga, 1935; Momongan *et al.*, 1964; Nagura, 1964). Differences in processing techniques, however, appear to result in differences in the nutritive value of copra meal for broilers with subsequent variable results (Panigrahi, 1992). The acceptance of copra meal in broiler diets in developing countries, therefore, has been slow.

The utilization of copra meal seems to be limited by its capacity to restrict food intake, including its nutrient density and high water-absorbing capacity relative to the limited gut volume of young birds (Panigrahi *et al.*, 1987; Panagrahi, 1991). Nevertheless, a high residual-oil copra cake (221 g/kg) was well tolerated in diets for broiler chickens at an inclusion rate of 250 g/kg (Panigrahi *et al.*, 1987), with some workers reporting that a properly supplemented starter diet for broilers could sustain an inclusion rate of 400 g/kg copra meal without any adverse effects on growth and feed conversion efficiency over a 7-week growing period (Thomas and Scott, 1962; Panagrahi, 1991).

An aspect of copra meal that has received surprisingly little attention, and one that is more likely to limit its widespread use, is the composition of its NSP. It is well recognized that the crude fiber, NDF and ADF determinations are now considered to be unsatisfactory techniques for 'fiber' determination as they fail to measure all of the soluble fiber (Mugford, 1993). A better indicator of the carbohydrate fraction, and one that is required if the use of copra meal is to increase in intensive animal production, is the NSP profile of copra meal, since knowledge of this may then allow for planned exogenous enzyme intervention.

The nature of the NSP in coconut kernel (Balasubramaniam, 1976) and copra meal (Saittagaroon *et al.*, 1983) is known. Balasubramaniam (1976) reported that galactomannan formed 61% of the total polysaccharides in the kernel followed by mannan (26%) and cellulose (13%). In contrast, Saittagaroon *et al.* (1983) reported that the major NSP in copra meal is mannan (mannose units joined through ß-1, 4 linkages), accounting for 61% of the total polysaccharides, followed by substantial amounts of cellulose, minor amounts of arabinoxylogalactan, galactomannan, and traces of arabinomannogalactan and galactoglucomannan.

Given this, a logical progression to enhance the nutritive value of copra meal would be to incorporate a mannanase preparation in diets containing copra meal, but surprisingly little research has been conducted. Teves *et al.* (1988) hydrolyzed copra meal in a mannanase preparation from *Streptomyces spp.* for 24 hours at room temperature, and reported a significant reduction in crude fiber in comparison to the untreated copra meal (6.3 vs. 12.7%, $P < 0.01$). Diets fed to broilers with enzyme-treated copra meal (at 5, 10 and 15% of the diet) showed significant improvements ($P < 0.05$) in apparent digestible dry matter (68.3 to 81.0%), crude fiber (25.5 to 61.9%), crude protein (49.7 to 61.2%), crude fat (83.3 to 91.5%), and AME (12.88 vs. 10.86 MJ/kg) in comparison to untreated copra meal. This was reflected in an increased weight gain (558.5 vs. 433.8 g, $P < 0.05$) and a better feed:gain (gram feed per gram live-weight gain) ratio (2.13 vs. 2.78, $P < 0.05$) of broilers between 2 and 5 weeks of age.

Additional research funded by Alltech Inc. at Gatton College, The University of Queensland, Australia, has focused on the use of supplementary mannanase in diets for broiler chickens containing 200 g/kg copra meal (Kumar and Dingle, unpublished data). Mannanase was used to hydrolyze the bonds in mannan to mannose which would then follow the glycolytic pathway into pyruvate, thus providing energy to the birds which may otherwise be unavailable.

In this experiment, a total of 320 birds were maintained in battery brooders and fed the following diets for 42 days:

1. Corn-soyabean meal (SBM),
2. Corn-SBM, with 200 g/kg copra meal,
3. Corn-SBM, with 200 g/kg copra meal plus Alltech mannanase (1 kg/tonne),
4. Corn-SBM, with 200 g/kg copra meal plus Alltech mannanase (2 kg/tonne).

Diets were fed both as a 'starter' and a 'finisher', and were formulated to contain 12.2 and 12.6 MJ ME/kg, 22 and 20% crude protein, 1.18 and 1.00% available lysine, and 0.90 and 0.68% methionine + cysteine, respectively.

Addition of Alltech mannanase increased liveweight at 42 days of age in comparison to birds fed either the corn/soya or the corn/soya plus 200 g/kg copra meal diets by, on average, 5.3 and 2.6%, respectively, ($P = 0.04$). The feed:gain ratio was improved with enzyme addition by, on average, 3.3 and 1.3%, respectively (Table 5) but this result was not statistically significant.

Although these increases were not dramatic, this is at least some evidence that mannanase addition to diets containing copra meal causes improvements in performance. However, the increases are not as marked as one may have expected given the high mannan content of copra meal. The mannanase-treated copra meal used in the study of Teves *et al.* (1988) had been treated for 24 hours with the mannanase preparation prior to its incorporation in broiler diets, supporting the concept that pre-incubation of feedstuffs with exogenous enzymes is necessary for maximum effect. Pre-feeding treatment of by-products with enzymes in an aqueous environment has been shown to have considerable commercial potential (S. Maughan, personal communication).

Table 5. Effects of Alltech Mannanase supplementation of copra-meal based diets on broiler performance at 42 days of age. Values are mean ± SEM*.

Item	Corn-SBM	Corn-SBM and: Copra meal	Copra meal+ enzyme (1 kg/t)	Copra meal+ enzyme (2 kg/t)	*P*-value
Liveweight, kg	1833 [a] ± 28.7	1880 [ab] ± 31.0	1931 [b] ± 16.9	1925 [b] ± 24.3	0.04
Total feed consumption, g/bird	3334 [a] ± 89.0	3344 [a] ± 64.8	3418 [a] ± 49.1	3355 [a] ± 68.9	0.82
FCR, g feed/g gain	1.82 [a] ± 0.030	1.78 [a] ± 0.030	1.77 [a] ± 0.020	1.74 [a] ± 0.020	0.35
Mortality, %	1.25 [a]	2.50 [a]	1.25 [a]	1.25 [a]	0.94

*From Kumar and Dingle (unpubl. data).

[a, b] Within rows, values not followed by the same superscript differ.

Alternatively, other factors such as the amount of residual oil and any damage to proteins that may have occurred during mechanical extraction, may have been limiting the performance of birds fed copra meal.

For example, copra meal can contain up to 16% residual oil (Table 4) which is rich in fats such as lauric acid (12:0; 45–55%) and in other short- and medium-chain saturated acids (C_8-C_{14}; 35–40%) (Gunstone, 1996). Is there potential for the use of lipase in diets containing copra meal? Panigrahi (1991) reported a study where copra meal containing 220 g/kg residual lipid was incorporated into diets for broilers, but performance was depressed relative to control diets devoid of copra meal. Is there potential for incorporation of lipase plus bile acids, especially in young birds up to 21 days of age, as it is known (Krogdahl, 1985) that utilization of saturated fats is relatively low? Alternatively, the mannanase used in this study may not have been specific for the copra meal incorporated into the diets and not cleaved sufficient mannan molecules to effect a larger response in performance. Nevertheless, it is evident that more work is required if copra meal is to be included reliably in diets for chickens, and biotechnology would appear to be an intrinsic aspect of this research.

Canola

Canola is the registered name for rapeseed containing less then 2% of the total fatty acids in the oil as erucic acid and less than 30 moles of alkenyl glucosinolates per gram of oil-free dry matter (Aherne and Bell, 1990). In recent years in both the pig and poultry industries interest has increased in including fats and oils in diets to reduce dust levels in barns and due to the greater use of genetically lean, high-producing animals whose appetites may not have kept pace with their potential for growth (Aherne and Bell, 1990). In some countries such as Canada, large quantities of canola meal (the meal remaining after canola oil is extracted) and canola seed are available for use in the animal industries.

Canola seed contains between 38 and 42% ether extract dependent upon variety and growing conditions, whereas canola meal varies more widely in the amount of residual oil it contains. For example, canola meal produced in Canada contains between 2 and 4% oil (much of this being in the form of gums which are added back to the meal), whereas that produced in Australia, where the extraction of oil is not as complete, can contain up to 11.2% residual lipid (Mullan and Pluske, unpublished data). In situations where the residual oil level is high and (or) full-fat canola seed is fed, the judicious use of lipase would appear to offer advantages as a means of increasing the amount of energy released. However, canola seed and canola meal may contain some undesirable constituents (glucosinolates, tannins, sinapine) which, if fed in high concentrations, can limit animal productivity. Indeed, another role for synthetically-produced enzymes may be to enhance the degradation of such anti-nutritional factors.

Conclusions

In this paper we have limited our discussion to increased utilization of rice bran, copra meal and canola using exogenous enzymes, and particularly that of lipase. Releasing energy from feed fats by the use of enzymes can be economically beneficial given the high energetic value of fat (35–39 MJ/kg). Only small increases in the efficiency of digestion may be required, therefore, to impart significant increases in energy availability, especially in young, rapidly growing birds (and pigs).

It is important that nutritional quality be defined for feedstuffs of these types to allow efficient and cost-effective feed formulation. It is of global importance given the predicted shortages of conventional raw materials for the compound feeds industries, and the increase in demand for meat products, particularly in the Asian region. Furthermore, the dependence on imported ingredients in some of these countries has resulted in a certain degree of neglect in the full utilization of local materials as substitutes (Teves *et al.*, 1988). Collectively, it is quite evident that increased utilization of by-products such as these will be necessary in the future, and in this regard the biotechnology industry has a major role to play. Finally, we have limited our discussion to broiler chickens, although the rationale for using enzymes in diets containing unconventional feedstuffs can be applied equally to other animal industries such as pigs, aquaculture, and other avian species.

References

Aboosadi, M.A., J.R. Scaife, I. Murray and M. Bedford. 1996. Effect of supplementation with cell wall degradation enzymes on the growth performance of broiler chickens fed diets containing rice bran. Br. Poult. Sci. 37: S41-S42.

Aherne, F.X. and J.M. Bell. 1990. Canola seed: Full-Fat. In: Non-traditional Feed Sources for Use in Swine Production. P.A. Thacker and R.N. Kirkwood (Eds.). Butterworths, MA, 79–90.

Annison, G., P.J. Moughan and D.V. Thomas. 1995. Nutritive activity of soluble rice bran arabinoxylans in broiler diets. Br. Poult. Sci. 36: 479–488.

Askbrant, S. and D.J. Farrell. 1987. Utilization of oil in seed meals determined with chickens at different ages. In: Recent Advances in Animal Nutrition in Australia 1987. D.J. Farrell (Ed.). University of New England, Armidale, Australia, 182–186.

Baidya, N., P. Biswas and L. Mandal. 1995. Metabolizable energy value of expeller copra (*Cocos nucifera*) cake in layers. Indian Vet. J. 72: 767–768.

Balasubramaniam, K. 1976. Polysaccharides of the kernel of maturing and matured coconuts. J. Food Sci. 41: 1370–1373.

Farrell, D.J. 1994. Utilization of rice bran in diets for domestic fowl and ducklings. World's Poult. Sci. J. 50: 115–131.

Fronda, F.M. and M.P. Mallonga. 1935. Protein supplements in poultry rations. V. Copra meal as a supplement in rations for growing chicks. Phillip. Agric. 24: 326–336.

Gomez, M.X. and D. Polin. 1974. Influence of cholic acid on the utilization of fats in the growing chicken. Poult. Sci. 53: 773–781.

Gunstone, F.D. 1996. Fatty acid and lipid chemistry. Blackie Academic and Professional, Glasgow, UK.

Houston, D.F. 1972. Rice bran and polish. In: Rice Chemistry and Technology. D.F. Houston (Ed.). American Association of Cereal Chemists Inc., St. Paul, MN, 272.

Ketels, E. 1994. The metabolizable energy values of fats in poultry diets. PhD Thesis, University of Ghent, Belgium.

Krogdahl, A. 1985. Digestion and absorption of lipids in poultry. J. Nutr. 115: 675–685.

Momongan, V.G., L.S. Castillo, A.R. Gatapia and R.S. Resurreccion. 1964. High levels of copra meal in poultry and livestock rations. I. Methionine and lysine supplementation in broiler rations. Phillip. Agric. 48: 163–180.

Mugford, D. 1993. Current methods for measurement of dietary fibre: choices and suitability. In: Dietary Fibre and Beyond. S. Samir and G. Annison (Eds.). Occasional Publications, Vol. 1. Nutrition Society of Australia, 19–36.

Nagura, D. 1964. The use of coconut oil meal in chick diets. Ceylon Vet. J. 12: 40–42.

Panigrahi, S. 1991. Metabolizable energy (ME) value of high residual lipid copra meal in formulating broiler chick diets. Trop. Sci. 31: 141–145.

Panigrahi, S. 1992. Effects of different copra meals and amino acid supplementation on broiler chick growth. Br. Poult. Sci 33: 683–687.

Panigrahi, S., D.H. Machin, W.H. Parr and J. Bainton. 1987. Responses of broiler chicks to dietary copra cake of high lipid content. Br. Poult. Sci 28: 589–600.

Polin, D., T.L. Wing, K. Ping and K.E. Pell. 1980. The effect of bile acids and lipase on absorption of tallow in young chicks. Poult. Sci. 59: 2738–2743.

Saittagaroon, S., S. Kawakishi and M. Namiki. 1983. Characterisation of polysaccharides of copra meal. J. Sci. Food Agric. 34: 855–860.

Saunders, R.M. 1986. Rice bran composition and potential food uses. Food Rev. Int. 1: 465–495.

Smits, C.H.M. 1996. Viscosity of dietary fibre in relation to lipid digestibility in broiler chickens. PhD Thesis, Wageningen Agricultural University, The Netherlands.

Teves, F.G., A.F. Zamora, M.R. Calapardo and E.S. Luis. 1988. Nutritional value of copra meal treated with bacterial mannanase in broiler diets. In: Recent Advances in Biotechnology and Applied Biology. Proceedings of 8th International Conference on Global Impacts of Applied Microbiology and International Conference on Applied Biology and Biotechnology. Chinese University Press, Hong Kong, 497–507.

Thomas, D.V., I.T. Kadim, P.J. Moughan and S. Bourne. 1997. Effect of lipase supplementation of rice bran on excreta energy content in adult cockerels. In: Proceedings of the Australian Poultry Symposium (In press).

Thomas, O.A. and M.L. Scott. 1962. Coconut oil meal as a protein supplement in practical poultry diets. Poult. Sci. 41: 477–485.

Thorne, P.J., J. Wiseman, D.J.A. Cole and D.H. Machin. 1988. Use of diets containing high levels of copra meal for growing/finishing pigs. Trop. Agric. (Trinidad) 65: 197–201.

Thorne, P.J., D.J.A. Cole and J. Wiseman. 1990. Copra meal. In: Non-traditional Feed Sources for Use in Swine Production. P.A. Thacker and R.N. Kirkwood (Eds.). Butterworths, MA, 123–129.

Warren, B.E. and D.J. Farrell. 1990a. The nutritive value of full-fat and defatted Australian rice bran. I. Chemical composition. Anim. Feed Sci. Tech. 27: 219–228.

Warren, B.E. and D.J. Farrell. 1990b. The nutritive value of full-fat and defatted Australian rice bran. II. Growth studies with chickens, rats and pigs. Anim. Feed Sci. Tech. 27: 229–246.

Warren, B.E. and D.J. Farrell. 1990c. The nutritive value of full-fat and defatted Australian rice bran. III. The apparent digestible energy content of defatted rice bran in rats and pigs and the metabolisability of energy and nutrients in defatted and full-fat bran in chickens and adult cockerels. Anim. Feed Sci. Tech. 27: 247–257.

Warren, B.E. and D.J. Farrell. 1990d. The nutritive value of full-fat and defatted Australian rice bran. IV. Egg production of hens on diets with defatted rice bran. Anim. Feed Sci. Tech. 27: 259–268.

THE PERFORMANCE OF BROILERS FED WITH DIETS CONTAINING ALLZYME VEGPRO®

M.J. SCHANG[1], J.O. AZCONA[1] and J.E. ARIAS[2]

[1] *INTA – Instituto Nacional de Tecnología Agropecuaria, Pergamino, Argentina,*
[2] *Alltech Inc., Nicholasville, Kentucky, USA*

Introduction

The high prices of cereal grains and oil seeds seen across the world may vary in the future, although it is estimated that prices will remain higher than in the past. This situation has not only brought about an increase in animal feeding costs but has also further challenged animal nutritionists to find new alternatives to improve efficiency of food utilization by animals (Wyatt and Graham, 1996). In this regard enzyme supplements for feed ingredients besides cereals are receiving increased attention. Enzyme supplements, particularly cocktails, are added to improve the digestibility of feed ingredients and reduce the detrimental effects of antinutritional factors present in plant protein sources (Table 1, Charlton, 1996).

Since 1958 most of the research efforts in the enzyme field were dedicated to improve the digestibility of some cereals that could replace corn usage in animal feeds (Fry *et al.*, 1958). Basically the results of this work showed that the addition of the enzyme ß-glucanase improved performance of growing birds when fed barley-based diets while the inclusion of pentosanases to either wheat or rye-based diets also resulted in better poultry performance (Leslie, 1995).

The presence of non-starch polysaccharide (NSP) compounds in feed grains causes an increase in digesta viscosity in the intestine. This increase in viscosity reduces nutrient digestibility and absorption with a resulting negative impact on performance of non-ruminant animals (Bedford, 1991).

As previously mentioned, the main approach to enzyme use in animal feeds has been to improve cereal utilization. However, recently it has been shown there is great potential in improving the digestibility of other components of the diet, in particular soybean meal (Pugh, 1995). Work by Bonino *et al.* (1991) and Azcona and Schang (1991, 1993) showed that the amount of gross energy that is metabolized in corn is larger than in soybean meal (89–91% vs 60–75%). The work conducted by McNab and Pugh (1993, cited by Charlton, 1996) indicated that addition of Allzyme Vegpro increased true metabolizable energy (nitrogen corrected) (TME) values of several plant protein sources by 4 to 14% (Figure 1).

Table 1. Some antinutritional factors present in several protein sources.

Soybean meal	Trypsin inhibitors, lectins, saponins, oligosaccharides (raffinose, stachyose)
Rapeseed meal	Glucosinolinates, tannins, phenolic acids, fiber
Sunflower meal	Fiber
Lupins	Alkaloids, fiber
Peas	Lectins, tannins, fiber, oligosaccharides

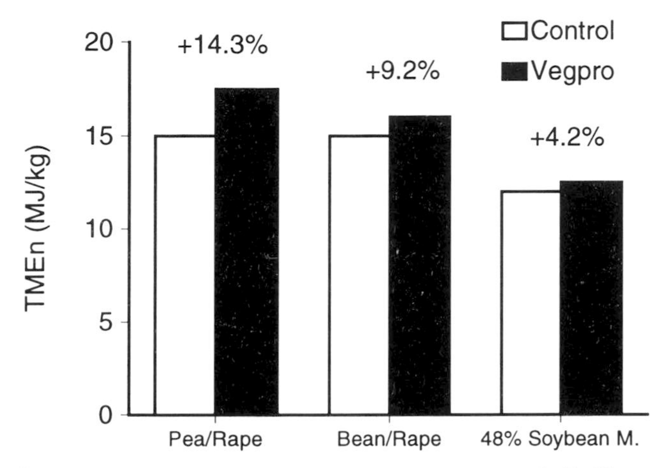

Figure 1. Effect of Vegpro addition on nitrogen-corrected true metabolizable energy (TMEn) of several plant protein sources (from Charlton, 1996).

Based on the information previously reported and considering that in many countries poultry diets are formulated to contain a combination of corn, soybean meal, full fat soybeans and wheat bran, an experiment was designed to evaluate effects of an enzyme complement designed primarily for soy (Allzyme Vegpro, Alltech Inc.) in diets containing different amounts of these ingredients.

Materials and methods

The experiment was conducted at the Poultry Section of the National Institute of Agricultural Technology (INTA) located at Pergamino, Buenos Aires, Argentina. There were eight experimental treatments with five replicates of 10 birds each in a 2 × 2 × 2 factorial arrangement (Table 2). Half of the diets were formulated to supply 100% of nutrient needs while the other half were calculated to supply approximately 90% of energy and available amino acid requirements. Diets of each group were formulated to contain either corn/soybean meal (C+SBM) or corn/soybean meal/full fat soy bean/wheat bran (C+SBM+FFS+WB). Finally, each of the diets was evaluated with and without Allzyme Vegpro addition.

Day-old Arbor Acres broiler males were placed in battery pens and given starter (days 1 to 21), finisher (days 22 to 42) and market (days 43 to 53) diets. Allzyme Vegpro was added at 300 and 200 g/tonne of feed for the starter

Table 2. Experimental treatments.

Treatment	Ingredient	Nutrient density	Allzyme Vegpro
1	C+SBM	High	Without
2	"	"	With
3	"	Low	Without
4	"	"	With
5	C+SBM+FFS+WB	High	Without
6	"	"	With
7	"	Low	Without
8	"	"	With

C+SBM=Corn+soybean meal; C+SBM+FFS+WB=Corn+soybean meal+full fat soy+wheat bran.

Table 3. Calculated nutrient content of experimental diets.

	Diet density	
Nutrient	High	Low
Starter (days 0 to 21)		
Protein (%)	21.5	19.4
TME (kcal/kg)	3350	3115
Met+cys (%)	0.85	0.76
Lys (%)	1.30	1.13
Finisher (days 22 to 42)		
Protein (%)	20.0	18.0
TME (kcal/kg)	3500	3255
Met+cys (%)	0.84	0.75
Lys (%)	1.19	1.04
Market (days 43 to 52)		
Protein (%)	18.0	16.2
TME (kcal/kg)	3550	3300
Met+cys (%)	0.69	0.62
Lys (%)	1.05	0.90

TME = True metabolizable energy.

and finisher/market periods, respectively. The composition of the experimental diets is presented in Table 3.

Once a week all birds were individually weighed and feed consumption was recorded by pen. For each period feed conversion was calculated on a pen basis. At day 53 of age, 10 birds from each treatment were slaughtered and the fat pad weight recorded. The results were subjected to analysis of variance.

Results and discussion

The 'ingredient' factor did not affect body weight gain (P>0.05), although broilers given the C+SBM+FFS+WB diets had higher (P<0.05) feed consumption and higher (P<0.05) feed conversion than those given C+SBM diets (Table 4).

Table 4. Performance results according to the source of variation at different days of age.

	Weight gain (kg)			Feed consumption (kg)			Feed conversion		
Age, days	21	42	53	21	42	53	21	42	53
Ingredient (A)									
C+SBM	0.67	2.11	2.74	1.06	3.85[a]	5.69[a]	1.58[a]	1.83[a]	2.09[a]
C+SBM+FFS+WB	0.67	2.10	2.71	1.08	4.02[b]	5.96[b]	1.63[b]	1.92[b]	2.20[b]
Diet density (B)									
High	0.71[a]	2.24[a]	2.86[a]	1.08	3.97	5.84	1.52[a]	1.77[a]	2.04[a]
Low	0.62[b]	1.98[b]	2.59[b]	1.05	3.90	5.81	1.70[b]	1.97[b]	2.24[b]
Allzyme Vegpro (C)									
Without	0.66	2.09[a]	2.70	1.05[a]	3.92	5.81	1.59[a]	1.88	2.15
With	0.67	2.12[b]	2.74	1.09[b]	3.95	5.84	1.62[b]	1.87	2.14
Interaction									
A × B	ns	ns	ns	ns	ns	ns	ns	ns	ns
A × C	**	**	ns	ns	ns	ns	***	***	ns
B × C	ns	**	ns	ns	ns	ns	ns	ns	ns

Means within the same column with different letters are significantly different $P<0.05$.
C+SBM = Corn + soybean meal; C+SBM+FFS+WB = Corn + soybean meal + full fat soy + wheat bran.
** $P<0.05$; *** $P<0.01$; ns, nonsignificant.

The differences in feed consumption may have originated in differences in the energy content of the diets. This hypothesis was confirmed in light of the effects on TME determined on experimental diets using the method of Sibbald (1976). These data showed that C+SBM+FFS+WB diets were 100 cal/g lower than in C+SBM rations. It is possible that an overestimation of the energy content of full fat soybeans could have been the origin of the observed TME differences.

As expected, a significant ($P<0.05$) effect of diet density on the performance of the birds was observed. Birds receiving a low density diet had a lower body weight gain and feed conversion than those given high density diets.

Response to Allzyme Vegpro addition depended on diet density. The enzyme effect also varied with diet composition (SBM alone or in combination with FFS+WB, Table 5). When birds were fed high density diets no effect of Allzyme Vegpro addition was observed in any of the evaluated parameters ($P>0.05$). These results would be in agreement with the comments of Charlton (1996) indicating that the benefits of using enzymes may be limited when added to diets formulated to exceed animal nutrient requirements.

On the contrary, the addition of Allzyme Vegpro to low density diets containing C+SBM+FFS+WB resulted in a significant ($P<0.05$) improvement in body weight gain. When the enzyme was added to C+SBM diets, however, no effect was observed on this parameter.

The larger abdominal fat content of birds receiving diets containing Allzyme Vegpro may be related to the capacity of the enzyme to increase energy rather

Table 5. Effect of diet density, composition and Vegpro supplementation on broiler performance at 53 days.

T	Weight gain (kg)	Δ (%)	Feed intake (kg)	Δ (%)	FCR	Δ (%)	Abdominal fat (g)	Δ (%)
1	2.84[a]		5.70[cd]		2.01[d]		3.0	
2	2.89[a]	1.7	5.76[bcd]	1.0	1.99[d]	−1.0	3.2	6.7
3	2.61[b]		5.65[d]		2.17[b]		2.9	
4	2.60[bc]	−0.3	5.65[d]	0.0	2.18[b]	0.5	3.8	31.0*
5	2.85[a]		5.99[ab]		2.10[c]		3.6	
6	2.84[a]	−0.4	5.91[abc]	−1.3	2.08[c]	−1.0	3.5	−2.8
7	2.52[c]		5.87[abcd]		2.33[a]	3.2		
8	2.64[b]	4.7 **	6.06[a]	3.2	2.30[a]	−1.3	3.8	19.0*
CV (%)	2.3		3.0		2.0		31.4	

Means within the same column with different letters are statistically different ($P<0.05$).
** ($P<0.05$); * ($P<0.10$).
T = treatment; Δ = difference; CV = coefficient of variation.

than the amino acid digestibility of the feed. This observation would be in agreement with the results previously found by McNab and Pugh in 1993 (cited by Charlton, 1996).

Under practical circumstances when formulating poultry diets there are two approaches to evaluate the economic impact of enzyme addition (Pugh, 1995; Wyatt and Graham, 1996). First, assign added energy and amino acid content to the diet ingredients; or second, reduce by certain amounts the requirement for energy and amino acids in the matrix. In this regard, prices for the two major ingredients used in Latin American poultry diets (corn and soybean meal) together with broiler retail prices were collected from several countries in October, 1996 (Figure 2). Considering that feed cost represents between

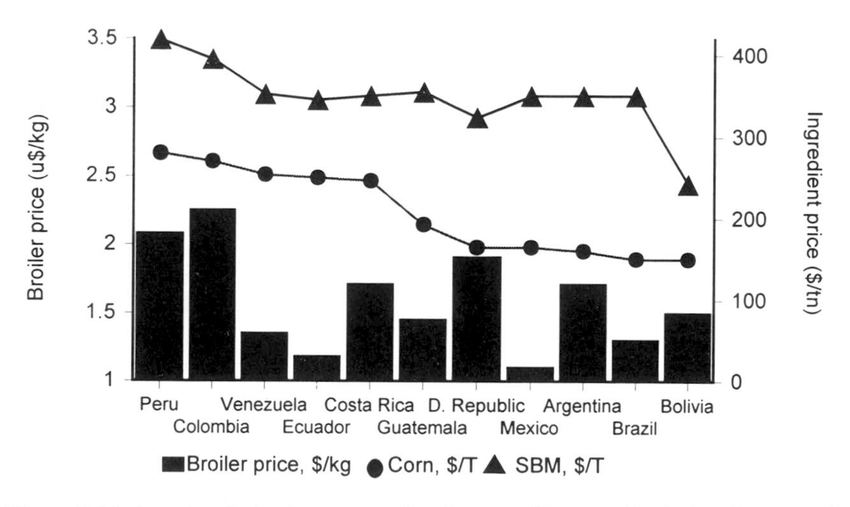

Figure 2. Market prices for broilers, corn and soybean meal in several Latin American countries.

60 and 80% of total production costs and taking into account that two major feedstuffs represent most of that feed cost, it is possible to define a range where the use of enzymes could result in some important cost reductions. This reduction in feeding costs would improve the actual benefit:cost ratio in many Latin American countries making broiler companies more competitive worldwide. The uses of any of the restrictions mentioned above in a least-cost computer program may help nutritionists in achieving that goal.

Conclusions

The results obtained in this experiment demonstrate that the addition of Allzyme Vegpro to C+SBM+FFS+WB low density diets improved broiler body weight gain and broiler feed conversion by 4.5 and 1.5%, respectively. These results may be due to the improvement in energy availability (Charlton, 1996) and amino acid digestibilities (Pugh, 1995). Additional studies might more specifically determine the economic impact of this enzyme in individual markets.

References

Azcona, J.O. and M.J. Schang. 1991. Poroto de soja, maices y sorgo, su uso eficiente en la alimentación de aves. Nutrición Animal Aplicada, 19:32–35.

Azcona, J.O. and M.J. Schang. 1993. Digestibilidad verdadera de aminoacidos en materias primas argentinas. Balanceados Argentinos, CAFAB, 73:24–28.

Bedford, M.R., H.L. Classen and G.L. Campbell. 1991. The effect of pelleting, salt and pentosanase on the viscosity of intestinal contents and the performance of broilers fed rye-based diets. Poultry Sci. 70:1571.

Bonino, M.F., M.J. Schang and J.O. Azcona. 1991. Tabla de composición de ingredientes argentinos. Balanceados Argentinos, CAFAB, 63:32–70.

Charlton, P. 1996. Expanding enzyme applications: Higher amino acid and energy values for vegetable proteins. Proceedings of the 12th Annual Symposium on Biotechnology in the Feed Industry. Nottingham University Press, Loughborough, Leics., Great Britain pp. 317–326.

Fry, R., J.B. Allred, L.O. Jensen and J. McGinnis. 1958. Effect of pearling barley and of different supplements to diets containing barley on chick growth and feed efficiency. Poultry Sci. 37:281–288.

Leslie, A.J. 1995. Future potential for cereal enzymes. Proceedings of the International Launch of Allzyme Vegpro. Alltech, Inc., USA.

Pugh, R. 1995. Allzyme Vegpro: Its history and application for soya and legumes. Implications for the poultry industry worldwide. Alltech, Inc., USA.

Sibbald, I.R. 1976. A bioassay for true metabolizable energy in feedingstuffs. Poultry Sci. 55:303–308.

Wyatt, C. and H. Graham. 1996. Enzymes to the rescue. Feed Management, 47:18–22.

ENZYME APPLICATIONS IN CORN/SOYA DIETS FED PIGS

DAVID I. KITCHEN

Amalgamated Farmers, plc, Kinross, New Hall Lane, Preston, Lancashire, UK

Introduction

Legislation in the UK has banned the use of mammalian meat and bone meal in animal feedstuffs. There is, therefore, an increase in the use of vegetable proteins of which extracted soyabean meal is a major contributor. Some maize by-products are used, but wheat and barley constitute the greater portion of the content of pig diets. It is probable that less maize by-products will be used in future if the 'anti-genetically modified corn' lobby are successful in their endeavors.

Against this background, nutritionists have to continue to provide feed specifications which maximize the biological and economic efficiency of UK pig production in an increasingly competitive worldwide pig industry.

Biotechnology now provides the nutritionist with a source of products which can be used to complement the pig's digestive and metabolic processes to improve both biological and economic efficiencies. Biosynthetically produced enzymes have been a significant factor in improving the economics of poultry production and there are strong indications that the economics of pig production can be improved by the incorporation of specific enzymes into pig feeds. Additionally, intensive livestock production is now critically evaluated by its impact on the environment. Enzymes may, therefore, have a significant role to play in ameliorating the output of animal waste nitrogen and phosphorus.

The recent high rate of genetic improvement in the pig also gives nutritionists a challenge to provide nutrients in specific ratios in order to meet increasingly higher levels of production. Responses to enzyme addition to feedstuffs will, therefore, not only depend on substrate availability, but also the physiological ability of the pig to use the resultant nutrients to meet its production potential. Such genotype-nutrition interactions are now being explored, particularly by UK pig breeding companies, who are issuing their own recommendations for feeding their particular genotype. It is, therefore, becoming increasingly important to understand the nutrient requirements of the modern pig, calculate how these can be met commercially, and then measure how the use of specific enzymes can improve performance and help the pig industry comply with environmental and consumer requirements.

Enzymes in young pig diets

Enzymes can be employed in dry feeding (meal, pellets), liquid feeding and wet feeding systems. Use of enzymes to pretreat feedstuffs offers much promise for the future. In the UK most enzymes are targeted at pigs between weaning and 30 kg liveweight because this is where the most consistent positive responses have been observed (Inborr, 1994). By definition, therefore, such enzymes are incorporated into pelleted feeds fed dry. Enzymes *per se* are only part of the total nutritional package and must be considered in the context of the dietary stresses to which pigs are exposed in modern feeding management and the nutritional requirements necessary to reach target growth rates.

The act of weaning has a profound effect on future growth rates. If this is managed badly it has a deleterious effect on lifetime performance. In 1985 Whittemore demonstrated that pigs growing at less than 300 g per day immediately post-weaning were losing body protein (Figure 1) and replacing this with body water, while liveweight remained static. Such pigs were subsequently shown to be disadvantaged for the rest of their productive life. Higher daily liveweight gains over a pig's productive life can be achieved by maximizing feed intake (Whittemore, 1985) and providing feed higher in digestibility. Unfortunately, these factors can be antagonistic. Increased feed intake can reduce digestibility through increasing the rate of passage and thereby limiting exposure of digesta to enzymatic hydrolysis and absorption. The opposite is also true. Reduced intake results in higher digestibility.

The digestibility of a feed substrate varies with the type and concentrations of enzymes present in the gastrointestinal tract. These change as the pig ages; and when a suckling pig is weaned off sow's milk onto solid feed (Aumaitre and Corring, 1978, Corring *et al.*, 1978). At the time of weaning (between 21 and 28 days of age) intestinal enzyme levels are at the lowest concentration for the digestion of both sow's milk and other diets (Figure 2). Similarly, Aumaitre (1972) demonstrated a seven-day time lag between the introduction of a new or increased level of substrate to the diet and the increase in enzyme activities to meet the increasing levels of substrate intake (Figure 3).

Thus at this point there is a window of opportunity to fortify the diet with supplementary enzymes to complement those being produced in the gut to improve diet digestibility. A similar observation was made by Dierick and Decuypere (1994). For some enzymes, such as amylase, this window of opportunity is limited by the rapid development of the pig's inherent digestive enzymes. For some enzymes, however, this window of opportunity is wider due to either the limited availability or absence of specific endogenous enzymes.

Viscosity of digesta does not appear to interfere with the process of digestion in the young pig to the same extent as in poultry (Dierick and Decuypere, 1994). The most probable explanations of the improved performance in young

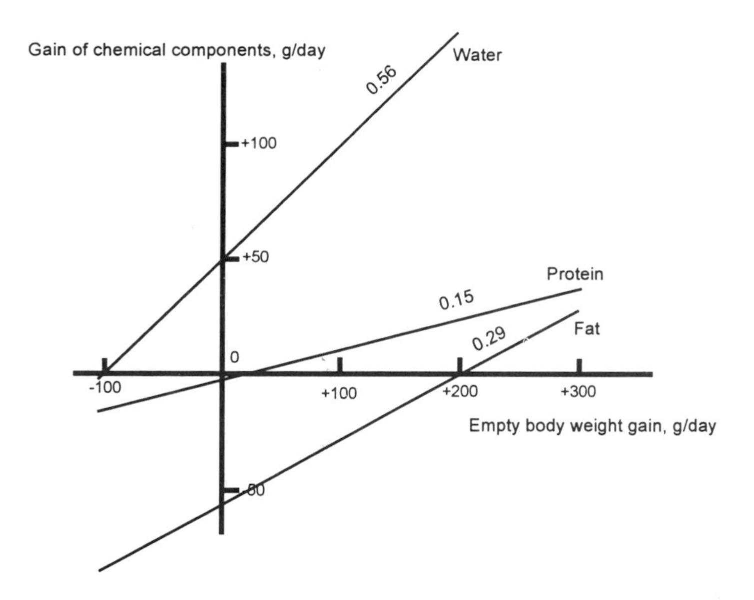

Figure 1. Chemical components of piglet gain in relation to rate of body weight gain (Whitemore, 1985). At 0 weight gain pigs lost about 50 g of fat but gained 50 g of water. Protein gain comprised about 15% total gain and occurred simultaneously with fatty tissue losses between 20 and 200 g daily body weight.

pigs due to enzyme addition are effects on factors such as feed transit time, endogenous losses, non-starch-polysaccharide (NSP) degradation and (or) changes in gut flora. These factors may be important, but Graham and Bedford (1993) suggested that the relative reduction in digesta viscosity due to enzyme addition is more important than the absolute value because reducing digesta viscosity means that less digestive enzyme is required to achieve the same rate of digestion than when digesta are more viscous. This would account for the improved weight gain and feed utilization reported by Inborr (1994). Viscosity of feed ingredients were reported to increase with pelleting temperature (Spring *et al.*, 1996). Cellulase reduced this viscosity by 10% where the pelleting temperature was 60°C and 18% where the pelleting temperature was 100°C. As most newly weaned pigs in the UK are fed dry extruded diets, pelleting temperature may have some implication in the variability of results reported when enzymes are included in pig diets.

Similarly, there are inherent dangers in using changes in fecal digestibilities as a measure of the effectiveness of exogenous enzymes. As enzymes mainly affect digestion in the small intestine, differences seen in fecal apparent digestibities are usually an underestimation of the real effect (Graham *et al.*, 1988). The nutrient requirement of the modern hybrid pig, together with developments in diet formulation and a better understanding of the nutritional qualities of dietary ingredients, have created opportunities to use specific supplementary enzymes in the diets of pigs after the immediate post-weaning period.

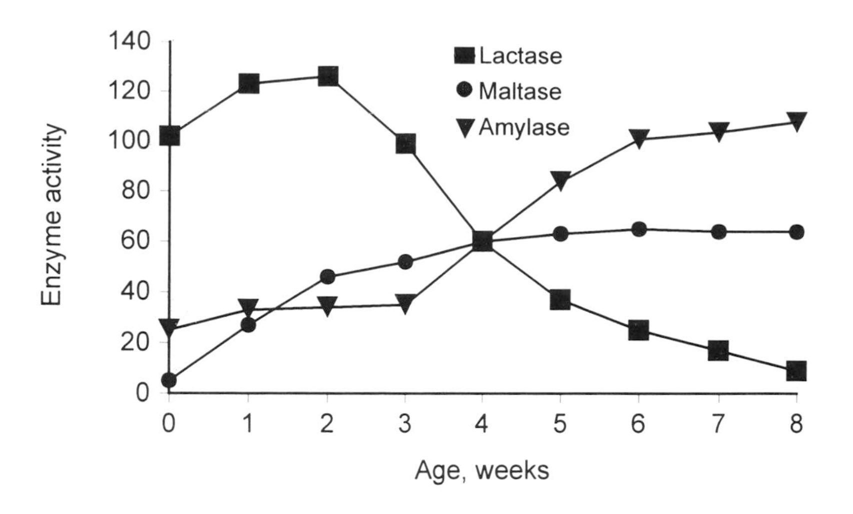

Figure 2. Effect of age on concentration of enzymes in the gut of the piglet (after Aumaitre and Corring, 1978).

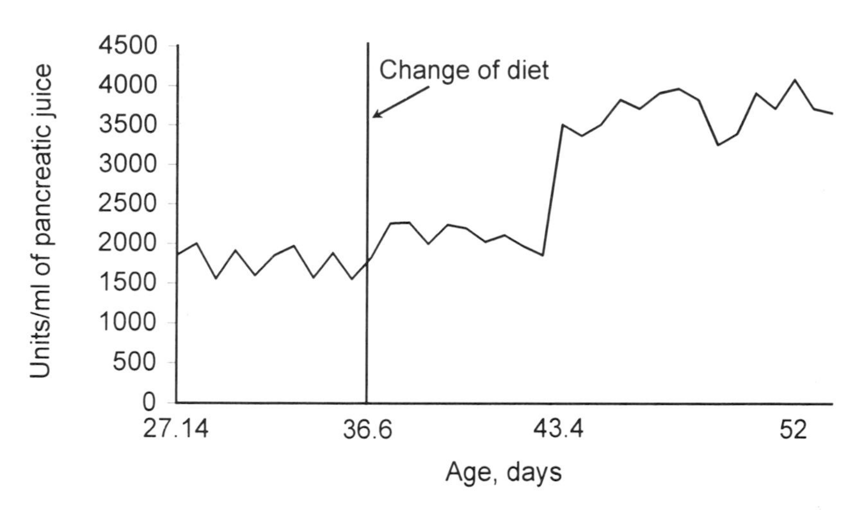

Figure 3. Pancreatic amylase activity in piglets following a change in diet (after Aumaitre, 1972).

ANTI-NUTRITIVE EFFECTS OF NSP

The digestible energy and total lysine requirements of growing pigs of 20 kg liveweight and above were calculated for the J.S.R. genotype (Close, 1994; Table 1). They can be applied, however, to most modern genotypes for the formulation of commercial feeds. Using the principles of 'Ideal Protein' (Cole,

1978; ARC, 1981) a diet was formulated to provide a total lysine content which was 7% of the total protein (Diet A, Table 2)

Table 1. Nutrient requirements of growing pigs*.

Liveweight (kg)	DE requirement, (MJ/day)	Lysine requirement (g/day)	Estimated intake (kg/day)
20	15.5	14.0	1.07
40	24.5	20.0	1.75
60	30.0	24.0	2.17
80	34.0	27.0	2.46
100	36.5	28.5	2.64

*After Close, 1994.

Table 2. Diet formulations to meet requirements of 20 kg growing pigs.

	Diet A	Diet B
Ingredients, %		
Wheat	49.7	39.1
Barley	15.6	10.0
Field peas	12.5	12.5
Full fat soya	4.2	7.4
Soybean meal (46%)	–	23.2
Extracted rape	5.0	2.2
Fish	7.5	–
Vegetable oil	2.5	2.5
Calcium carbonate	0.9	0.8
Salt	0.22	0.33
Dicalcium phosphate	0.90	1.39
L-lysine	0.37	0.06
DL-methionine	0.02	0.02
L-threonine	0.06	–
Vitamin/mineral supplement	0.50	0.50
Calculated analysis		
Oil, %	6.08	5.93
Crude protein, %	18.51	22.50
Crude fiber, %	3.60	3.69
Digestible energy, MJ/kg	14.80	14.80
Total lysine, g/kg	13.10	12.90
Lysine:DE	0.89	0.87
Lysine:crude protein	0.07	0.06
Non-starch polysaccharide, g/kg DM	126.0	156.0

The NSP content of diet A was 126 g/kg of dry matter as calculated from the data of Longland and Low (1995). The cost of such a diet is prohibitive, being £50 (January, 1997) more than diet B, which is formulated to more realistic commercial specifications. In this case the increased level of crude protein excluded fishmeal and replaced it with extracted soyabean meal (48%). The NSP content of diet B was 24% higher than diet A, at 156 g/kg of dry matter. NSP consist of a wide range of plant polymers which include cellulose, hemicellulose, pectic substances, ß-glucans, pentosans (arabinoxylans) and α-galactosidases (raffinose, stachyose, vebusose and ajucose) (Dierick and Decuypere, 1994). These complex substances influence many of the chemical processes which make up the digestive process of the pig. This in turn influences the digestion of the feed. Some of these effects have been reviewed by Low (1985) and Dierick (1989).

Longland *et al.* (1993) demonstrated that cereal diets containing higher levels of NSP have a lower digestibility and need a longer period of adaptation to maximize digestibility than diets with lower levels of NSP. This period of adaptation will result in, at best, poor feed conversion efficiency and higher cost per kilogram of liveweight gain, and at worst poor feed conversion together with reduced daily liveweight gain. Thus diets for pigs between weaning and 35 kg liveweight which contain high levels of NSP will have negative effects on production. Adding specific enzymes will improve such diets by reducing the anti-nutritive effects (Dierick and Decuypere, 1994).

Arabinoxylanase, ß-glucanase and α-galactosidase enzymes have been reported to increase the total metabolizable energy of barley by 10%, wheat by 6% (Anonymous, 1996) and extracted soyabean meal by 7% for broiler chickens (Charlton, 1996). They may exhibit a similar effect on digestible energy when included in pig diets.

In conclusion, efficiency of pig lifetime growth rate is maximized when pigs grow at a minimum rate of 300 g per day throughout the weaning process. The use of exogenous enzymes to complement those of the pig at this time appears to help achieve this target. Similarly, commercial pressures on feed costs dictate the use of feed ingredients which may contain anti-nutritive factors such as NSP. Once again, added enzymes appear to ameliorate anti-nutritive effects and encourage cost-efficient growth.

FARM TRIALS USING ALLZYME VEGPRO

The effectiveness of adding Allzyme Vegpro, an enzyme formulated to increase digestibility of soya, was tested in growing pig diets during autumn and winter of 1995/96 in a series of on-farm trials. A control diet was manufactured from cooked and uncooked cereals, cereal grain by-products, oil seed products and by-products including extracted soyabean meal, fish meal, whole heat-treated oil seeds, milk products, vegetable oils and vitamin and trace minerals. The test diet was manufactured by adding Allzyme Vegpro to the control diet at the rate of 500 g per tonne. Vegpro provided a mixture of enzymes, including arabinoxylanase, ß-glucanase and α-galactosidase. Five groups of pigs were fed the control diet and four groups the treatment diet.

Pigs fed the diet containing the Vegpro consumed more feed; but converted this feed better than did pigs fed diet A (Table 3). As a consequence they had, on average, a 60 g per day higher liveweight gain than the pigs fed the control diet. The treatment diet over control diet ratios for daily liveweight gain and feed conversion ratios were 1.1 and 0.95, respectively. These were of the same magnitude as those calculated by Dierick and Decuypere (1994) for similar blends of enzymes. Although the mean differences in performance between treatments were not statistically significant ($P>0.05$), when the economic principle of pig liveweight produced per tonne of feed was applied the diet containing Vegpro produced an extra 27.88 kg per tonne of feed. This is equivalent to £33.46 extra income at a liveweight price of £1.20 per kilogram (Gadd, personal communication).

Subsequently an additional series of trials were conducted on commercial farms as part of a competition for the best performance of pigs fed diets containing Vegpro. A small sample of the results are given in Table 4. It is worthy of note that product temperature varied between 30°C and 60°C during the manufacture of the diets used. Hence the conditions recorded by Spring *et al.* (1996) were infrequently reached.

Comments from the farms about the appearance of the pigs fed the diet containing Vegpro were very complimentary. The pigs had less 'non-specific scour' as a result of which they were cleaner and had more 'bloom'. This effect was also described by Inborr and Ogle (1988). The 'bloom' had a positive aesthetic effect on the stockmen which indirectly lead to an increase in feed sales.

In a recent trial reported from a commercial testing facility in the US the addition of 1 kg Alltech Vegpro to diets fed to 5 kg liveweight pigs gave a daily liveweight gain ratio of 1.03:1 over the diet without enzyme (Karnezoes, personal communication; Table 5). Once again this ratio agrees with those reported by Dierick and Decuypere (1994). In this trial an extra positive economic value of £42 was calculated for the extra liveweight produced by a tonne of feed containing Vegpro (Gadd, personal communication).

Enzymes and the environment

The environmental effect of enzymes included in pig diets cannot be ignored. The European public are becoming more aware and vociferous about environmental pollution from intensive livestock production. The main concern on mainland Europe is due to increasing nitrogen and phosphorus pollution of water supplies. In the UK phosphorus levels in water do not appear to be a problem; however nitrogen pollution as nitrate and ammonia in water and the atmosphere is regarded with concern. It was estimated that 80% of ammonia emissions contribute to the 'acid rain' effect (Pig Industry, 1996) which is having a deleterious effect on some forestry plantations. Though use of enzymes to reduce nitrogen excretion is comparatively new, the use of phytase to reduce phosphorus excretion offers more immediate promise. Jongbloed *et al.* (1993) demonstrated increased digestion of phosphorus from plant phytate when phytase was incorporated into the diet of pigs between 11 kg and 35 kg liveweight. They also reported that without microbial phytase only approximately 16% of phosphorus in maize and approximately 38% of phosphorus in soyabean meal is digested by pigs. The remaining dietary phosphorus is excreted via the urine and feces.

In conclusion, data from both the UK and USA have demonstrated that in commercial trials young pigs fed diets containing Allzyme Vegpro grow faster and return a greater income per tonne of feed than those fed a diet without the enzyme. It is also demonstrated that diets containing phytase reduce

Table 3. Performance of pigs fed diets with and without Allzyme Vegpro.

	Trial No.										
	1 Control	2 Control	3 Control	4 Control	5 Control	ALL Control	ALL Vegpro	6 Vegpro	7 Vegpro	8 Vegpro	9 Vegpro
Number of pigs	19	18	19	18	18	92	74	18	19	18	19
Start weight, total kg	307	337	378	405	409			297	390	378	462
Average start weight, kg	16.2	18.7	19.9	22.5	22.7	20.0	20.6	16.5	20.5	21.0	24.3
Days on trial	7	6	7	7	7			7	6	7	7
Average end weight, kg	19.9	22.5	25.1	26.7	26.3	24.1	25.1	20.9	24.3	26.3	28.8
Total end weight, kg	378	405	476	481	474			377	462	474	547
Total feed, kg	112	100	150	133	117			111	118	142	146
Daily gain, g	534	630	737	603	516	604	667	635	632	762	639
Daily feed per pig, g	842	926	1128	1056	929	978	1035	881	1035	1127	1098
Feed Conversion	1.58	1.47	1.53	1.75	1.80	1.62	1.55	1.39	1.64	1.48	1.72
Liveweight/tonne of feed, kg*						617.28	645.16				
Extra liveweight, kg*						–	27.88				

* Gadd, personal communication.
Courtesy of AF plc.

Table 4. Performance results of feeding control diet plus enzyme in a series of farm trials.

	Trial				
	1	2	3	4	5
Number of pigs	40	48	17	29	15
Mean start weight, kg	7.15	10.02	9.41	8.45	13.60
Mean end weight, kg	13.88	15.50	15.00	18.17	25.07
Mean daily intake, kg	327.0	465.0	504.2	575.0	853.3
Days on test	21	13	14	21	20
Mean daily gain, g	320.24	421.50	399.20	463.00	573.30
Mean FCR	1.02	1.10	1.26	1.24	1.49

Courtesy of AF plc.

Table 5. Response of pigs to inclusion of Vegpro*.

	Control	Vegpro
Total food eaten, kg 16–35 days	4.908	4.772
Weight gain, kg 16–35 days	3.66	3.78
Pigs per tonne of feed, kg	203.7	209.5
Pig weight per tonne of feed, kg	750	792
Extra liveweight/tonne of feed, kg	–	+42
Value at £1.20/kg at 35 days	–	50 ($84)

*Vegpro added at 1kg/tonne.

phosphorus pollution from pig enterprises. This, however, appears only to be of economic value at present where phosphorus pollution is perceived as a problem.

Enzymes addition to pig diets during the growing/finishing phase

In the older pig, data in the literature would suggest that enzymes are most effectively used either as a pretreatment of the diet, or in liquid feeding systems (Dierick and Decuypere, 1994; Brooks, 1995). Indeed, Brooks (1995) reported that Alltech's Allzyme Phytase was very effective at releasing phosphorus from feed ingredients which have little or no natural phytase such as soyabean meal when added to a single ingredient in a liquid medium (Figures 4 and 5). This improvement in the bioavailability of phosphorus means that a smaller dietary inclusion of phosphorus from inorganic sources is necessary and the levels in effluent can be reduced.

From practical experience, the inclusion of an enzyme mixture containing xylanase and ß-glucanase in a wet fed wheat, barley, soyabean, wheat feed and liquid whey diet, improved uniformity of growth and cleanliness of pigs between 35 and 85 kg liveweight. This resulted in increased sales of other pig feeds to this group of farmers! The work of Professor Brooks and his team at the University of Plymouth in England, where they are exploring possible techniques for modifying feed ingredients in liquid feeding systems by

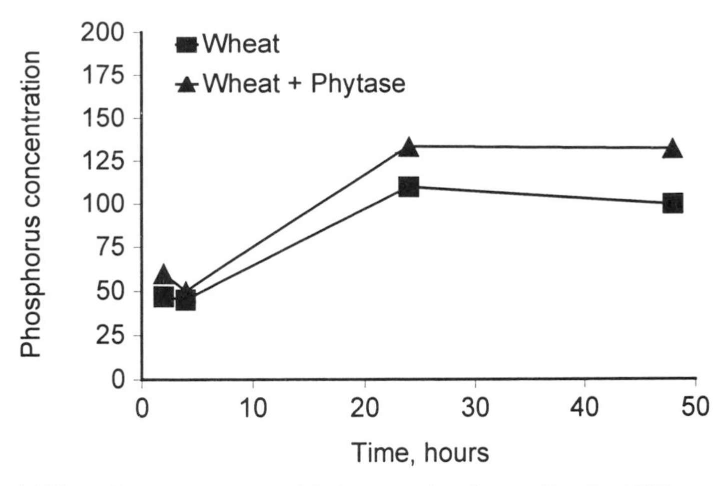

Figure 4. Effect of steeping raw materials in water plus phytase (Brooks, 1995).

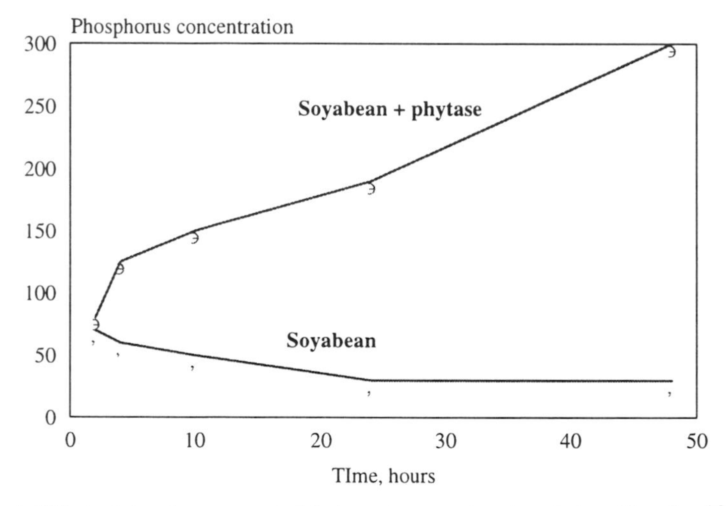

Figure 5. Effect of steeping raw materials in water or water plus phytase (Brooks, 1995).

biotechnical processes, could have a significant effect on how the pig is fed in future.

In a recent experiment conducted by Lindemann at the University of Kentucky it was demonstrated that Allzyme Vegpro may have a greater effect on the performance of pigs of less than 60 kg liveweight than in heavier pigs. There was an effect of lesser magnitude between 60 and 109 kg liveweight. Pigs between 26 kg and 63 kg liveweight were fed either a high density, corn/soya diet (CS) or a lower density corn/soya/midds (CSM) diet with or without Vegpro. The enzyme treatement resulted in higher daily liveweight gains in the CS diet ($P \leq 0.0$), but not with the CSM diet (Table 6). Similarly, pigs fed the CS diet with Vegpro had a more effecient feed conversion ($P \leq 0.093$). Similar, but smaller responses were recorded in the CSM diet

Table 6. Effect of Vegpro addition to diets for growing/finishing pigs on performance.

	Corn/soya		Corn/soya/midds		Specific comparisons *P* Values[a] Vegpro in corn/soya	Vegpro in corn/soya /midds
Vegpro, –/+	–	+	–	+		
Vegpro						
26–63 kg liveweight						
Daily liveweight gain, g	771	838	723	711	0.046	NS
Daily feed intake, g	1982	1917	2099	1964	NS	NS
FCR	2.57	2.29	2.91	2.76	0.093	NS
63–109 kg liveweight						
Daily liveweight gain, g	823	882	788	776	0.81	NS
Daily feed intake, g	2703	2773	3076	3026	NS	NS
FCR	3.28	3.16	3.91	3.90	NS	NS
26–109 kg liveweight						
Daily liveweight gain, g	799	858	770	747	0.041	NS
Daily feed intake, g	2370	2356	2678	2550	NS	–
FCR	2.96	2.75	3.48	3.41	NS	NS

with and without Vegpro addition. The responses in the finishing period (63–109 kg) were in the same direction but of a smaller magnitude as those in the growing period. Thus the overall performance (26–109 kg) was significantly improved ($P<0.05$) for daily liveweight gain and FCR where Vegpro was included in the CS diet with no significant differences being recorded in either of the CSM diets.

As with most trials involving enzymes, the ratios of the performance of pigs on the enzyme treated diet over that of the pigs fed the diet without enzymes was between 0.9 and 1.09:1. These ratios are similar to those reported above. In the CS diet in particular this indicates a potential economic benefit to many pig producers when Vegpro is used.

Enzymes into the twenty-first century

To date most of the data on enzymes in corn/soya diets have been produced using the younger pig and focused on production parameters or phosphorus excretion. In the UK more research is now aimed at solving the environmental and welfare problems of pig production. Between 25 and 30% of sows are now housed outdoors, many in nitrogen pollution-sensitive areas where the land conditions are most suitable for year round outdoor pig production. In the UK, ADAS and CEDAR workers have demonstrated that stocking density and vegetation type have significant effects on nitrate leaching from land on which outdoor sows are kept. In fact, only a stocking density of 12 sows to the hectare on established grassland produced a sufficiently low level of nitrate leaching over the autumn and winter periods of a two-year trial to meet anticipated EU maximum limits (Pig Industry, 1996) (Figure 6).

As with phosphorus and phytase, there is a role for enzymes to improve

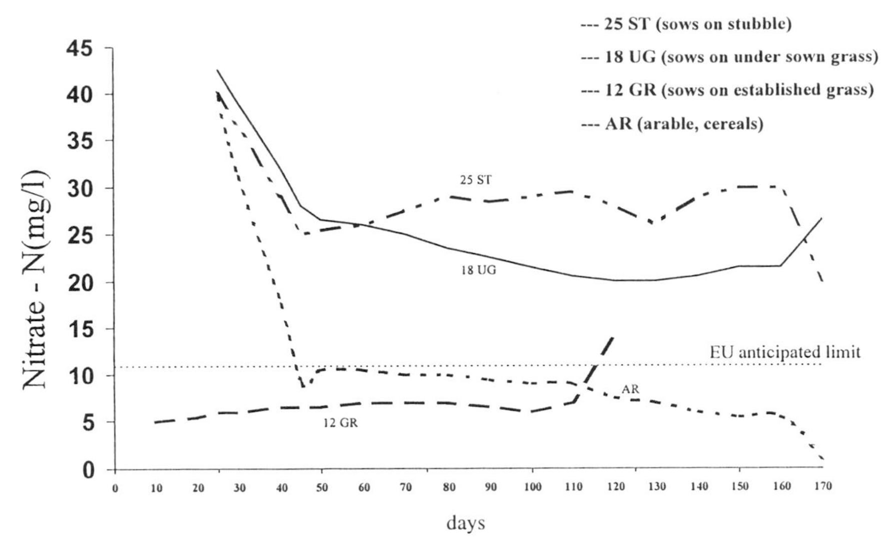

Figure 6. Nitrate concentrations in drainage water 1995/96 (adapted from Pig Industry, 1996).

nitrogen digestibility in pig diets, to reduce both the nitrogen levels of the diet and, as a consequence, the quantities of nitrogen excreted. Such enzymes may not consist only of proteases, but may consist of cellulases, xylanases and α-galactosidases. These would act together to release plant cell nitrogen before the ileal-cecal junction and hence improve nitrogen utilization and reduce environmental pollution.

References

Anonymous, 1996. Various commercial sources.

ARC. 1981. Agricultural Reserch Council. The Nutrient Requirements of Pigs. Commonwealth Agricultural Bureaux. Slough: England.

Aumaitre, A. 1972. Development of enzyme activity in the digestive tract of the suckling pig. Nutritional Significance and Implications for Weaning. In: World Reviews of Animal Production 8:54–68.

Aumaitre, A. and T. Corring. 1978. Development of digestive enzymes in the piglets from birth to 8 weeks. In: Nutritional Metabolism 22:231–255.

Brooks, P. 1995. Development in liquid feeding systems: The role of the compounder. In: Alltech's European Lecture Tour 1995, Alltech Inc.

Cole, DJA. 1978. Amino Acid Nutrition of the Pig. In: Recent Advances in Animal Nutrition. W. Haresign and D. Lewis (eds). Butterworths, London. 59–72.

Charlton, P. 1996. Expanding enzyme applications: higher amino acid and energy values for vegetable proteins. In: Biotecnology in the Feed Industry. Proceedings of the 12th Annual Symposium. (T.P. Lyons and K.A. Jacques, eds), Nottingham University Press, Loughborough, Leics. UK pp. 317–326.

Close, W.H. 1994. The Nutritional Requirements of the J.S.R. Genotype. J.S.R. Healthbred Limited, Southburn, Driffield, East Yorkshire, England.

Corring, T., A. Aumaitre and G. Durand. 1978. Nutrition and Metabolism. 22:24–255.
Dierick, N.A. 1989. Archives of Animal Nutrition: 39:241–261.
Dierick, N.A. and J.A. Decuypere. 1994. Supplementary enzymes to improve utilization of pig diets. Proceedings 45th Annual Meeting of EAAP, Edinburgh.
Graham, H. and M. Bedford. 1993. Feed enzymes, effect of heat treatment and enzymes in poultry feed. Amandus Kahl Seminar, Eurtatier, Hannover 12, p. 11.
Graham, H., W. Lowgren, D. Petterson and P. Aman. 1988. As cited by Graham, H. Enzymes in monogastric feeds. International Milling Flour and Feed. July, 1991. pp. 33–36.
Inborr, J. 1994. Supplementation of pig starter diets with carbohydrate degrading enzymes – stability, activity, and mode of action. Agricultural Science Finland 3(Suppl. 2):8.
Inborr, J. and R.D. Ogle. 1988. Effect of enzyme treatment of piglet feed on performance and post-weaning diarrhoea. Swedish Journal of Agricultural Research 18:129–133.
Jongbloed, A.W., P.A. Kemme, A. Mroz and Z. 1993. The role of microbial phytases in pig production. In: Enzymes in Animal Production. Proceedings of 1st Symposium, Kartause, Ittingen, Switzerland. pp. 173–179.
Longland, A.C. and A.G. Low. 1995. Prediction of the energy values of alternative feeds for pigs. In: Recent Advances in Animal Nutrition. (P.C. Garnsworthy and D.J.A. Cole, eds.) Nottingham University Press : 187–209.
Longland, A.C., A.G. Low, D.B. Quelch, and S.P. Bray. 1993. Adaptation to the digestion of non-starch-polysaccharides in growing pigs fed cereal and semi-purified based diets. British Journal of Nutrition 70:557–566.
Low, A. 1985. In: Digestive Physiology in the Pig. (Just *et al.* eds). Copenhagen National Institute of Animal Science . 157–179.
Pig Industry. 1996. Pollution's a 'hot potato'. The official journal of the British Pig Association. December/January 1996/1997:16–17. B.C. Publications, England.
Spring, P., K. Newman, C. Wenk, R. Messikommer and M. Vukic-Vranjes, 1996. Effect of pelleting temperature on the activity of different enzymes. Poultry Science . 75:285–438.
Whittemore, C.T. 1985. Nutrition of sow and weaner. The Feed Compounder. January, 1985. pp. 42–49.

THE USE OF ENZYMES IN RUMINANT DIETS

GEOFFREY ANNISON

Australian Food Council, Kingston, Australia

Introduction

Almost since the first appreciation that the digestive processes of all animals involved the breakdown of the macro-nutrients by endogenous enzymes, animal nutritionists have attempted to enhance the process by application of exogenous enzymes. This has been used most successfully in monogastrics and, particularly, in broiler chickens. The science behind the use of enzymes in broiler chickens is well reviewed (Annison and Choct, 1991; Annison, 1993). The reasons for the success of the application of exogenous enzymes in this species are:

1. advanced analytical techniques have been applied defining target substrates in the diets and identifying required activities of enzymes;
2. the relatively simple digestive tract of the chicken is particularly sensitive to nutritional perturbations which can subsequently be overcome by enzymes; and
3. the suitability of chickens as inexpensive laboratory animals allows large scale, statistically well designed experiments to elucidate the chemistry, biochemistry and physiology behind the nutritional observations. Subsequent product development is also greatly enhanced by the ease of experimentation with chickens.

In light of this, the use of enzymes to improve the quality of ruminants diets can be seen to be a far greater challenge. Although the same analytical techniques can be applied to ruminant feedstuffs, the digestive tract of ruminants is far more complex than that of chickens. Much is known, of course, about the rumen, but nevertheless attempts to modify its function with enzymes have been much less spectacular than seen in broiler chickens.

Notwithstanding the difficulties in using enzymes in ruminant diets stemming from the complexity of the rumen, and the expense of using sheep or cattle as experimental animals, opportunities for use of exogenous enzymes as feed additives exist. These are founded on two observations:

1. the digestibility of organic matter in ruminants rarely approaches 100% and is frequently considerably less; and
2. new feed raw materials, many of low quality, are being promoted for ruminants and enzymes may assist in their breakdown in the rumen or elsewhere in the digestive tract.

The use of enzymes in ruminants has recently been extensively reviewed (Beauchemin and Rode, 1996; Hristov *et al.*, 1996). This paper looks past the recent developments to some potential uses of enzymes in the future.

The use of enzymes in silage versus the use in diets

Enzymes may find application in ruminant nutrition in two ways:

1. as silage improving agents; or
2. as feed additives expected to act immediately prior to ingestion or at some point during the passage through the gastrointestinal tract of the animal.

In this paper only the application of enzymes in feeds will be discussed. The use of enzymes as silage improving agents holds great potential but it could be argued that this is more of a feed technology or biochemical engineering problem and less attention needs to be paid to the physiology, biochemistry and nutrition of the ruminant. The constraints on developing successful enzyme uses are quite different for the two situations.

Examples of promising results from enzyme use in ruminant diets

FORAGE-BASED DIETS

Beauchemin and Rode (1996) make the point well that experiments investigating the effects of enzymes in ruminant diets have recorded both positive and negative results since the 1960s. There are, however, several recent studies demonstrating positive effects of feed enzymes in diets fed to lactating and growing cattle. A fibrolytic enzyme preparation has been shown to improve the nutritive value of an alfalfa hay/alfalfa silage/barley mixed diet (Stokes and Zheng, 1995). Dry matter intake was increased by 10.7% compared to controls while milk yield was up by 14.8%.

Lewis *et al.* (1995) also demonstrate effects of enzymes on this type of diet. Cows fed the diet treated with enzyme produced 1.3 kg/day more milk than cows fed control diets. The feed intake of these cattle was also increased significantly by 2 kg/day.

Alfalfa hay diets were also shown to be amenable to enzyme treatment by Beauchemin and Rode (1996). An enzyme solution containing xylanase activity sprayed onto the forage resulted in an increase in average daily gain (ADG) of 13% without significant changes in dry matter (DM) intake. A second enzyme preparation containing xylanase and cellulase activities tested in the same

Table 1. Positive effects of fibrolytic enzymes in corn silage fed to cattle.*

	Enzyme level (×1)			
	0	15×	22.5×	30×
ADG[†], kg/day	1.06	1.07	1.12	1.23
% of control	100	101	106	116
DM[†] intake, kg/day	6.6	6.3	6.3	6.2
% of control	100	95	95	94
FCR[†], kg DM/kg gain	6.22	5.88	5.63	5.05
% of control	100	95	91	81

* From Beauchemin and Rode (1996).
[†] ADG, average daily gain; DM, dry matter; FCR, feed conversion ratio.

Table 2. Performance data for cows fed forage treated with two levels of enzyme.*

	Enzyme			
	Untreated	1.25 l/ton	2.5 l/ton	5 l/ton
Dry matter intake, kg/day	24.3[a†]	26.2[b]	26.1[b]	26.6[b]
Energy corrected milk yield, kg/day	41.1[a]	42.1[a]	48.1[b]	41.9[a]
Condition score, 1–5	2.6[a]	3.0[b]	2.6[a]	3.0[b]

* Sanchez *et al.* (1996). [†] Values with unlike letters are significantly different, $P<0.05$.

study at three different levels increased ADG and DM intake at the lowest level, depressed ADG and DM intake at the intermediate level and increased ADG, resulting in an improved feed conversion ratio (FCR), at the highest level.

A similar enzyme preparation has also been effective at raising the nutritional value of corn-silage diets when sprayed on before feeding. In this case high levels of enzyme were required but a convincing dose-response relationship for ADG and DM intake was observed (Table 1).

Less convincing dose-response data were reported by Sanchez *et al.* (1996) in an experiment in which dairy cattle were fed an alfalfa haylage mixed ration with three levels of addition of enzyme. The intermediate level of enzyme addition was seen to be more effective than the higher level compared to control animals which were fed an untreated diet (Table 2). A return of the energy corrected milk yield to levels similar to those of the controls at high levels of enzyme addition is surprising given the maintained increase in DM intake. The authors suggested a partitioning of energy towards improved body condition at the higher levels of enzyme of addition but the mechanism by which this could possibly happen remains obscure.

CEREAL-BASED DIETS

As with the studies of exogenous enzyme use in forage-based diets, experiments examining the effect of enzymes when included in cereal-based diets have provided less than dramatic results.

Beauchemin and Rode (1996) reviewed a number of their experiments.

These authors postulated that barley as compared to corn had higher levels of fiber and lower levels of starch. Thus, enzyme preparations high in xylanase and cellulase activity might be expected to significantly improve the nutritive quality of the barley compared to the effects on corn. Enzyme preparations high in xylanase and cellulase activity, as measured by reducing sugar release, are more effective applied to barley compared to corn *in vitro*.

Addition of enzyme preparations high in xylanase and low in cellulase activity and the opposite (i.e. high in cellulase but low in xylanase) failed to produced substantial performance responses when used on barley or corn diets and subsequently fed to feedlot cattle (Table 3). The barley was clearly shown to be better than corn as judged by ADG.

This study is of interest because it highlights some of the problems faced by the experimenter when studying the effects of enzymes. Firstly, it should be noted that although barley contains high levels of xylans there are also high levels of β-glucans which contribute significantly to the fiber fraction. In fact, it is this component of the barley which has been shown to be primarily responsible for the poor nutritive value of the cereal for poultry. The β-glucan polysaccharide has subsequently been the target of β-glucanase enzymes in poultry diets with great commercial success (Annison, 1993). Xylanases, on the other hand are much less potent against barley and are ineffective against maize (corn) which contains low levels of both types of polysaccharide. Secondly, neither cereal grain contains appreciable amounts of cellulose and so one would not expect to see a great effect of cellulase enzymes in these diets. Thus, in the experiment described by Beauchemin and Rode (1996) a great response to either enzyme combination on barley or corn in highly unlikely and the results confirm this. There is a possibility that side activities in the enzymes may have effects, but this undescores the importance of defining carefully the enzyme activities of preparations used in feeding studies.

The authors suggest that the FCR was significantly improved in cattle fed the barley diet with the high xylanase activity, but from the data this is more likely to be a statistical anomaly rather than a true biological effect.

Other studies with cereals have yielded conflicting results. Boyles *et al.* (1992) showed that treatment of steam-flaked as well as dry-rolled sorghum with a commercial enzyme mix containing cellulolytic, amylolytic and proteolytic activities resulted in improved ADG and FCR for steers. A similar preparation used by Chen *et al.* (1995) failed to demonstrate any effect in milk production or composition in cattle fed dry-rolled or steam flaked sorghum.

Table 3. Effect of fibrolytic enzymes in corn and barley-based diets for feedlot cattle.*

	Barley			Corn		
	Control	LX:HC[†]	HX:LC[†]	Control	LX:HC[†]	HX:LC[†]
ADG, kg/day	1.43[a‡]	1.40[a]	1.52[a]	1.33[b]	1.33[b]	1.19[b]
DMI, kg/day	9.99	9.86	9.53	9.55	9.10	9.29
FCR	7.11[f]	7.13[f]	6.33[e]	7.26[ab]	6.95[a]	7.83[b]

* From Beauchemin and Rode (1996).
[†] LX – low xylanase, HC – high cellulase, HX – high xylanase, LC – low cellulase.
[‡] Values with unlike letters are significantly different, $P<0.05$.

The challenge for ruminant nutritionists in the use of enzymes

In looking ahead at possible directions in which enzymes for use in ruminants should be developed it is important to consider the basic function of the rumen and nutritional requirement of the ruminant. The biggest challenge for nutritionists is to supply adequate amounts of protein and energy to the animal; and it is in meeting this demand that enzyme technology is likely to yield the most profitable results. The provision of micro-nutrients is also important, of course, but enzyme technology is unlikely to play a role in this area in the short term.

The rumen has two basic functions:

1. to release energy from carbohydrate sources which would otherwise be unavailable to the animal; and
2. to fix non-protein nitrogen in a form which can be utilized by the animal, i.e. as microbial protein.

These functions are carried out by the rumen microflora and fauna which consist of bacteria, fungi and protozoa. The microflora produce a vast array of enzymes which 'process' the feed material. The complexity of this system cannot be overstated. The organisms function synergistically and competitively and, due to their great diversity, can adapt to a wide variety of feeds converting them to energy substrates and proteins suitable for the host animal. Many of these feedstuffs are completely intractable as far as monogastrics are concerned. Thus ruminants represent a remarkable means of transforming feedstuffs into a form suitable for human use (i.e. meat, milk and fibers).

To optimize the performance of ruminants, nutritional strategies with the following objectives are required:

1. maximized fermentation of carbohydrates which are unable to be digested and absorbed in the small intestine;
2. miniminized fermentation of carbohydrate which is digested and absorbed in the small intestine;
3. maximum synthesis of microbial protein from non-protein nitrogen; and
4. minimal breakdown of dietary protein.

Each of these objectives will be discussed below in terms of how they may be addressed by the application of enzymes.

Carbohydrate fermentation in the rumen

As with monogastrics, carbohydrates in ruminants can be classified into two types: those which are digested and absorbed in the small intestine (available carbohydrate) and those which are not. Essentially available carbohydrates are the monosaccharides glucose, fructose and galactose, disaccharides such as lactose and the maltose series [unlike monogastrics, ruminants appear not to have a sucrase function in their small intestine (Kreikemeier *et al.*, 1990)], and starches. The unavailable carbohydrates are essentially the remainder which may occur in plant feedstuffs. They are the non-starch polysaccharides (NSPs) and also the oligosaccharides such as those of the raffinose series and fructo-oligosaccharides. These unavailable carbohydrates are found in the 'fiber' component of feedstuffs.

Unlike the nutritionist, the rumen microflora do not distinguish between these two types of carbohydrates. All may be fermented in the rumen leading to the common metabolic products – primarily short chain fatty acids (SCFAs) and methane and heat (Preston and Leng, 1987, Figure 1). The SCFAs are absorbed directly from the rumen and can be used for both catabolic and anabolic (i.e. gluconeogenesis) processes. The process of fermentation results in significant losses of energy in the form of methane, hydrogen and heat. Thus when dietary glucose, for example, bypasses the rumen and is absorbed in the small intestine, the efficiency of energy utilization is increased by around 30%.

The rumen microflora degrade and ferment NSPs very efficiently. Whilst there is great diversity in the types of NSP found in feedstuffs, there is also a very large number of carbohydrases produced by the rumen microflora to degrade them (Table 4).

The list of enzymes in Table 4 is by no means complete and for each type many different individual enzymes with slightly different optimal activities, binding sites and products have been identified.

For the degradation of NSP to constituent sugars, which is the prerequisite for fermentation, many polysaccharidases acting in concert are required. For example, the breakdown of arabinoxylans, a structural polysaccharide found in cell walls of forages and the in the endosperm of cereals, requires a range of enzymes working sequentially. Essentially enzymes which remove the side chains of arabinose, acetyl groups, ferulic acid and glucuronic acid act first followed by xylanases which can cleave the xylan main chain (Figure 2). The breakdown of cellulose also requires a series of enzymes which includes endo-1,4β–D-glucanases, 1,4β–D-glucan cellobiohydrolases and β-glucosidases.

The breakdown of NSPs to fermentable sugars is therefore a complex system of cooperation between microorganisms and their enzymes. To match or better this with the application of exogenous enzymes as feed additives is a tall order. Nevertheless, digestion of fibrous material is rarely complete in the

Table 4. Enzyme activities found in the rumen.*

Cellulase	α-L-arabinofuranosidase
Endoglucanase	Ferulic acid esterase
Exoglucanase	β-D-glucuronidase
β-glucosidase	Amylase
Xylanase	Arabinase
Xylosidase	Laminarinase
Acetyl xylan esterase	Lichenase
Acetyl esterase	Pectinase

* From Cheng *et al.* (1989).

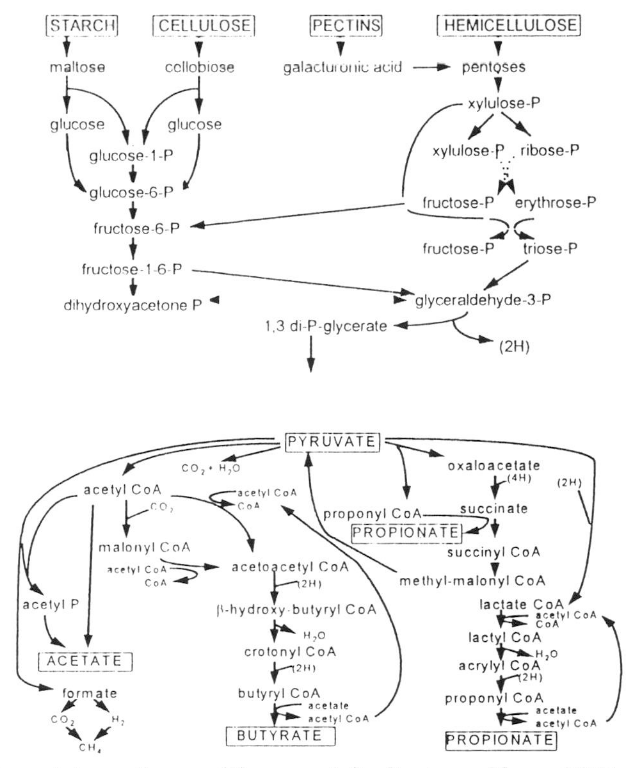

Figure 1. Fermentation pathways of the rumen (after Preston and Leng, 1987).

rumen and examining the factors which may limit the breakdown of the NSPs provides direction for future enzyme research.

Factors limiting NSP degradation in the rumen – possible enzyme targets

Assuming a competent microflora and no extrinsic factors which may constrain or redirect their activity, the rate and extent of NSP degradation is determined broadly by two factors:

1. the molecular structure and 'order' of polysaccharides in the feed material; and
2. the presence of inhibiting structures (such as lignin).

INFLUENCE OF POLYSACCHARIDE STRUCTURE

Cellulose, the main NSP in forage, occurs as a highly ordered insoluble crystalline structure which is very resistant to enzymic attack. In the rumen, the diverse microflora employ a wide range of cellulase complexes, some with

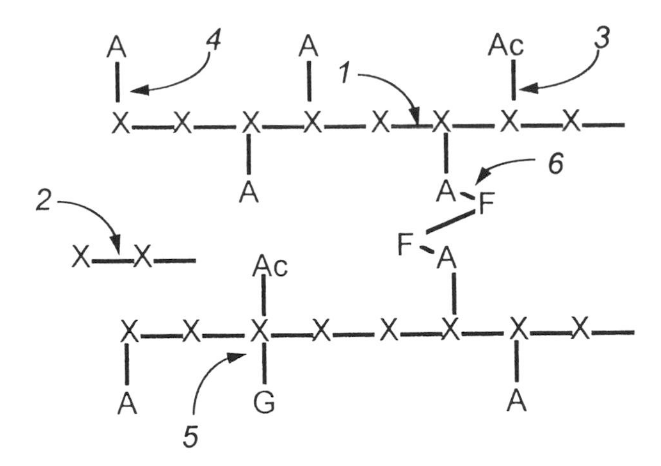

Figure 2. The structure of arabinoxylans of cereals and grasses and the enzymes necessary to break it down in the rumen. A – arabinose, X – xylose, G – glucuronic acid, Ac – acetate, F – ferulic acid. *1* – xylanase, *2* – xylobiase, *3* – acetyl esterase, *4* - arabinofuranosidase, *5* – glucuronidase, *6* – ferulic acid esterase.

as many as five distinct enzyme activities. Enzymic attack necessarily involves attachment of microflora (Cheng *et al.*, 1989) which allows the cell surface-associated cellulase complexes to come to close contact with the substrate. Adhesive forces are very large at the micro level and this probably results in a localized disruption of the ordered cellulose structure and depolymerization proceeds.

The cellulase complexes are found mainly associated with the cell walls of bacteria and other microflora and very little free cellulase activity is found in the rumen liquor. This would suggest that any application of exogenous enzymes to provide fibrolytic activity may require application of live cellulolytic organisms. This option of using 'probiotic' organisms to deliver enzymic activity is certainly feasible, but falls outside the scope of this paper.

An encouraging development is that a highly active cell-free cellulase complex has been isolated from the rumen fungus *Neocalleomastix frontalis* (Wood *et al.*, 1986). Production of the cellulase complex as a feed additive may be possible. Another promising observation is that although the thermophilic bacterium *Clostridium thermocellum* possesses 15 unique endoglucanases, only one of these, along with a non-catalytic protein, which might have a role in bacterial cell adhesion to cellulose, is necessary for the hydrolysis of a crystalline cellulose (Wu *et al.*, 1988). Clearly this type of enzyme/protein combination may also provide opportunities for cellulolytic exogenous enzyme development.

As far as other NSP are concerned, they are thought to be present as less ordered structures than cellulose. They tend to be more soluble and certainly more fermentable, but there is evidence that subtle changes in their structures can affect their digestibility. Li *et al.* (1992) examined the chemical composition and nutritive value of the annual legumes *Trifolium subterraneum* cv.

Junee and *Trifolium resupinatum* cv. Kyambro. In spite of the levels of most components being similar, including the sugars arabinose and xylose, the digestibility of xylose was lower in Kyambro than Junee (Table 5). One interpretation of this is that the structure of the arabinoxylans in the Kyambro is different, i.e. the pattern of substitution of the arabinose sugars along the xylan backbone in Kyambro may be less vulnerable to arabinase and xylanase attack. Another possibility is that the higher levels of galactose and mannose-containing polysaccharides in the Kyambro may protect the arabinoxylan polysaccharide. Observations such as these suggest a role for specific highly active arabinoxylanases or galactomannanases as additives for the animals feeding on these types of forage.

The presence of substituent groups on polysaccharide structures may inhibit the activity of carbohydrases. Common non-carbohydrate substituents are acetyl groups and ferulic acid groups such as those found on arabinoxylans. Both of these have been identified as possible inhibitors to the degradation of forage in the rumen (Richards, 1976). These groups are both attached through ester linkages, thus the development of potent esterases as feed additives may enhance the degradation of feed materials, particularly the hemicellulose portion.

ENZYMIC REMOVAL OF NSP-PROTECTING STRUCTURES

Lignin is the major component of feedstuffs directly inhibiting the degradation of polysaccharides in the rumen. It is overcome during the colonization of feed particles by the rumen microflora, particularly the protozoa and fungi. Lignin is a three dimensional network built up from phenylpropane units. The precursors are coniferyl, sinapyl and p-coumaryl alcohols. Lignin may protect the NSP of plant feedstuffs by creating a physical barrier, by creating a hydrophobic microenvironment close to the substrate or through direct covalent linkage to the substrate polysaccharides (Theander, 1988).

Lignin is degraded to some extent by the action of rumen microflora, primarily the fungi. The lignin does not, however, appear to be utilized by the rumen microflora to any extent. It is dissolved and can be recovered from the rumen liquor as lignin-polysaccharide complexes (Orpin, 1988).

Table 5. Composition of mature annual legumes *Trifolium subterraneum* cv. Junee and *Trifolium resupinatum* cv. Kyambro and the *in vivo* digestibilities of some components.*

	Junee		Kyambro	
	Level (g/kg)	Digestibility	Level (g/kg)	Digestibility
Hemicellulose	114	65.6	115	46.9
Cellulose	294		287	
Lignin	98.2		88.9	
Arabinose	31.4	74.9	30.9	72.6
Xylose	49.9	52.2	50.5	36.7
Mannose	8.0	85.8	19.3	45.8
Galactose	21.7	77.3	32.2	53.3

* From Li *et al.* (1992).

The characteristic feature of low quality roughages, such as straw, is the refractory nature of their highly lignified cell walls. Lignin degradation is very slow and it has been recognized that acceleration of this process in the rumen may be beneficial in increasing both the rate and extent of release of fermentable carbohydrate. A single lignase produced by the soft-rot fungus *Phanerochaete chrysosporium* is capable of causing a high degree of lignin depolymerization (Tien and Kirk, 1983). The enzyme acts like a peroxidase causing a free-radical-mediated cleavage of carbon-carbon bonds, which requires hydrogen peroxide or oxygen. The requirement for such oxidative conditions may limit the usefulness of this enzyme as a feed additive for ruminants as most of the rumen contents are highly anaerobic. Nevertheless, some oxygen does find its way into the rumen from the intestinal contact surfaces and this may be enough to allow this enzyme to operate. It is also possible that with the application of recombinant-DNA technology the lignase gene may be modified to allow the enzyme to operate in more anaerobic conditions.

The cuticular layer of forage materials also represents a considerable barrier to the invasion of rumen microorganisms. Microscopic examination of the feed particles taken from the rumen shows the cuticle to be virtually devoid of attached microorganisms. Cutin is a polyester of C16 and C18 hydroxy and hydroxyepoxy fatty acids. The ester linkage is known to be hydrolyzed by some pathogenic bacteria and some aerobic bacteria, but there are no reports of rumen bacteria exhibiting cutinase activity (Cheng *et al.*, 1989). Cutinase therefore represents an opportunity as a feed enzyme additive for cattle fed low quality forage.

Starch digestion in the ruminant – possibility for enzyme additives

The major component of cereal grains is starch, a polymer of glucose. The digestion of starch in the rumen is usually high, although the rate of digestion can vary considerably among cereal types. Maize starch, for example, is digested slowly as the starch granules are surrounded by a protein matrix which acts a barrier to amylase enzymes. This has been demonstrated dramatically in *in vitro* rumen fluid incubation experiments where starch from maize has been shown to degrade more slowly than wheat starch (Opatpatanakit *et al.*, 1994, Figure 3). Rapid fermentation of starch in the rumen is not considered desirable as it can lead to acidosis, bloat and liver abscess (Rowe and Pethick, 1994). The use of diets high in cereal are associated with a depression of digestion of fiber from forage components. This is thought to be due primarily to an inhibition of the cellulolytic microflora by the low pH which accompanies rapid fermentation of the cereal starch. Clearly there is no role for amylases which would exacerbate this problem and enhance a process which, as far as the host animal is concerned, is energetically unfavorable. Under these circumstances however, exogenous cellulases and hemicellulases, particularly those active at low pH, may be of value.

The studies of Rowe and Pethick (1994) have highlighted another problem.

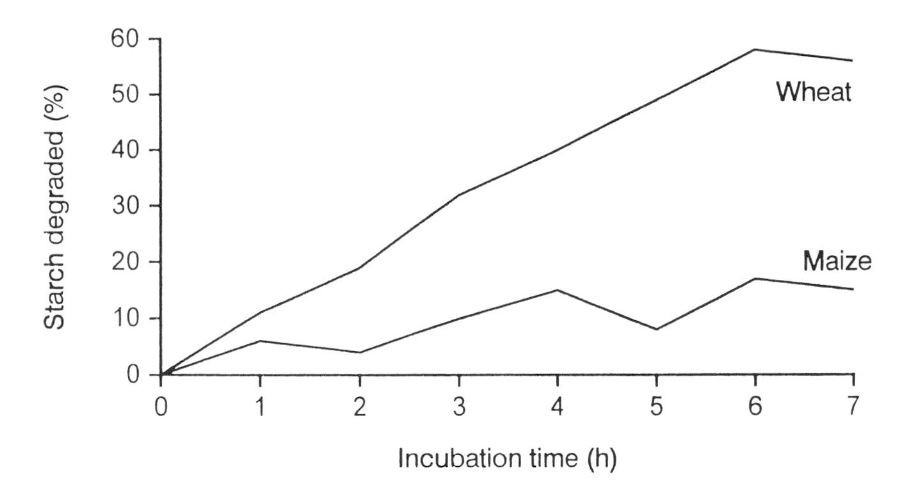

Figure 3. Degradation of wheat and maize starch during incubation in rumen fluid (after Opatpatanakit *et al.*, 1994).

When starch escapes digestion in the rumen with cereals such as maize and sorghum it may also pass through the small intestine to be fermented in the large bowel. Extensive fermentation in this organ may cause severe scouring. Thus acidosis may also occur in the hindgut in the ruminant. Rowe and Pethick (1994) counsel the use of antibiotic feed additives such as virginiamycin to control this problem but the judicious use of amylase enzymes to enhance the small intestinal breakdown of starch and absorbtion of glucose may also be a solution. The challenge here, of course, is to get the enzyme past the rumen to the site of preferred activity – the small intestine. Encapsulating technology has already been developed for the delivery of methionine and lysine to the small intestine and it should be possible to adapt it to enzyme delivery.

Protein digestion in the ruminant

A major concern for ruminant nutritionists is that on low quality forage diets the release of energy from carbohydrates by fermentation is so slow that the microflora start to break down dietary protein and use it as an energy source. The only solution to this problem is to speed up the carbohydrate fermentative processes and possible enzyme strategies to achieve this have been described above. Clearly, the uses of exogenous proteases targeted at the rumen are inappropriate. The small intestine is the preferred site for the breakdown of proteins from the diet and is also the site for rumen-synthesized microbial protein digestion. As small intestinal digestibility of dietary and microbial protein is only in the region of 60 to 70% (Harmon, 1993) an opportunity may exist here for the use of exogenous proteases as feed additives. Again, as described above, delivery of the enzyme to the site of activity is a concern but this may be achieved with the rumen bypass encapsulating technology.

The ruminant as a monogastric

When young ruminants are suckling, milk bypasses the rumen through the esophageal groove and goes straight to the abomasum and, hence, to the small intestine where digestion occurs. The reflex responsible for this usually diminishes in adult life, but can be maintained indefinitely if animals are provided daily, but not necessarily exclusively, liquid feed (Orskov, 1983). One can envisage feeding regimes being developed in the feedlot situations where feed concentrates high in cereal and protein are fed in the liquid form so that they bypass the rumen. Lower quality forage may also be fed which would be digested in the rumen.

In this situation the NSPs from the cereals (i.e. wheat, barley, rye) may display anti-nutritive activity as has been described in poultry and pigs. These would be appropriate targets for the exogenous enzymes. As ruminants seem to be sensitive to rapid fermentation in the large bowel if starch is digested incompletely exogenous amylases may also prove to be beneficial.

Enzymes targeted against naturally occurring toxins

Many plant feedstuffs, particularly those eaten by grazing ruminants, contain toxins which inflict stock losses. The tropical legume *Leucaena leucocephala* has caused problems because Australian sheep are sensitive to poisoning by the compound mimosine, which this plant produces, and its degradation product hydroxypyridone. To combat this problem, microorganisms responsible for the resistance of Indonesian goats to mimosine were transferred to ruminants in Australia.

Further work has examined the possibility of using microorganisms to detoxify flouroacetate which occurs naturally in the tissues of many plants such as *Acacia, Gastrolobium* and *Oxylobium.* In ruminants, fluoroacetate has a lethal dose between 0.25 and 0.5 mg/kg. A bacterium *Moraxalla* sp. possesses an enzyme which hydrolyzes fluoroacetate and attempts have been made to transfer this into the rumen species *Butyrivibrio fibrisolvens* and *Bacteriodes ruminicola* (Gregg and Sharpe, 1989).

A major problem with introducing new detoxifying bacteria to the rumen is that they must colonize and remain present for extended periods in the rumen. In order to do this they must compete with the other rumen organisms.

An alternative approach to providing live organisms would be to simply provide to the rumen the enzymes responsible for the detoxification. In some cases this may require several different enzyme activities to ensure the final degradation products are innocuous.

Maintaining enzyme activity in the rumen

The loss of exogenous enzyme activity from the rumen may occur through degradation by microbial proteases or through washout in the rumen liquor. The evidence is good, however, that degradation of exogenous enzymes in the rumen may not be a significant problem. Chesson (1993) has reported that extensive glycosylation of fungal enzymes protects them from proteolytic attack in monogastric animals. In addition, Annison (1992), has demonstrated that β-glucanase and xylanase activities can be detected in the terminal ileum of chickens following dietary supplementation with commercial feed enzyme preparations indicating that they survive the proteolytic environment of the chicken digestive tract. Achieving stabilty of exogenous enzymes in the rumen therefore does not seem to be a major problem and indeed, Hristov *et al.* (1996) have demonstrated that exogenous enzyme activity is maintained over several hours in the rumen and small intestine of the dairy cow.

As long as the exogenous enzyme partitions into the aqueous phase there must be some washout. This may not be a problem in the feedlot situation or where regular feeding occurs. It does, however, present a problem in the grazing situation. A possible solution to this problem is to immobilize the enzyme within the rumen. Large, solid pellets which remain in the rumen have been successfully used to provide slow release of trace elements to the animal. This idea may be extended to providing exogenous enzyme activity. Immobilized enzymes, attached to inert supports, are already used extensively in some industries. It may be that highly active enzymes attached to supports which remain intact for an extended period of time may enhance the rumen function considerably. Wide arrays of enzyme activities capable of breaking down a range of feed components could greatly enhance the efficiency of digestion and improve productivity. Interestingly, the notion borrows also from nature, where, as described above, the cellulase complexes of many rumen microorganisms are associated with the cell surface.

Conclusion

The introduction of exogenous enzyme activity to improve the performance of ruminants provides considerable challenge. With the judicious application of microbiological, biochemical and nutritional science, directions for promising research and product development are becoming clear. Whilst the highly complex nature of the rumen environment presents considerable impediments to the application of simple enzyme systems, it also provides opportunities for the application of ingenious biotechnologies which use the rumen to their advantage.

References

Annison, G. 1992. Commercial enzyme supplementation of wheat-based diets raises ileal glycanase activities and improves apparent metabolisable energy, starch and pentosan digestibilities in broiler chickens. Anim. Fd. Sci. Technol. 38:105.

Annison, G. 1993. The role of wheat non-starch polysaccharides in broiler nutrition. Aust. J. Agric. Res. 44:405.

Annison G. and M. Choct. 1991. Anti-nutritive activities of cereal non-starch polysaccharides in broiler diets and strategies mimimizing their effects. World's Poult. Sci. J. 47:232.

Beauchemin, K.A. and L.M. Rode. 1996. Use of feed enzymes in ruminant nutrition. *In:* Animal Science Research and Development. Meeting Future Challenges. Proceedings of the Canadian Society of Animal Science Meeting. Ed. L.M. Rode. Lethbridge, Alberta, pp. 103–130.

Boyles, D.W., C.R. Richardson, K.D. Robinson and C.W. Cobb. 1992. Feedlot performance of steers fed steam-flaked grain sorghum with added enzymes. *In:* Proceedings Western Section American Society Animal Science 43:502.

Chen, K.H., J.T. Huber, J. Simas, C.B. Theurer, P. Yu, S.C. Chan, F. Santos, Z. Wu, R.S. Swingle and E.J. DePeters. 1995. Effect of enzyme treatment or steam-flaking on sorghum grain on lactation and digestion in dairy cows. J. Dairy Sci. 78:1721.

Cheng K.-J., C.W. Forsberg, H. Minao and J.W Costerton. 1989. Microbial ecology and physiology of feed degradation within the rumen. *In:* Physiological Aspects of Digestion and Metabolism in Ruminants. Proceedings of the Seventh International Symposium on Ruminant Physiology. Eds T.Tsuda, Y. Sasaki and R. Kawashima. Academic Press.

Chesson, A. 1993. Feed enzymes. Anim. Fd. Sci. Tech. 45:65.

Gregg, K. and H. Sharpe. 1989. Enhancement of rumen microbial detoxification by gene transfer. *In:* Physiological Aspects of Digestion and Metabolism in Ruminants. Proceedings of the Seventh International Symposium on Ruminant Physiology. Eds T.Tsuda, Y. Sasaki and R. Kawashima. Academic Press. pp. 179–735.

Harmon, D.L. 1993. Nutritional regulation of postruminal digestive enzymes in ruminants. J. Dairy Sci. 76:2102.

Hristov, A., T.A. McAllister and K.-J. Cheng. 1996. Exogenous enzymes for ruminants: Modes of action and potential applications. Proc. 17th Western Nutrition Conference, Edmonton, Alberta. Sept. 1996.

Kreikemeier, K.K., D.L. Harmon, J.P. Peters, K.L. Gross, C.K. Armendariz and C.R. Krehbiel. 1990. Influence of dietary forage and feed intake on carbohydrase activities and small intestinal morphology. J. Anim. Sci. 68:2916.

Lewis, G.E., W.K. Sanchez, R. Treacher, C.W. Hunt and G.T. Pritchard. 1995. Effect of direct-fed fibrolytic enzymes on lactational performance of midlactation Holstein cows. *In:* Proceedings of Western Section American Society Animal Science. 46:310.

Li, X., R.C. Kellaway, R.L. Ison and G. Annison. 1992. Chemical composition and nutritive value of mature annual legumes for sheep. Anim. Feed. Sci. Technol. 37:221.

Opatpatanakit, Y., R.C. Kellaway, I.J. Lean, G. Annison and A. Kirby. 1994. Microbial fermentation of cereal grains *in vitro*. Aust. J. Agric. Res. 45:1247.

Orpin, C.G. 1988. Ecology of rumen anaerobic fungi in relation to the nutrition of the host animal. *In:* The Roles of Protozoa and Fungi in Ruminant Digestion. Eds J.V. Nolan, R.A. Leng and D.I. Demeyer. Penambul Books. Armidale. pp. 29–38.

Orskov, E.R. 1983. The oesophageal grove reflex and its practical implications in the nutrition of young ruminants. *In:* Maximum Livestock Production from Minimum Land. Eds C.H. Davis, T.R. Preston, M. Haque and M. Saadullah. Bangladesh Agricultural University, Myemensingh. pp. 47–53.

Preston, T.R and R.A. Leng. 1987. Matching Ruminінant Production Systems with Available Resources in the Tropics and Sub-Tropics. Penambul Books, Armidale, Australia.

Richards, G.N. 1976. Search for factors other than 'lignin shielding' in protection of cell-wall polysaccharides from digestion in the rumen. *In:* Carbohydrates in Plants and Animals Misc. papers 12 pp. 129–135. Landbauwhogeschool Wageningen (Veeman and Zonen:Wageningnen).

Rowe, J.B. and D.W. Pethick. 1994. Starch digestion in ruminants – problems, solutions and opportunities. *In:* Proceedings of the Nutrition Society of Australia. pp. 40–52.

Sanchez, W.K., C.W. Hunt, M.A. Guy, G.T. Pritchard, B. Swanson, T. Warner and R.J. Treacher. 1996. Effect of fibrolytic enzymes on lactational performance of dairy cows. *In:* Proceedings of American Dairy Science Association. Crovallis, Oregon. July 14–17. (poster).

Stokes, M.R. and S. Zheng. 1995. The use of carbohydrolase enzymes as feed additives for early lactation cows. 23rd Biennial Conference on Rumen Function, Chicago, IL p. 35 (Abstr).

Theander, O. 1988. Plant cell walls – their chemical properties and rumen degradation. *In:* The Roles of Protozoa and Fungi in Ruminant Digestion. Eds J.V. Nolan, R.A. Leng and D.I. Demeyer. Penambul Books. Armidale. pp. 1–12.

Tien, M. and K.T. Kirk. 1983. Lignin degrading enzyme from the hymenomycete *Phenerochaete chrysosporium* burds. Science 221:661.

Wood, T.M., C.A. Wilson, S.I. McCrae and K.N. Joblin. 1986. A highly active extracellular cellulase from the anaerobic rumen fungus *Neocallimastix frontalis*. FEMS Microbiol. Lett. 34:37.

Wu, J.H.D., W.H. Orme-Johnson and A.L. Demain. 1988. Two components of an extracellular protein aggregate from *Clostridium thermocellum* together degrade crystalline cellulose. Biochemistry. 27:1703.

ENZYME SUPPLEMENTATION OF BROILER AND TURKEY DIETS TO ENHANCE WHEAT UTILIZATION

J.L. GRIMES, P.R FERKET and A.N.CROUCH

Department of Poultry Science, North Carolina State University, Raleigh, North Carolina, USA

Introduction

The poultry industry has been challenged recently by high corn prices. In some cases during 1996, corn was difficult to find regardless of price. Even though wheat has also at times been in shorter supply than usual, many poultry producers, especially turkey producers, have considered wheat and wheat products or screenings as alternatives to corn for poultry rations (Wyatt and Graham, 1996). Wheat production has increased in the last several decades on a worldwide basis, and in conjunction with the North American Free Trade Agreement, General Agreement on Tariffs & Trade and other trade agreements, wheat may become competitive with corn on a routine basis (Ward, 1995). The National Research Council (1994) list wheat as having 7–13% less metabolizable energy (ME) than corn (1523 kcal/lb for corn versus 1313 and 1418 kcal/lb for hard red winter and soft white winter wheat, respectively). However, wheat ME can vary greatly due to many factors such as genetics, location of wheat production, environmental growing conditions, and amount of wheat kernels or foreign material (such as straw or weed seeds) in a particular lot of wheat. Wheat has a higher crude protein content than corn (14.1 and 11.5 versus 8.5%) and also has higher levels of the essential amino acids lysine and methionine than corn. Crude protein content, however, can vary for both grains. The superior pelleting characteristics of wheat are well known in the poultry industry. Many producers utilize wheat at 10 to 20% to improve pellet quality.

Poultry vary in their response to dietary wheat especially when wheat is a major part of the ration, i.e. greater than 20 to 30%. Wheat, as well as other grains such as barley and rye, contains a class of poorly digested substances called non-starch polysaccharides (NSP) which are associated with the endosperm cell wall of the grain (Ward, 1995; Morgan *et al.*, 1995). In wheat, arabinoxylans are the major NSP, and consist of a main chain of ß-(1,4)-linked xylose to which sidechains of arabinose are attached. Increased intestinal viscosity is attributed to the water-soluble NSP fraction, and it is the arabinose sidechains that impart a water soluble nature to this NSP. The long polymers entangle nutrients and have a high water-holding capacity, resulting in an increase in intestinal viscosity (Ward, 1996). Arabinoxylans are soluble in the gut of the bird yet they are indigestible and form a highly viscous gel-like material in the gut of the animal (Morgan *et al.*, 1995). Studies have shown that these viscous indigestible polysaccharides cause an enlargement of digestive organs, including the pancreas, as they produce more intestinal enzymes to compensate for the digestive impediment (Ikegami *et al.*, 1983). The negative relationship between the NSP and nutrient digestibility and ME provides evidence of its anti-nutritive character, while fecal recoveries of 86% attest to its poor digestibility. This viscous gel interferes with digestion and absorption of nutrients in wheat as well as with digestion and absorption of nutrients found in the other ingredients of the ration. Poor digestibility of rations means that nutrients can be less available to the bird. This in turn can lead to poor performance such as reduced growth and decreased feed efficiency (i.e. higher feed:gain ratios). Wheat-fed birds can also have increased incidence of pasty beaks and vents and wet litter (Ward, 1995). Wet litter can, in turn, lead to other performance problems such as increased disease, poor air quality, leg problems, and decreased growth or feed efficiency.

Enzymes have been developed to reduce the negative effects of NSP and improve the feeding value of wheat in poultry diets. Pentosanases are the class of enzymes most effective for supplementing dietary wheat. By use of pentosanases, arabinoxylans are hydrolyzed into smaller polymers and basic components of arabinose and xylose. Once hydrolyzed, the pentoses precipitate from solution and cease to form the viscous gel-like material. This reduces the viscosity in the gut and normalizes digestion in the intestinal tract (Wyatt and Graham, 1996; Ward, 1996). The end result is better digestion and absorption of nutrients from wheat and other feed ingredients, as well as improving litter quality (Pijsel, 1996). Even though the arabinoxylans are hydrolyzed, the basal pentoses are not digested or absorbed very well and probably contribute very little energy to the bird. The result of the enzyme action is to release nutrients trapped by the highly viscous gut contents resulting from NSP. Enzymes can be made to withstand pelleting temperatures in the 85–90°C range, but higher temperatures can denature enzymes and destroy their activity. Under conditions of feed manufacturing and milling which require higher temperatures, liquid enzymes can be sprayed on the feed after pelleting.

Studies at North Carolina State University

Work at North Carolina State University supports much of what others have reported (Crouch *et al.*, 1996). Two experiments were conducted using 450 broiler chicks (Experiment 1) and 450 Large White turkey poults (Experiment 2). The chicks and poults were randomly assigned to 45 pens in four Petersime batteries on day of hatch. There were nine replicates per treatment with 10 chicks or poults per replicate. During the experimental period the birds were offered feed *ad libitum* in trough feeders and water by nipple drinkers. Once the birds were distributed among pens, each pen was assigned to one of five feed treatments formulated to be isocaloric and isonitrogenous (Table 1). The five dietary treatments consisted of a corn/soybean meal based ration, and four other rations containing 40% wheat from two locations, North Carolina (NC) and western Canada, supplemented with and without enzyme (Allzyme Pentosanase). The enzyme used in this study

Table 1. Composition of the treatment diets (%) used in Experiments 1 and 2.

	Chick diets		Poult diets	
	Corn	Wheat	Corn	Wheat
Ingredient				
Corn	53.0	16.0	42.8	5.4
Wheat	–	40.0	–	40.0
Soybean meal (48.5%)	36.0	32.8	50.5	47.8
Poultry Fat	6.5	6.8	1.3	1.7
Dical (18.5% P)	1.9	1.7	2.6	2.4
Limestone	1.4	1.4	1.4	1.5
Salt	0.4	0.4	0.3	0.3
Minerals*	0.2	0.2	0.2	0.2
Choline Cl 60%	0.2	0.2	0.2	0.2
DL-methionine	0.1	0.1	0.1	0.1
Vitamins[†]	0.05	0.05	0.1	0.1
Enzyme[‡]	–	0.1	–	0.1
Calculated analysis				
Crude protein[§], %	23.0	23.0	28.0	28.0
ME, kcal/lb	3395	3395	2797	2797
Crude fat, %	9.18	8.53	3.44	3.44
Methionine, %	0.52	0.50	0.60	0.60
Meth. + Cys., %	0.90	0.90	1.05	1.05
Lysine, %	1.33	1.33	1.66	1.66
Calcium, %	1.00	1.00	1.20	1.20
Available phosphorus, %	0.45	0.45	0.60	0.60
Sodium, %	0.22	0.22	0.17	0.17
NSP[¶]				

* Mineral mix supplied the following per kilogram of chick and poult diet: 120 mg Zn as $ZnSO_4 \cdot H_2O$; 120 mg Mn as $MnSO_4 \cdot H_20$; 80 mg Fe as $FeSO_4 \cdot H_2O$; 10 mg Cu as $CuSO_4$; 2.5 mg I as $Cu(IO_3)_2$; 1.0 mg Co as $CoSO_4$.

[†] Vitamins supplied the following per kg of diet when added at 0.05%: vitamin A, 6600 IU; vitamin D_3, 2000 ICU; vitamin E, 33 IU; vitamin B_{12}, 19.8 µg; riboflavin, 6.6 mg; niacin, 55 mg; d-pantothenate, 11 mg; menadione, 2 mg; folic acid, 1.1 mg; thiamine, 2 mg; pyridoxine, 4 mg; d-biotin, 126 µg; 0.15 mg Se as Na_2SeO_3; 50 mg ethoxyquin; and 100 g wheat midds.

‡ Allzyme PT added to wheat diets contained pentosanase at 1000 xylanase units/g.

§ Crude protein (%) determined by AOAC method 4.2.08. Chicks: corn, 22.8; North Carolina (NC) wheat, 22.8; NC wheat + enzyme, 23.5; Canadian (CN) wheat, 24.0; CN wheat + enzyme, 22.9; poults: corn, 26.9; NC wheat, 28.5; NC wheat + enzyme, 28.9; CN wheat, 28.5; CN wheat + enzyme, 28.7.

¶ NSP levels determined by the method of Choct and Annison (1992). Chicks: corn, 3.9; NC wheat, 2.8; NC wheat + enzyme, 2.7; CN wheat, 3.5; CN wheat + enzyme, 3.6; poults: corn, 2.5; NC wheat, 2.4; NC wheat + enzyme, 3.0; CN wheat, 3.7; CN wheat + enzyme, 3.4.

contained 1000 xylanase units per gram and was added to the applicable treatment rations at a rate of 1 kg per ton. The feed treatments were prepared in mash form and fed throughout the two experiments.

In Experiment 1, individual chick body weight (BW) and feed consumption were determined on days 0, 7, 14, and 21; while in Experiment 2, individual poult BW and feed consumption were determined on days 0, 6, 12, and 18. BW, weight gain, feed consumption, and adjusted feed conversion were determined for each of these periods. On the termination dates for the chick and poult experiments; days 21 and 18, respectively, one bird per pen (nine per treatment) was used to determine gut content supernatant viscosity. Birds were selected with feed present to each pen. The data were subjected to regression analyses using the General Linear Model program of SAS (SAS, 1992). Significant differences between treatment means were determined by the least significant difference test at $P \leq 0.05$, unless stated otherwise.

Rations made with wheat from either source without the enzyme resulted in high gut viscosity for both chicks and poults when compared to those fed the corn-based rations. However, the addition of the pentosanase enzyme to the wheat-based rations resulted in gut viscosity values the same as for corn based rations for both sources of wheat fed to both chicks and poults (Table 2). Growth performance and feed to gain ratio responses were more variable (Tables 3, 4, and 5). The chicks and poults fed Canadian wheat-based rations with or without enzyme performed as well as those fed the corn-based ration. However, the chicks and poults fed North Carolina wheat-based rations without the enzyme had reduced growth rates and increased feed to gain ratios compared to chicks or poults fed corn or Canadian wheat-based rations. When the North Carolina wheat-based rations were fed with the pentosanase enzyme to both chicks and poults, their performance was equal to those fed corn or Canadian wheat.

Table 2. Effect of diet and dietary enzyme supplementation of wheat on intestinal viscosity (centipoise*) for chicks at 21 days of age (Experiment 1) and poults at 18 days of age (Experiment 2).

	Speed (rpm[†]) for			
	Chicks		Poults	
Treatment	1.5	6.0	1.5	6.0
Corn/soy	4.08[a]	3.15[a]	4.92[b]	3.61[bc]
NC wheat	6.19[c]	5.26[c]	7.32[a]	5.37[a]
NC wheat + Allzyme PT	4.49[a]	3.35[a]	3.93[b]	3.43[b]
CN wheat	6.52[b]	5.16[b]	5.48[b]	4.07[b]
CN wheat + Allzyme PT	4.57[a]	3.52[a]	3.74[b]	2.93[c]
SEM	0.39	0.31	0.71	0.41

[a,b,c] means within a column with difference in superscripts are significantly different ($P \leq 0.05$).

* Centipoise (cP). A centimeter-gram-second unit of dynamic viscosity equal to one dyne-second per square centimeter.

[†] Revolutions per minute.

Table 3. Effect of dietary enzyme supplementation of wheat on body weights (g) of chicks at days 7, 14, and 21 (Experiment 1) and poults at days 6, 12, 18 (Experiment 2).

	Body weight at days					
	Chicks			Poults		
Treatment	7	14	21	6	12	18
Corn/soy	109.5	316.3	612.1[b]	103.1	222.2	356.0
NC wheat	113.4	306.3	555.5[a]	101.5	216.1	356.0
NC wheat + Allzyme PT	117.5	317.3	590.8[b]	104.5	237.7	388.5
CN wheat	118.3	320.4	615.0[b]	100.3	222.7	375.3
CN wheat + Allzyme PT	110.4	316.1	617.5[b]	103.1	229.6	374.0
SEM	2.94	6.48	9.07	2.73	6.09	9.04

[a,b] means with different superscripts within a column are significantly different ($P \leq 0.05$). There were no significant differences ($P \geq 0.05$) for beginning body weights of chicks (39.1 g) or poults (50.0 g).

Table 4. Effect of dietary enzyme supplementation with wheat on body weight gains (g) of chicks and poults for weekly periods until three weeks of age (Experiments 1 and 2).

	Body weight gain for weekly periods					
	Chicks			Poults		
Treatment	0–7	7–14	14–21	0–6	6–12	12–18
Corn/soy	70.6	206.8	292.9[a]	48.3	119.4[b]	133.8[b]
NC wheat	74.7	192.9	248.9[b]	46.5	114.5[b]	139.8[b]
NC wheat + Allzyme PT	78.5	200.1	274.2[ac]	49.7	133.2[a]	150.7[a]
CN wheat	78.9	202.0	293.7[a]	44.7	121.5[ab]	151.3[a]
CN wheat + Allzyme PT	70.9	20.54	297.9[a]	48.2	126.4[a]	144.3[ab]
SEM	2.88	4.65	8.98	2.57	4.39	4.18

[a,b,c] means within a column with different superscripts are significantly different ($P \leq 0.05$).

Table 5. Effect of dietary enzyme supplementation with wheat on feed:gain conversions for chicks and poults (Experiment 1 and 2).

	Feed:grain conversion at days							
	Chicks				Poults			
Treatment	0–7	7–14	14–21	0–21	0–6	6–12	12–18	0–18
Corn/soy	1.09	1.00[c]	1.46	1.25[b]	1.27	1.60[a]	1.86	1.66
NC wheat	1.11	1.35[a]	1.77	1.49[a]	1.38	1.75[b]	1.87	1.74
NC wheat + Allzyme PT	1.08	1.06[bc]	1.63	1.33[b]	1.27	1.58[a]	1.85	1.65
CN wheat	1.08	1.10[b]	1.53	1.31[b]	1.37	1.59[a]	1.79	1.64
CN wheat + Allzyme PT	1.00	1.07[bc]	1.57	1.30[b]	1.29	1.64a	1.86	1.68
SEM	0.028	0.029	0.076	0.036	0.046	0.033	0.041	0.029

[a,b,c] means within a column with difference in superscripts are significantly different ($P \leq 0.05$).

Discussion

In both experiments, supplementation of diets containing either North Carolina or Canadian wheat with a pentosanase enzyme resulted in a significant improvement in bird performance. Birds fed Canadian wheat without pentosanase performed comparably to those fed the corn-based diet, suggesting that this particular source of wheat did not need an enzyme. Birds fed North Carolina wheat without pentosanase did not perform comparably to those fed the corn-based diet. However, with the addition of pentosanase, birds fed North Carolina wheat performed equally or better than those fed the corn or Canadian wheat-based diets. The variable performance observed in this study of birds fed two sources of wheat with and without enzyme supplementation agrees with other reports.

In general, as NSP levels increase in grains, ME decreases. For example, corn is relatively low in NSP and high in ME; whereas barley and rye are relatively high in NSP and low in ME. However, wheat can vary in NSP level and poultry seem to vary in their response to wheat-based rations. Other researchers have noted the variable response of both broilers and turkeys to wheat-based rations, especially when wheat has been incorporated at high levels. Zumbado and Dale (1994) reported that a wide variation exists in the proximate composition of various wheat products, while Summers *et al.* (1968) reported that ME and metabolizable dry matter can vary among wheat sources and by-products. Hulet *et al.* (1993) reported that the substitution of wheat for corn in starting diets fed to male turkeys had no effect on BW gain. However, in the early 1990s, a considerable amount of low-priced wheat arrived from western Canada. When used to completely replace corn in diets this wheat usually resulted in decreased performance (Ward, 1995). There were indications that some of this wheat had been subjected to frost-damage. Wheat often contains high levels of pentosans when exposed to early freezing temperatures or grown under dry conditions (Ward, 1995; Friesen *et al.*, 1992).

Researchers have also reported variable performance of birds fed wheat diets supplemented with enzyme designed to enhance utilization of NSP. Brenes *et al.* (1993) reported chick performance (weight gain 0 to 42 days) was significantly improved when the wheat-based diet was supplemented with an enzyme (wheat = 1772 g gain versus wheat + enzyme = 1996 g gain). However, Ward (1996) reported the effect of enzyme supplementation on the performance of broilers fed 60% wheat diets varied, depending on the age of the bird. Performance to day 21 was not significantly improved when an enzyme was added, while performance from day 22 to 42 was significantly improved with an enzyme added at various levels. Feed conversion was significantly improved from day 22 to 42 when an enzyme was added at 0.02 to 0.1% of the diet. Leeson *et al.* (1996) conducted three separate experiments with the addition of an enzyme into wheat fed to broilers and turkeys. They reported that enzyme supplementation in the first experiment had a positive effect on the growth rate of turkeys to 72 days of age, with a significant improvement seen in the 28 to 56 day period. In a second experiment with turkeys no significant difference was seen in performance when an enzyme was added to the diet beginning at 28 days of age. In the third experiment, which involved broiler chicks, the addition of an enzyme resulted in improved early growth rate from 0 to 28 days of age. In addition, Wyatt and Graham (1996) reported an improved feed:gain ratio when an enzyme was added to a wheat-based broiler diet, but actual 49 day final weight was not significantly improved. Pijsel (1996) reported that dietary enzymes may also be affected by the addition of an antibiotic. Testing several diets containing wheat, Pijsel (1996) found that some antibiotics tended to enhance the enzyme response while other antibiotics had little effect.

Lowered intestinal viscosity has been suggested as one possible explanation for the improvement in performance of birds fed enzyme-supplemented wheat diets, especially when feeding proportionally high levels of wheat (Ward, 1996). Increased intestinal viscosity is also considered to be an important constraint to digestion by interfering with the diffusion of pancreatic enzymes (Morgan *et al.*, 1995). Enzyme supplementation is the most economical and effective means available to reduce intestinal viscosity (Ward, 1996). Morgan *et al.* (1995) reported that birds fed wheat-based diets which result in low intestinal viscosity can have intestinal viscosity further lowered when an enzyme is added to the ration. However, certain high molecular weight NSP molecules can be responsible for most of the highly viscous material forming in the intestinal tracts of birds. Therefore, the total NSP level and the NSP level responsible for high gut viscosity can vary somewhat independently. As demonstrated by Crouch *et al.* (1996), NSP levels of wheat or resulting changes in gut viscosity cannot always be used to reliably predict wheat ME. Crouch *et al.* (1996) reported that birds fed Canadian wheat-based rations had higher gut viscosity than birds fed a corn-based ration. Addition of a pentosanase enzyme to the ration reduced gut viscosity while performance was unaffected.

Hughes *et al.* (1994) fed four varieties of wheat, with and without enzymes, to broilers. Gut viscosity was reduced by addition of an enzyme in three of the four wheat varieties. Similarly, apparent ME was increased by the addition of an enzyme in the same three varieties where viscosity was reduced. For the wheat variety where gut viscosity was not greatly affected, addition of the enzyme also had no effect on apparent ME. Furthermore, Hickling (1994) reported that addition of an enzyme to five cargo samples of wheat from western Canada reduced the variation in ME from 225 kcal/kg to 80 kcal/kg. The samples with the highest original ME were affected the least while the samples with the lowest original ME were affected the most. All of these reports support observations of variation of bird performance when fed wheat-based rations. They also suggest that some of this variation might be caused by differences in digestion and or utilization of different sources of wheat by poultry. These reports also indicate that some of this variation can be reduced by utilization of proper enzyme supplementation.

Enzymes must be matched to the grain source in the ration. Wheat-based rations would require enzyme mixtures with predominantly pentosanase activity while other grains such as barley or triticale would require enzyme mixtures with predominantly ß-glucanase activity. The decision of what to feed poultry usually comes down to the most economical choice for production of the final product. Addition of enzymes to poultry rations will add to the diet cost. Poultry producers should then examine records for increased performance on the same amount of feed or the same performance on less feed. Also, producers should look for a decrease in variation in bird performance which might be just as important in some cases as economics of performance.

Summary

Increased intestinal viscosity is generally associated with reduced growth performance. Supplementation of wheat diets with a pentosanase enzyme may reduce gut content supernatant viscosity for both chicks and poults. This lowered viscosity is generally but not always accompanied by improved performance of chicks and poults fed wheat-based rations. Considerable variation can exist in wheat and in response of poultry fed wheat. Both chick and poult performance have been reported to be affected by wheat source. Addition of enzymes to wheat diets can reduce the variation in bird performance due to source of dietary wheat. Wheat is an excellent ingredient for poultry diets. However, dietary modifications might be needed when substituting certain or multiple sources of wheat for corn.

References

Brenes, A., M. Smith, W. Guenter and R. Marquardt. 1993. Effect of enzyme supplementation on the performance and digestive tract size of broiler chickens fed wheat and barley based diets. Poultry Sci. 72:1731–1739.

Choct, M. and G. Annison. 1992. The inhibition of nutrient digestion by wheat pentosans. Br. J. Nutr. 67:123–132.

Crouch, A.N., J.L. Grimes, P.R. Ferket and L.N. Thomas. 1996. Effect of enzyme supplementation of starter diets for broilers and turkeys to enhance wheat utilization. Poultry Sci. 75(Suppl. 1):14.

Friesen, O.D., W. Guenter, R. Marquardt and B. Rotter. 1992. The effect of enzyme supplementation on the apparent metabolizable energy and nutrient digestibilities of wheat, barley, oats, and rye for the young broiler chick. Poultry Sci. 71:1710–1721.

Hickling, D. 1994. Varying bushel weights. Proceedings of the Fifteenth Western Nutrition Conference. Winnipeg, Manitoba.

Hughes, R., G. Annison and M. Choct. 1994. Effect of Allzyme pentosanase and wheat variety on performance, apparent metabolisable energy and ileal digesta viscosity of broilers. Poster presented at the tenth Alltech Symposium, Alltech Inc., Nicholasville, KY.

Hulet, R.M., D.M. Denbow and L.M. Potter. 1993. Effects of lighting and dietary energy source on male market turkeys. Poultry Sci. 72:1459–1466.

Ikegami, S., N. Tsuchihashi, S. Nagayama, H. Harada, E. Nishide and S. Innami. 1983. Effect of indigestible polysaccharides on function of digestion and absorption in rats. J. Jpn. Soc. Nutrition Food Sci. 36:163–168.

Leeson, S., L. Caston and D. Yungblut. 1996. Adding Roxazyme to wheat diets of chicken and turkey broilers. J. Appl. Poultry Res. 5:167–172.

Morgan, A., M. Bedford, A. Wilo, M. Nurminen, K. Autio, K. Poutanen and T. Parkkonen. 1995. How enzymes improve the nutritional value of wheat. Zootecnica International 18(4):44–50.

National Research Council. 1994. Nutrient Requirements of Poultry. ninth rev. ed. National Academy Press, Washington, DC.

Pijsel, C. 1996. Is there an interaction between antibiotics and enzymes? World Poultry 12(2):44–45.

SAS Institute. 1992. Version 6.08. SAS Institute, Inc., Cary, NC.

Summers, J.D., S.J. Slinger, W.F. Pepper and E.T. Moran, Jr. 1968. Biological evaluation of selected wheat fractions from nine different wheat samples for energy and protein quality. Poultry Sci. 47:1753–1760.

Ward, N.E. 1995. With dietary modification, wheat can be used for poultry. Feedstuffs 67(33):14–16.

Ward, N.E. 1996. Intestinal viscosity, broiler performance. Poultry Digest 4:12–16.

Wyatt, C. and H. Graham. 1996. Enzymes to the rescue. Feed Management 47(6):18–22.

Zumbado, M. and N. Dale. 1994. True metabolizable energy of wheat byproducts. Poultry Sci. 73(Suppl.1):8.

ORGANIC SELENIUM

NON-GLUTATHIONE PEROXIDASE FUNCTIONS OF SELENIUM

JOHN R. ARTHUR

Rowett Research Institute Bucksburn, Aberdeen, UK

Introduction

The trace element selenium has been associated with a wide range of clinical problems and metabolic activities in animals and man (Arthur and Beckett, 1994). Many of these effects are due to proteins which contain selenium as selenocysteine, a very effective redox catalyst at physiological pH (Arthur *et al.*, 1996). The biochemical and metabolic basis for some of the functions of selenium are understood, although many associations between micronutrient and disease remain anecdotal and have yet to be explained (Arthur and Beckett, 1994). There is, however, justification for connecting selenium with a wide range of functions and diseases as the micronutrient occurs as a component of, at least, 30 and perhaps as many as 100 proteins. This review will consider some of the more recently discovered functions of selenium and effects of selenium deficiency and how these may have implications for animal and human health.

Selenium deficiency and disease

As with micronutrient deficiencies such as those of copper, cobalt, iodine and zinc, the effects of severe selenium deficiency in animals have been investigated and are clearly described (see Combs and Combs, 1986). Consequently, in clinical veterinary practice severe selenium deficiency is usually recognized and methods of treatment have been established (McDowell *et al.*, 1996). Thus further occurrence of selenium deficiency disease is usually prevented by the appropriate supplementation of animals or their diets (Combs and Combs, 1986).

Selenium deficiency, generally combined with vitamin E deficiency, is the cause of several degenerative diseases in animals. The most widely recognized problems are skeletal and cardiac myopathies, and liver necrosis. Additionally, in both farm and laboratory animals selenium deficiency can cause both male and female infertility, poor growth and exudative diathesis (Combs and

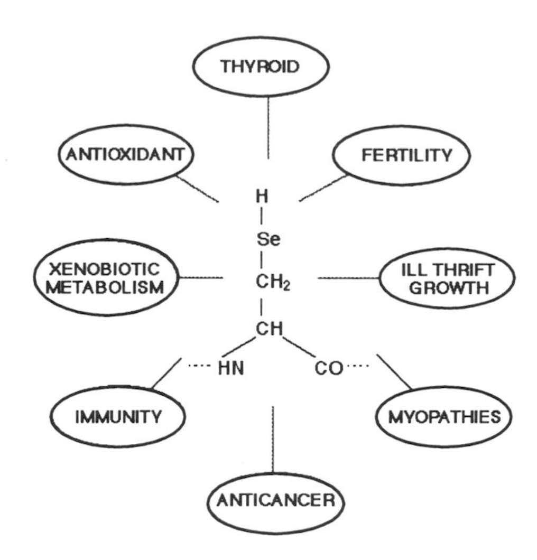

Figure 1. Metabolic pathways and diseases associated with selenium in animals. Selenium (Se) is shown as selenocysteine in the primary structure of a protein. At physiological pH over 99% of the selenium is in the form Se^- which can act as an efficient redox catalyst.

Combs, 1986) (Figure 1). The essentiality of selenium for human health has also been established. In China selenium deficiency is associated with an endemic cardiomyopathy known as Keshan disease (Levander and Whanger, 1996). However, selenium deficiency alone does not cause this condition and other stresses such as viral infection or vitamin E deficiency may also be necessary to initiate the disease process (Hill *et al.*, 1996; Beck *et al.*, 1994, 1995). Selenium supplementation has nonetheless greatly decreased the incidence of Keshan disease so the micronutrient deficiency is probably the major factor underlying the cardiomyopathy. Also in China, selenium deficiency has been associated with Kaschin-Beck disease which causes damage to cartilage in growing bones. The exact relationship between this disease and selenium has not been established and other factors must also be involved in its etiology (Combs and Combs, 1986). Along with these diseases, selenium has been associated with fertility, immune function, thyroid hormone metabolism, drug metabolizing enzymes, heart disease, cancer and effects on mood and well-being (see Arthur and Beckett, 1994) (Figure 1). These latter diseases and potential biological functions have been the topics of much current selenium-related research.

Biological functions of selenium

INORGANIC SELENIUM

Although most biological effects of selenium are probably mediated by selenocysteine-containing proteins, inorganic selenium compounds such as selenite will react with low molecular weight thiols to form covalent selenium sulfur compounds (Ip and Ganther, 1992). Some of these products are cytotoxic and may be the mediators of anti-cancer effects of selenium (Wu *et al.*, 1995; Lanfear *et al.*, 1994). Reaction of inorganic selenium with biological molecules may divert selenium from a form which can participate in selenocysteine synthesis. Alternatively, since inorganic selenium compounds are required for selenocysteine synthesis, these may provide a more available source of selenium for repletion of deficient animals (Daniels, 1996). The cytotoxic effects of selenium-thiol compounds such as selenodiglutathione can be described as 'non-glutathione peroxidase' functions for the micronutrient. However the majority of 'non-glutathione peroxidase' functions are likely to be those dependent on the increasing number of selenoproteins that are being characterized.

SELENOPROTEINS

The first functional selenoprotein to be characterized in mammals was cytosolic glutathione peroxidase (cGSHPx) (Rotruck *et al.*, 1973). Subsequently, many attempts were made to associate selenium deficiency disease with loss of cGSHPx activity. In particular, when combined with vitamin E deficiency, loss of cGSHPx activity would allow increased production of reactive oxygen species, with consequent damage to polyunsaturated fatty acids in membranes and essential proteins throughout the cell (Hoekstra, 1975). Although loss of cGSHPx activity and oxidative mechanisms could explain some effects of severe selenium deficiency, antioxidant systems did not fully explain all the biochemical changes in deficiency. For example, some drug-metabolizing enzymes in rat liver could be induced by selenium deficiency, apparently independent of changes in cGSHPx activity, and selenium-deficient rats could grow normally with liver cGSHPx activities of less than 10% of control (Reiter and Wendel, 1983, 1984; Burk, 1983; Arthur *et al.*, 1987a,b). Along with the discovery of cGSHPx-independent functions of selenium, only approximately 30% of selenium in the rat was found in the peroxidase, the remainder being distributed in other proteins (Beilstein and Whanger, 1988). *In vivo* labeling with ^{75}Se has been used to indicate the number of selenoproteins which may be present in mammalian systems. There are predicted to be between 20 and 30 selenoproteins based on radioactive bands which appear on separation of ^{75}Se-labeled tissue on SDS PAGE acrylamide gels (Bansal *et al.*, 1991; Behne *et al.*, 1988, 1996; Evenson and Sunde, 1988). Using the ratio of abundant to non-abundant genes in the genome, Burk and Hill (1993) have estimated that there may be up to 100 selenoproteins in mammalian systems. Thus, to better explain the functions of selenium, efforts have to be made to further characterize selenoproteins and associate them with particular biochemical or physiological activities. To date 14 selenoproteins have been further characterized by purification and (or) cloning. Additionally, and of particular importance in the understanding of the physiological roles of selenium, is investigation of the control of selenoprotein expression by selenium intake (Bermano *et al.*, 1995; Lei *et al.*, 1995). Selenoproteins which vary in concentration within the normal range of dietary selenium intake are those which are most likely to be associated with diseases caused by selenium deficiency. Additionally the response of individual selenoproteins to selenium deficiency can indicate which are the most important for maintenance of cell function. For example, in rats type I iodothyronine deiodinase (IDI) activity in the thyroid gland is maintained or may increase in selenium deficiency. In contrast, cGSHPx activity in liver may decrease by over 90% (Bermano *et al.*, 1995). Thus, maintenance of normal thyroid hormone metabolism is more crucial to the survival of the rat than hydrogen peroxide metabolism in the liver. However in the selenium-deficient rat, cGSHPx activity decreases very little in brain indicating that within an animal the selenoproteins may have 'organ-specific' functions of importance to overall health (Buckman *et al.*, 1993).

DIETARY FORMS OF SELENIUM

When investigating control of selenium function in animals it is important to recognize that the selenium in selenocysteine is incorporated as selenophosphate, an inorganic chemical (Low *et al.*, 1995). However, most selenium in natural diets occurs as selenomethionine, an organic compound (Daniels, 1996). Thus, any selenium absorbed from the diet must be converted from this organic form, into an inorganic form, then back to an organic form to fulfill its biochemical functions. The pathway for selenocysteine formation in animals therefore provides several possible stages for control of selenoprotein synthesis. This is particularly significant when selenium supplies are limiting and the most essential selenoproteins have to be synthesized at the expense of those which may be lost with few adverse effects on health.

Selenoproteins and their functions

GLUTATHIONE PEROXIDASES

Of selenoproteins which have been characterized (Figure 2), four are glutathione peroxidases. These enzymes metabolize hydrogen peroxide and lipid peroxides in different compartments of the cell. Thus the essential role of selenium in antioxidant systems can be explained. For example, phospholipid hydroperoxide glutathione peroxidase, which protects cell membranes from lipid hydroperoxides, is likely to be the cause of interactions between selenium and vitamin E deficiency (reviewed Arthur *et al.*, 1996; Arthur and Beckett, 1994). A major proportion of research into selenium biochemistry has concerned the function of cGSHPx, since this was the first selenoenzyme to be identified and it is the most abundant selenoprotein in most cells.

IODOTHYRONINE DEIODINASES

The next largest group of selenoproteins are three iodothyronine deiodinases (ID) which regulate the conversion of thyroxine (T4) to 3,3',5-triiodothyronine (T3), the active thyroid hormone or reverse triiodothyronine

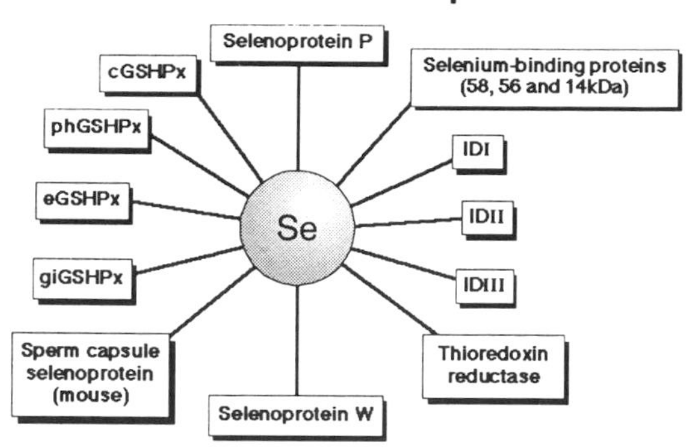

Figure 2. Selenoproteins which have been purified from mammalian cells. A more detailed discussion of the properties and functions of these proteins can be found in Arthur and Beckett (1994); Arthur *et al.* (1996); Burk and Hill (1993); Larsen and Berry (1995); Salvatore *et al.* (1996b) and Vendeland *et al.* (1995).
Key: cGSHPx, cystolic glutathione peroxidase; phGSHPx, phospolipid hydroperoxide glutathione peroxidase; eGSHPx, extracellular glutathione peroxidase; giGSHPx, gastrointestinal gluthione peroxidase; IDI, IDII and IDIII, type I type II and type III iodothyronine deiodinases.

(rT3), the inactive thyroid hormone (Salvatore *et al.*, 1995, 1996b; Larsen and Berry, 1995; Arthur *et al.*, 1990a). Type I ID (IDI) is expressed in liver, kidney, brain, pituitary and brown adipose tissue of ruminants. Type II ID (IDII) occurs in brain and pituitary of all species examined and brown adipose tissue of humans. Additionally, there is some evidence that IDII activity is expressed in skeletal muscle and in thyroid of humans (Salvatore *et al.*, 1996a). IDII catalyzes conversion of T4 to T3 within tissues which can not utilize circulating T3. Type III ID (IDIII) converts T4 to rT3 and T3 to diiodothyronine, is found in brain, skin and placenta and functions to deactivate thyroid hormones (Larsen and Berry, 1995). IDI is more sensitive than IDII or IDIII to the effects of selenium deficiency. Thus, in selenium-deficient rats, IDI activity may decrease by 90%, whereas IDII in brain will only decrease by 50% (Beckett *et al.*, 1987, 1989). A characteristic effect of selenium deficiency on plasma thyroid hormones caused by decreases in IDI activity is an acute increase (up to two-fold) in T4 concentration and a smaller decrease in T3 concentration (Beckett *et al.*, 1987). However, in more prolonged selenium deficiency plasma thyroid hormone concentrations may return to control levels despite large decreases in IDI activity in liver and kidney (Mitchell *et al.*, 1996).

OTHER SELENOPROTEINS

Other selenoproteins that have been identified are selenoprotein P, selenoprotein W, thioredoxin reductase, selenium binding proteins (58, 56 and 14 kDa), sperm capsule selenoprotein and a protein in the glandular epithelial cells of the rat prostate. Selenoprotein P contains 10 selenocysteine molecules and accounts for 60 to 80% of plasma selenium (Burk and Hill, 1994). The function of selenoprotein P remains uncertain, although it was originally thought to have a transport function distributing selenium to different organs via the circulation. However it seems unlikely that a complex, covalent molecule such as selenoprotein P would be synthesized purely as a transport form of selenium. Preferential synthesis of selenoprotein P in selenium-deficient rats, associated with increased protection against diquat toxicity, has led to the proposal of an antioxidant or cell membrane stabilizing function for the protein (Burk *et al.*, 1995).

Selenoprotein W is expressed in skeletal muscle with a subunit molecular weight of 9.5 kDa. The protein has been isolated with and without glutathione molecules bound to the amino acid backbone. The function of selenoprotein W has not been defined and it has been postulated to have antioxidant properties consistent with its ability to bind reduced glutathione. Additionally, loss of selenoprotein W may be associated with white muscle disease in sheep with combined selenium and vitamin E deficiencies (Levander and Whanger, 1996; Beilstein *et al.*, 1996; Vendeland *et al.*, 1995).

Thioredoxin reductase purified from a human lung cancer cell line contains selenium as selenocysteine at the penultimate amino acid at the C-terminus of the protein. The selenocysteine is encoded by a TGA stop codon in the thioredoxin reductase cDNA isolated from human placenta (Gladyshev *et al.*, 1996). Rat liver thioredoxin reductase contains selenium and its activity decreases by 75% in selenium deficiency (F. Nicol, J.R. Arthur, A.F. Howie and G.J. Beckett, unpublished). Thioredoxin reductase is involved in redox regulation of exposed disulfide groups on a range of enzymes and transcription factors. Thus through disruption of this regulation, selenium deficiency could influence the expression of a wide range of proteins

A sperm capsule selenoprotein been identified, which is consistent with the role of selenium in maintaining normal fertility. The nature of selenium incorporation into this protein is a subject of controversy since the translation start site in its mRNA is downstream of UGA codons which could specify selenocysteine (Cataldo *et al.*, 1996). The selenoprotein is a major component of the sperm capsule and has been postulated to have a structural function (Hansen and Deguchi, 1996). Similarly the role of the selenoprotein detected in rat prostate has not been determined (Kalcklosch *et al.*, 1995).

Less is known about the functions of selenium in the 14, 56 and 58 kDa binding proteins. For instance, the 14 kDa protein is fatty acid binding protein and it is uncertain whether the bound selenium has any physiological function (Sinha *et al.*, 1993).

The variety of selenoproteins identified emphasizes the wide range of biochemical pathways and thus physiological functions that can be modulated by changes in selenium status. Selenoprotein expression which varies within the normal range of dietary selenium intake is likely to underlie the injurious effects of selenium deficiency. Thus characterization of 'newer' selenoproteins may identify clinical problems which have hitherto not been linked to selenium deficiency.

Consequences of selenium deficiency

THYROID HORMONE METABOLISM

The role of selenium in IDIs implies that some of the consequences of selenium deficiency may be directly attributed to disturbances in thyroid hormone metabolism. In rats decreased pituitary growth hormone is consistent with lower IDII activity and T3 levels within the gland (Arthur *et al.*, 1990b). However in cattle, increases in growth on selenium supplementation are associated with changes in thyroid hormone metabolism but not changes in growth hormone (Wichtel *et al.*, 1996). Impairment of IDII activity in brown adipose tissue may decrease the ability of animals to respond to cold stress, as T3 synthesis within the tissue is required for uncoupling protein synthesis and thermogenesis (Arthur *et al.*, 1991). Consistent with this hypothesis, uncoupling protein concentrations are lower in combined selenium and iodine deficiency in rats (Geloen *et al.*, 1990). However, trials with sheep have not revealed interactions between selenium and iodine supplementation or selenium and thyroid hormone supplementation despite growth responses to selenium (Donald *et al.*, 1994). In rats selenium deficiency can either exacerbate or ameliorate the effects of a concurrent iodine deficiency (Mitchell *et al.*, 1996). Similar differences in the interactions between selenium and iodine deficiencies also occur in humans. These contrasting interactions are likely to be caused by compensatory changes in thyroid hormone metabolism which vary in efficiency with age and the onset and duration of micronutrient deficiency (Mitchell *et al.*, 1996). For example, increased plasma T4 concentrations, which are characteristic of *acute*, severe selenium deficiency, do not occur in more prolonged deficiency. This is due to induction of pathways for T4 metabolism other than selenium-dependent liver and kidney IDI activities, which predominate in normal nutritional conditions (Chanoine *et al.*, 1993). Thus the effects of selenium deficiency on thyroid hormone metabolism are variable but should not be underestimated since thyroid hormones are central to normal growth and development.

IMMUNITY AND INFECTION

Effects of selenium deficiency on the immune system and on the toxic properties of viruses require further research to clarify their underlying cause. Selenium deficiency adversely impairs several aspects of immunity ranging from neutrophil function to resistance to infection (Levander *et al.*, 1995; Turner and Finch, 1991; Boyne and Arthur, 1986). Decreased cGSHPx activity would allow peroxides to kill the neutrophil as well as the cell it is ingesting (Boyne and Arthur, 1986). However, it is less easy to relate other immune changes, such as impaired antibody production, to low cGSHPx activity in selenium deficiency. As well as impairing the host immune system, selenium deficiency can bring about changes in infective organisms. A non-cardiotoxic coxsackievirus B3 becomes cardiotoxic after passing through selenium-deficient mice. This change in phenotype is associated with specific changes in bases in the viral genome although the mechanism by which this occurs is unknown (Beck, 1996).

Conclusions

The identification of at least 30 selenoproteins and further characterization of 14 of these proteins is consistent with important roles for selenium in maintaining normal metabolism. Selenoproteins are involved in antioxidant systems, thyroid hormone metabolism, redox regulation and possible structural functions all of which are essential for optimal health. Clinical problems such as myopathies are associated with selenium deficiency and impairment of cell antioxidant systems. However, less is known about adverse effects of selenium deficiency caused by impairment of other biochemical pathways which are dependent on selenium. These latter reactions are likely to underlie some of the decreases in immune response and increases in cancer incidence which are being linked to low selenium status. Thus the range of clinical problems that might be related to selenium deficiency are potentially much wider than those which have been associated with functions of glutathione peroxidase. A major challenge is to determine which of these hitherto unrecognized clinical problems are of importance for economic or public heath reasons.

Acknowledgments

Work in the author's laboratory is supported by the Scottish Office Agriculture, Environment and Fisheries Department (SOAEFD).

References

Arthur, J.R. and G.J. Beckett. 1994. New metabolic roles for selenium. Proc. Nutr. Soc. 53:615–624.

Arthur, J.R., P.C. Morrice, F. Nicol, S.E. Beddows, R. Boyd, J.D. Hayes and G.J. Beckett. 1987a. The effects of selenium and copper deficiencies on glutathione S-transferase and glutathione peroxidase in rat liver. Biochem. J. 248:539–544.

Arthur, J.R., F. Nicol, R. Boyne, K.G.D. Allen, J.D. Hayes and G.J. Beckett. 1987b. Old and new roles for selenium. In: Trace Substances in Environmental Health XXI Ed D.D. Hemphill, Univ Missouri. pp. 487–498.

Arthur, J.R., F. Nicol and G.J. Beckett. 1990a. Hepatic iodothyronine deiodinase: The role of selenium. Biochem. J. 272:537–540.

Arthur, J.R., F. Nicol, P.W.H. Rae and G.J. Beckett. 1990b. Effects of selenium deficiency on the thyroid gland and on plasma and pituitary thyrotrophin and growth hormone concentrations in the rat. Clin. Chem. Enzym. Commun. 3:209–214.

Arthur, J.R., F. Nicol, G.J. Beckett and P. Trayhurn. 1991. Impairment of iodothyronine 5′-deiodinase activity in brown adipose tissue and its acute stimulation by cold in selenium deficiency. Can. J. Physiol. Pharmacol. 69:782–785.

Arthur, J.R., G. Bermano, J.H. Mitchell and J.E. Hesketh. 1996. Regulation of selenoprotein gene expression and thyroid hormone metabolism. Biochem. Soc. Trans. 24:384–388.

Bansal, M.P., C. Ip and D. Medina. 1991. Levels and Se-75-labeling of specific proteins as a consequence of dietary selenium concentration in mice and rats. Proc. Soc. Exp. Biol. Med. 196:147–154.

Beck, M.A. 1996. The role of nutrition in viral disease. J. Nutr. Biochem. 7:683–690.

Beck, M.A., P.C. Kolbeck, L.H. Rohr, Q. Shi, V.C. Morris and O.A. Levander. 1994. Benign human enterovirus becomes virulent in selenium-deficient mice. J. Med. Virol. 43:166–170.

Beck, M.A., Q. Shi, V.C. Morris and O.A. Levander. 1995. Rapid genomic evolution of a non-virulent Coxsackievirus B3 in selenium-deficient mice results in selection of identical virulent isolates. Nature Medicine 1:433–436.

Beckett, G.J., S.E. Beddows, P.C. Morrice, F. Nicol and J.R. Arthur. (1987). Inhibition of hepatic deiodination of thyroxine caused by selenium deficiency in rats. Biochem. J. 248:443–447.

Beckett, G.J., D.A. MacDougall, F. Nicol and J.R. Arthur. 1989. Inhibition of type I and type II iodothyronine deiodinase activity in rat liver, kidney and brain produced by selenium deficiency. Biochem. J. 259:887–892.

Behne, D., H. Hilmert, S. Scheid, H. Gessner and W. Elger. 1988. Evidence for specific selenium target tissues and new biologically important selenoproteins. Biochim. Biophys. Acta 966:12–21.

Behne, D., A. Kyriakopoeulos, C. Weissnowak, M. Kalckloesch, C. Westphal and H. Gessner. 1996. Newly found selenium-containing proteins in the tissues of the rat. Biol. Tr. Elem. Res. 55:99–110.

Beilstein, M.A. and P.D. Whanger. 1988. Glutathione peroxidase activity and chemical forms of selenium in tissues of rats given selenite or selenomethionine. J. Inorg. Biochem. 33:31–46.

Beilstein, M.A., S.C. Vendeland, E. Barofsky, O.N. Jensen and P.D. Whanger. 1996. Selenoprotein W of rat muscle binds glutathione and an unknown small molecular weight moiety. J. Inorg. Biochem. 61:117–124.

Bermano, G., F. Nicol, J.A. Dyer, R.A. Sunde, G.J. Beckett, J.R. Arthur and J.E. Hesketh. 1995. Tissue-specific regulation of selenoenzyme gene expression during selenium deficiency in rats. Biochem. J. 311:425–430.

Boyne, R. and J.R. Arthur. 1986. The response of selenium deficient mice to *Candida albicans* infection. J. Nutr. 116:816–822.

Buckman, T.D., M.S. Sutphin and C.D. Eckhert. 1993. A comparison of the effects of dietary selenium on selenoprotein expression in rat brain and liver. Biochim. Biophys. Acta 1163:176–184.

Burk, R.F. 1983. Biological activity of selenium. Ann. Rev. Nutr. 3:53–70.

Burk, R.F. and K.E. Hill. 1993. Regulation of selenoproteins. Ann. Rev. Nutr. 13:65–81.

Burk, R.F. and K.E. Hill. 1994. Selenoprotein P. A selenium-rich extracellular glycoprotein. J. Nutr. 124:891–1897.
Burk, R.F., K.E. Hill, J.A. Awad, J.D. Morrow, T. Kato, K.A. Cockell and P.R. Lyons. 1995. Pathogenesis of diquat-induced liver necrosis in selenium-deficient rats: Assessment of the roles of lipid peroxidation and selenoprotein P. Hepatology 21:561–569.
Cataldo, L., K. Baig, R. Oko, M.A. Mastrangelo and K.C. Kleene. 1996. Developmental expression, intracellular localization, and selenium content of the cysteine-rich protein associated with the mitochondrial capsules of mouse sperm. Mol. Reprod. Dev. 45:320–331.
Chanoine, J.P., S. Alex, S. Stone, S.L. Fang, I. Veronikis, J.L. Leonard and L.E. Braverman. 1993. Placental 5-deiodinase activity and fetal thyroid hormone economy are unaffected by selenium deficiency in the rat. Pediatric Res. 34: 88–292.
Combs, G.F. and S.B. Combs. 1986. The Role of Selenium in Nutrition. New York: Academic Press Inc.
Daniels, L.A. 1996. Selenium metabolism and bioavailability. Biol. Tr. Elem. Res. 54:185–199.
Donald, G.E., J.P. Langlands, J.E. Bowles and A.J. Smith. 1994. Subclinical selenium insufficiency 6. Thermoregulatory ability of perinatal lambs born to ewes supplemented with selenium and iodine. Austr. J. Exp. Agric. 34:19–24.
Evenson, J.K. and R.A. Sunde. 1988. Selenium incorporation into selenoproteins in the Se-adequate and Se-deficient rat. Proc. Soc. Exp. Biol. Med. 187:169–180.
Geloen, A., J.R. Arthur, G.J. Beckett and P. Trayhurn. 1990. Effect of selenium and iodine deficiency on the level of uncoupling protein in brown adipose tissue of rats. Biochem. Soc. Trans. 18:1269–1270.
Gladyshev, V.N., K.T. Jeang and T.C. Stadtman. 1996. Selenocysteine, identified as the penultimate C-terminal residue in human T-cell thioredoxin reductase, corresponds to TGA in the human placental gene. Proc. Natl. Acad. Sci. USA 93:6146–6151.
Hansen, J.C. and Y. Deguchi. 1996. Selenium and fertility in animals and man – a review. Acta Vet. Scand. 37:19–30.
Hill, K.E., Y.M. Xia, B. Akesson, M.E. Boeglin and R.F. Burk. 1996. Selenoprotein P concentration in plasma is an index of selenium status in selenium-deficient and selenium-supplemented Chinese subjects. J. Nutr. 126:138–145.
Hoekstra, W.G. 1975. Biochemical function of selenium and its relation to vitamin E. Fed. Proc. 34:2083–2089.
Ip, C. and H.E. Ganther. 1992. Comparison of selenium and sulfur analogs in cancer prevention. Carcinogenesis 13:1167–1170.
Kalcklosch, M., A. Kyriakopoulos, C. Hammel and D. Behne. 1995. A new selenoprotein found in the glandular epithelial cells of the rat prostate. Biochem. Biophys. Res. Commun. 217:162–170.
Lanfear, J., J. Fleming, L. Wu, G. Webster and P.R. Harrison. 1994. The selenium metabolite selenodiglutathione induces p53 and apoptosis:

Relevance to the chemopreventive effects of selenium? Carcinogenesis 15:1387–1392.

Larsen, P.R. and M.J. Berry. 1995. Nutritional and hormonal regulation of thyroid hormone deiodinases. Ann. Rev. Nutr. 15:323–352.

Lei, X.G., J.K. Evenson, K.M. Thompson and R.A. Sunde. 1995. Glutathione peroxidase and phospholipid hydroperoxide glutathione peroxidase are differentially regulated in rats by dietary selenium. J. Nutr. 125:1438–1446.

Levander, O.A. and P.D. Whanger. 1996. Deliberations and evaluations of the approaches, endpoints and paradigms for selenium and iodine dietary recommendations. J. Nutr. 126:S2427-S2434.

Levander, O.A., A.L. Ager and M.A. Beck. 1995. Vitamin E and selenium: Contrasting and interacting nutritional determinants of host resistance to parasitic and viral infections. Proc. Nutr. Soc. 54:475–487.

Low, S.C., J.W. Harney and M.J. Berry. 1995. Cloning and functional characterization of human selenophosphate synthetase, an essential component of selenoprotein synthesis. J. Biol. Chem. 270:21659–21664.

McDowell, L.R., S.N. Williams, N. Hidiroglou, C.A. Njeru, G.M. Hill, L. Ochoa and N.S. Wilkinson. 1996. Vitamin E supplementation for the ruminant. Anim. Feed. Sci. Tech. 60:273–296.

Mitchell, J.H., F. Nicol, G.J. Beckett and J.R. Arthur. 1996. Selenoenzyme expression in thyroid and liver of second generation selenium- and iodine-deficient rats. J. Mol. Endocrinol. 16:259–267.

Reiter, R. and A. Wendel. 1983. Selenium and drug metabolism-I, Multiple modulations of mouse liver enzymes. Biochem. Pharmacol. 32:3063–3067.

Reiter, R. and A. Wendel. 1984. Selenium and drug metabolism-II, Independence of glutathione peroxidase and reversibility of hepatic enzyme modulations in deficient mice. Biochem. Pharmacol. 33:1923–1928.

Rotruck, J.T., A.L. Pope, H.E. Ganther, A.B. Swanson, D.G. Hafeman and W.G. Hoekstra. 1973. Selenium: biochemical role as a component of glutathione peroxidase, Science. 179:588–590.

Salvatore, D., S.C. Low, M. Berry, A.L. Maia, J.W. Harney, W. Croteau, D.L. St Germain and P.R. Larsen. 1995. Type 3 iodothyronine deiodinase: Cloning, in vitro expression, and functional analysis of the placental selenoenzyme. J. Clin. Invest. 96:2421–2430.

Salvatore, D., T. Bartha, J.W. Harney and P.R. Larsen. 1996a. Molecular biological and biochemical characterization of the human type 2 selenodeiodinase. Endocrinology, 137:3308–3315.

Salvatore, D., H. Tu, J.W. Harney and P.R. Larsen. 1996b. Type 2 iodothyronine deiodinase is highly expressed in human thyroid. J. Clin. Invest. 98:962–968.

Sinha, R., M.P. Bansal, H.E. Ganther and D. Medina. 1993. Significance of selenium-labelled proteins for selenium's chemopreventive functions. Carcinogenesis, 14:1895–1900.

Turner, R.J. and J.M. Finch. 1991. Selenium and the immune response. Proc. Nutr. Soc. 50:275–285.

Vendeland, S.C., M.A. Beilstein, J.Y. Yeh, W. Ream and P.D. Whanger. 1995. Rat skeletal muscle selenoprotein W: cDNA clone and mRNA modulation by dietary selenium. Proc. Natl. Acad. Sci. USA. 92:8749–8753.

Wichtel, J.J., A.L. Craigie, D.A. Freeman, H. Varela Alvarez and N.B. Williamson. 1996. Effect of selenium and iodine supplementation on growth rate and on thyroid and somatotropic function in dairy calves at pasture. J. Dairy Sci. 79:1865–1872.

Wu, L., J. Lanfear and P.R. Harrison. 1995. The selenium metabolite selenodiglutathione induces cell death by a mechanism distinct from H_2O_2 toxicity. Carcinogenesis 16:1579–1584.

THE ROLE OF SELENIUM IN POULTRY NUTRITION

AUSTIN H. CANTOR

Department of Animal Sciences, University of Kentucky, Lexington, Kentucky, USA

Introduction

Selenium (Se) was reported to be an essential dietary nutrient 40 years ago. Schwarz and Foltz (1957) demonstrated that Se was the elusive Factor 3 that prevented liver necrosis in rats. In the same year there were two reports showing that Se was the factor that prevented exudative diathesis in chicks (Patterson *et al.*, 1957; Schwarz *et al.*, 1957). Ten years later, Scott *et al.* (1967) observed that Se could prevent gizzard and cardiac myopathies in young turkey poults. Normally, these deficiencies are manifested in the young bird, e.g., under four weeks of age. Therefore, the Se status of the young bird is affected by its diet as well as the carryover of nutrients from the hen.

Many symptoms of Se deficiency can be alleviated by vitamin E. The effectiveness of vitamin E, however, depends on the degree of the Se deficiency. In the case of a very severe Se deficiency, the chick develops pancreatic fibrosis (Thompson and Scott, 1970), a condition that reduces secretions of pancreatic digestive enzymes. This, in turn, reduces absorption of a number of nutrients including fat and fat-soluble vitamins. Therefore, dietary vitamin E is ineffective in preventing pancreatic fibrosis.

A deficiency of Se can also affect older birds. While Se deficiency has been shown to cause depressed egg production in both chickens and turkeys, it has a much greater impact on hatchability (Cantor and Scott, 1974; Cantor *et al.*, 1978; Latshaw and Osman, 1974). When turkey breeder hens were given diets deficient in both Se and vitamin E, the poults that did hatch had classical signs of gizzard myopathy (Cantor *et al.*, 1978).

The requirements of poultry for dietary selenium are listed in Table 1. According to the National Research Council (1994), the dietary needs range from 0.06 ppm Se in the laying hen to 0.2 ppm for the turkey. Meeting the dietary requirement for Se depends on a number of factors, including the amount of naturally occurring Se in the feed ingredients, the amount of supplemental Se used, and the availability of Se in the ingredients (Cantor *et al.*, 1975a,b).

Table 1. Dietary selenium requirements for poultry*.

Species	mg Se/kg diet
Chicken, broiler	0.15
Chicken, Leghorn type, to 6 weeks	0.15
Laying hen (consuming 100 g feed/day)	0.06
Turkey	0.20
Duck	0.20

* From National Research Council, 1994.

Concentration of selenium in corn and soybean meal in the USA

The concentration of Se in the soil and its availability to crops vary greatly from one geographic region to another. As a result the concentration of Se in crops grown on those soils also show great variation. For example, according to several reports (Scott *et al.*, 1982; Mikkelsen *et al.*, 1989; National Research Council, 1983; Scott and Thompson, 1971), the Se concentration in soybean meal has been shown to vary from 0.06 to over 1 ppm. The same reports indicated variations from less than 0.02 ppm to more than 1 ppm Se in corn and from 0.01 to 3 ppm Se in wheat.

Kubota *et al.* (1967) published a map of the United States showing regions where the Se concentrations in forages and crops were likely to be high, moderate or low. Data used to generate this map included the Se concentrations in alfalfa harvested from various geographic regions. Subsequent maps (e.g., National Research Council, 1983) included data from numerous other reports which also contained concentrations of Se in a number of plants. Most of the data were generated in the early 1980s or before. In general, these maps show that the regions with low Se concentrations in crops include a number of Midwestern states (Illinois, Indiana, Ohio) which are important sources of both corn and soybean meal used in the poultry industry. A number of the southeastern states, where there are heavy concentrations of poultry production, are classified as having either variable or low concentrations of Se in forages and crops. On the other hand, there is a broad area consisting of mostly Central Plains and western states that is considered to be a high Se area. The references cited above indicate that within each of these areas there can be considerable variation in the Se concentration of crops. In view of this variation, we were interested in conducting a survey of selenium concentrations in feed ingredients to provide feed manufacturers and poultry producers with sufficient information to prevent Se deficiencies in commercial poultry production.

A total of 254 samples of corn from the 1994 harvest, which originated from 10 different states in the USA, were analyzed for Se. Samples of soybean meal were obtained from 42 soybean processing plants located in 18 different states. Each processing plant was requested to submit three samples over a period of about 18 months. The samples were produced from the 1994 and 1995 harvests. The geographic origin (i.e., where the corn or soybeans were grown) were identified whenever possible. The fluorometric procedure of Olson *et al.* (1975), as modified by Cantor and Tarino (1982), was used to measure Se in all samples.

Table 2 shows the concentrations of Se in corn samples from 10 states. Values were very low for four states (Delaware, Maryland, Ohio and Virginia), low for two states (Illinois and Indiana), moderate for two states (Kentucky and Minnesota) and high for the remaining two states (Nebraska and Iowa). The highest state average (0.428 ppm for Minnesota) was almost 18 times that of the lowest value (0.024 ppm for Delaware). It was also noted that there was considerable variation within the various states, especially in states that had moderate to high values.

Results of analyses on high protein soybean meal (referred to as 48% soybean meal) and 44% soybean meal are presented separately. Table 3 shows the Se levels for 48% soybean meal obtained from 18 different states. States with very low averages (<0.10 ppm) included Delaware, Kentucky, Maryland, Ohio and South Carolina. Alabama, Illinois, Indiana, North Carolina and Virginia had low averages (0.10 to 0.20 ppm), while Georgia had a moderate value (0.25 ppm). High mean concentrations (0.3 to 0.5 ppm) were found in samples from Arkansas, Kansas and Mississippi, while very high mean concentrations (>0.5 ppm) were found in samples from Iowa, Minnesota, Missouri and Nebraska. The highest average (0.737 ppm for Missouri) was approximately 12 times the lowest state average (0.06 ppm for Delaware, Maryland and Virginia). The highest individual value (1.189 ppm for Iowa) was almost 20 times the lowest values (0.06 ppm for Delaware, Maryland and Virginia).

Table 4 shows the results of analyses for 44% soybean meal samples. In general, they are similar to the values seen for the 48% soybean meal samples. Within most, but not all, of the states found in both tables, the values in Table 3 are slightly lower. This would be expected because Se in plant tissues is associated with the protein fraction. Ohio and South Carolina had the lowest state means (0.075 and 0.073 ppm, respectively), while Missouri had the highest state mean (0.77 ppm). The individual sample

Table 2. Concentration of selenium in samples of corn obtained from 10 States. Values are in mg/kg, as-is basis.

State	N*	Mean	SD†	Minimum	Maximum
Delaware	5	0.024	0.004	0.019	0.028
Illinois	27	0.054	0.018	0.025	0.085
Indiana	23	0.058	0.055	0.012	0.268
Iowa	78	0.210	0.162	0.009	0.575
Kentucky	44	0.126	0.076	0.033	0.448
Maryland	14	0.026	0.015	0.010	0.059
Minnesota	20	0.147	0.086	0.006	0.340
Nebraska	11	0.428	0.141	0.225	0.644
Ohio	19	0.027	0.012	0.011	0.054
Virgina	13	0.037	0.017	0.011	0.072

* Number of samples.
† Standard deviation.

Table 3. Concentration of selenium in samples of 48% soybean meal obtained from 18 states. Values are in mg/kg, as-is basis

State	N*	Mean	SD†	Minimum	Maximum
Alabama	6	0.170	0.049	0.118	0.238
Arkansas	1	0.466	–	0.466	0.466
Deleware	3	0.062	0.019	0.049	0.084
Georgia	3	0.213	0.157	0.111	0.395
Illinois	7	0.144	0.057	0.091	0.246
Indiana	3	0.120	0.029	0.087	0.138
Iowa	26	0.798	0.200	0.467	1.318
Kansas	6	0.500	0.106	0.350	0.623
Kentucky	3	0.099	0.035	0.071	0.138
Maryland	1	0.056	–	0.056	0.056
Minnesota	4	0.578	0.034	0.544	0.623
Mississippi	6	0.588	0.327	0.246	1.158
Missouri	5	0.744	0.079	0.651	0.822
Nebraska	1	0.721	–	0.721	0.721
North Carolina	3	0.186	0.133	0.087	0.336
Ohio	6	0.088	0.025	0.068	0.121
South Carolina	3	0.061	0.001	0.060	0.062
Virginia	3	0.102	0.019	0.080	0.114

* Number of samples.
† Standard deviation.

Table 4. Concentration of selenium in samples of 44% soybean meal obtained from 13 states. Values are in mg/kg, as-is basis

State	N*	Mean	SD†	Minimum	Maximum
Aransas	1	0.419	–	0.419	0.419
Georgia	3	0.180	0.236	0.091	0.340
Illinois	3	0.123	0.019	0.102	0.136
Indiana	3	0.101	0.014	0.087	0.116
Iowa	23	0.686	0.232	0.219	1.174
Kansas	3	0.538	0.013	0.526	0.551
Minnesota	4	0.515	0.024	0.498	0.550
Mississippi	3	0.552	0.274	0.318	0.854
Missouri	4	0.772	0.093	0.648	0.845
Nebraska	1	0.617	–	0.617	0.617
Ohio	6	0.075	0.015	0.062	0.102
South Carolina	1	0.073	–	0.073	0.073
Virginia	2	0.120	0.008	0.113	0.126

* Number of samples.
† Standard deviation.

with the lowest Se concentration was obtained from Ohio (0.06 ppm) while that with the highest Se concentration was obtained from Iowa (1.17 ppm). The highest state average was approximately 11 times the lowest state average, while the highest individual value was approximately 19 times the lowest single value.

Availability of selenium in feedstuffs and selenium compounds

In addition to variations in concentrations of Se in feedstuffs, there can be considerable differences in Se availability. A number of feed ingredients were assayed for biological availability of Se using prevention of exudative diathesis in young chicks as the criterion (Cantor *et al.*, 1975b). To conduct these studies, chicks were obtained from breeder hens fed diets low in Se and vitamin E. This permitted a uniform development of deficiency symptoms over many trials. The chicks were fed semi-purified diets without vitamin E supplementation and with graded levels of the test ingredients and sodium selenite, which was used as a standard. Availability values ranged from 210% for dehydrated alfalfa meal (a result that was not explained) to 9% for fish solubles (Table 5). While the fish products assayed had very high Se concentrations, their availability values were low (< 25%). Other animal products also had low availability values. In contrast, most feedstuffs of plant origin had values from 60 to 90%. The availability of Se in a variety of compounds was also evaluated using the same assay procedure (Table 6). Of particular interest was the finding that Se in seleno-DL-cystine and sodium selenate were fairly available (73% and 74% respectively) while the average value for seleno-DL-methionine was only 37%, compared with sodium selenite. Much of the selenium in a variety of feedstuffs is thought to be in the form of selenium analogs of sulfur amino acids. Thus, the availability values for selenium compounds could partially explain Se availability in feedstuffs. Most animal by-products are generally considered to have equal or greater digestibility than plant products. This would suggest that digestibility is not responsible for the differences in availability between animal and plant feedstuffs.

The discovery reported by Rotruck *et al.* (1973) that Se is a constituent of the enzyme glutathione peroxidase was a major step in understanding the role

Table 5. Biological availability of selenium in feedstuffs for prevention of exudative diathesis in chicks (sodium selenite used as standard)*.

Feedstuff	Biological availability, %
Wheat	71
Brewer's yeast	89
Brewer's grains	80
Corn	86
Soybean meal	60
Cottonseed meal	86
Dehydrated alfalfa meal	210
Distiller's dried grains and solubles	65
Tuna meal	22
Poultry by-product meal	18
Menhaden fish meal	16
Fish solubles	9
Herring meal	25
Meat and bone meal	15

* Summarized from Cantor *et al.* (1975b)

Table 6. Biological availability of selenium in selenium compounds for prevention of exudative diathesis in chicks (sodium selenite used as standard)*.

Compound	Biological availability, %
Seleno-DL-methionine	37
Seleno-DL-cystine	73
6-Seleno-purine	20
Seleno-DL-methionine	44
Sodium selenate	74
Sodium selenide	42
Elemental selenium (gray)	7

* Summarized from Cantor *et al.* (1975b)

of Se in preventing a variety of deficiency symptoms. Noguchi *et al.* (1973) demonstrated how a decrease in plasma glutathione peroxidase was closely related to the onset of exudative diathesis in chicks fed Se-deficient diets. It was then shown that the availability of Se in selenomethionine for prevention of exudative diathesis was closely related to its ability to increase plasma glutathione peroxidase (Cantor *et al.*, 1975b). Plasma glutathione peroxidase was shown to be highly correlated with plasma Se concentrations when selenite, but not selenomethionine, was used as a dietary Se supplement. This suggested that Se from selenite could more directly be incorporated into glutathione peroxidase that Se from selenomethionine. Other studies with turkey poults did not show differences in utilization of Se from selenite and selenomethionine for increasing plasma glutathione peroxidase (Cantor *et al.*, 1982; Cantor and Tarino, 1982). However, menhaden fish meal, used as a source of Se, was shown to be significantly less effective than selenite for increasing plasma glutathione peroxidase in the poult (Cantor and Tarino, 1982).

In studies evaluating the efficacy of Se compounds for prevention of pancreatic fibrosis, selenomethionine was shown to be four times as effective as sodium selenite (Cantor *et al.*, 1975a). Supplementing a purified diet with 0.01 ppm Se as selenomethionine was as effective as adding 0.04 ppm Se as either sodium selenite or selenocystine for prevention of histological lesions as well as for increasing pancreatic Se concentrations. However, there were no differences in the effects of the three compounds on plasma Se concentrations and the activity of glutathione peroxidase in plasma and pancreatic tissue. The behavior of selenomethionine in this system is likely related to the fact that there is a high amount of protein synthesis in the pancreas. Thus, there is considerable transport of methionine as well as selenomethionine to that organ.

Comparative availability of selenium sources

Recently, there has been considerable interest in the use of organic forms of Se for supplementing animal diets. Mahan and Parrett (1996) reported greater retention of Se in muscle tissue of pigs when selenized yeast (Sel-Plex 50) was used as a Se supplement compared with sodium selenite.

However, the selenite appeared to be more available for enhancing the activity of glutathione peroxidase.

Edens *et al.* (1996a) compared the effects of feeding sodium selenite and selenized yeast (Sel-Plex 50) at 0.1 and 0.3 ppm upon body feathering of slow feathering male broilers in two trials. Use of selenized yeast resulted in better feathering measured at 3, 5 and 6 weeks of age and higher feather weight. In addition, under certain conditions (season and litter type) the use of selenized yeast resulted in decreased drip loss of breast meat (Edens *et al.*, 1996b).

Studies in our laboratory were conducted to compare the biological availability of selenized yeast (Sel-Plex 50, Alltech Inc.) with sodium selenite in young broiler chickens (Collins *et al.*, 1993). Following a 9-day depletion period, day-old males were fed (for 19 days) a low-Se basal diet alone, or with graded levels of Se (0.04, 0.08 or 0.12 ppm) provided by Sel-Plex 50 or selenite. Linear regression equations were determined for tissue Se vs dietary Se for each source. Using slope ratios, relative availability values for selenized yeast were calculated. Based on data for plasma, whole blood, liver and muscle Se (Table 7), it was concluded that Se in selenized yeast was about 67% available, compared with selenite. These results were in agreement with previous observations using plasma Se and glutathione peroxidase activity as criteria. However, in a recent trial we noted that selenized yeast was approximately 90% as available as selenite based on liver Se concentrations measured after six weeks.

Table 7. Biological availability of Sel-Plex 50 for broiler chickens based on tissue selenium concentrations (sodium selenite used as standard)*.

Criterion	Biological availability, %
Plasma Se	77
Blood Se	66
Liver Se	62
Muscle Se	65

* From Collins *et al.*, 1993

Sel-Plex 50 was also evaluated for its effect on egg Se concentrations following 3 and 6 weeks of feeding to laying hens (Cantor *et al.*, 1996). A basal diet was fed alone or with 0.3 ppm as either selenite or Sel-Plex 50 yeast. Egg Se concentrations were 0.05, 0.18 and 0.24 ppm after 3 weeks for the basal, selenite and Sel-Plex 50 treatments, respectively (Table 8). Egg Se for the Sel-Plex 50 treatment was significantly greater (+29%) than that for the selenite treatment. Similar results were obtained at 6 weeks. Thus, it appears that selenized yeast can be useful for enhancing egg Se levels.

Summary

Dietary deficiencies of Se have been shown to cause a variety of symptoms in various species of poultry. In the absence of adequate supplementation, these deficiencies can result in considerable economic loss for poultry producers.

Table 8. Effect of Sel-Plex 50 and sodium selenite on egg selenium content*.

	Basal	Selenite	Sel-Plex 50[2,3]
Daily Se intake, mg/hen	5	34	36
Egg Se, Week 3, ppm	0.05[c]	0.18[b]	0.24[a] (+29%)
Egg Se, Week 6, ppm	0.06[c]	0.20[b]	0.24[a] (19%)
Total Se/egg, Week 3, mg	2.6[c]	9.9[b]	13.0[a] (+31%)
Total Se/egg, Week 6, mg	3.3[c]	10.9[b]	13.5[a] (24%)

* From Cantor *et al.*, 1996

There is considerable variation in the Se levels found in two of the major ingredients used for poultry diets in the USA. In some cases, poultry producers and feed formulators may know that their major ingredients are coming from low-Se areas. In other cases they may not know the geographic origin of their feedstuffs. In both of these cases, it is highly recommended that the maximum allowable level of supplemental Se be used. Unfortunately, it is very expensive to monitor the Se levels in all batches of ingredients as they arrive at the feed mill. The values presented here may be useful in making some predictions of the expected Se values. However, it is important to keep in mind that there is considerable variation within states.

The data on Se levels in corn and soybean meal illustrate the possibility of encountering Se deficiencies in commercial poultry if inadequate levels of supplemental Se are used. In addition to use of adequate Se supplements, other measures that could be used to prevent deficiencies include saving the ingredients with higher Se levels for starter and breeder diets, using higher levels of vitamin E, and using ingredients with especially high levels of Se (e.g., fish meal, soybean meal with especially high Se concentrations). Unfortunately, all of these measures can be extremely costly.

In addition to variation in the naturally occurring levels of Se in feedstuffs, biological availability is another important factor influencing Se status in animals. A number of studies suggest that organic selenium sources (e.g., selenoamino acids, selenized yeast) can be effective in increasing tissue Se levels but may be less available than selenite for increasing the Se-dependent enzyme glutathione peroxidase. This suggests that the organic Se (e.g., selenomethionine) is deposited in tissues along with its sulfur analog and that it may only become available for glutathione peroxidase activity upon catabolism of protein and amino acids. This would explain why some organic sources of Se are less available for this enzyme activity when assayed in short feeding trials with very young chicks.

Feeding Sel-Plex 50 to broilers for 6 weeks results in liver Se levels that are slightly lower than levels obtained with selenite. In contrast, the use of Sel-Plex 50 in layer diets significantly increased the deposition of Se in the egg. Thus, selenium yeast may be especially useful in breeder diets for improving the Se status of the embryo and the newly hatched chick. It may also have an application for the production of nutritionally enhanced 'designer' eggs for human consumption.

References

Cantor, A.H. and M.L. Scott. 1974. The effect of selenium in the hen's diet on egg production, hatchability, performance of progeny and selenium concentrations in eggs. Poultry Sci. 53:1870.

Cantor, A.H. and J.Z. Tarino. 1982. Comparative effects of inorganic and organic dietary sources of selenium upon selenium levels and selenium-dependent glutathione peroxidase activity in blood of young turkeys. J. Nutr. 112:2187.

Cantor, A.H., M.L. Langevin, T. Noguchi and M.L. Scott. 1975a. Efficacy of selenium compounds and feedstuffs for prevention of pancreatic fibrosis in chicks. J. Nutr. 105: 106.

Cantor, A.H., M.L. Scott and T. Noguchi. 1975b. Biological availability of selenium in feedstuffs and selenium compounds for prevention of exudative diathesis in chicks. J. Nutr. 105: 96.

Cantor, A.H., P.D. Moorhead and K.I. Brown. 1978. Influence of dietary selenium upon reproductive performance of male and female breeder turkeys. Poultry Sci. 57:1337.

Cantor, A.H., P.D. Moorhead and M.A. Musser. 1982. Comparative effects of sodium selenite and selenomethionine upon nutritional muscular dystrophy, selenium-dependent glutathione peroxidase, and tissue selenium concentrations of turkey poults. Poultry Sci. 61:478.

Cantor, A.H., A.J. Pescatore, M.L. Straw and M.J. Ford. 1996. Effect of selenium yeast (Sel-Plex 50) on egg selenium concentrations. Poster, 12th Annual Symposium Biotechnology in the Feed Industry, Alltech Technical Publications, Nicholasville, KY.

Collins, V.C., A.H. Cantor, M.J. Ford and M.L. Straw. 1993. Bioavailability of selenium in selenized yeast for broiler chickens. Poultry Sci. 72 (Suppl. 1): 85.

Edens, F.W., T.A. Carter and A.E. Sefton. 1996a. Improved feathering with dietary organic selenium and its modification due to litter material. Poultry Sci. 75 (Suppl. 1): 114.

Edens, F.W., T.A. Carter and A.E. Sefton. 1996b. Influence of dietary selenium sources on post-mortem drip loss from breast meat of broilers grown on different litters. Poultry Sci. 75 (Suppl. 1): 60.

Kubota, J. W.H. Allaway, D.L. Carter, E.E. Cary and V.A. Lazar. 1967. Selenium in crops in the United States in relation to selenium-responsive diseases of animals. J. Agric. Food Chem. 15:448.

Latshaw, J.D. and M. Osman. 1974. A selenium and vitamin E responsive condition in the laying hen. Poultry Sci. 53:1704.

Mahan, D.C. and N.A. Parrett. 1996. Evaluating the efficacy of selenium-enriched yeast and sodium selenite on tissue selenium retention and serum glutathione peroxidase activity in grower and finisher swine. J. Anim. Sci. 74: 2967.

Mikkelsen, R.L., A.L. Page and F.T. Bingham. 1989. Factors affecting selenium accumulation by agricultural crops. In: Selenium in Agriculture and the Environment. Soil Sci. Soc. America, special publ. no. 23.

National Research Council, 1983. Selenium in Nutrition. Revised ed. National Academy Press, Washington, DC.

National Research Council, 1994. Nutrient Requirements of Poultry. 9th revised ed. National Academy Press, Washington, DC.

Noguchi, T., A.H. Cantor and M.L. Scott. 1973. Mode of action of selenium and vitamin E in prevention of exudative diathesis in chicks. J. Nutr. 103: 1502.

Olson, O.E., I.S. Palmer and E.E. Cary. 1975. Modification of the official method for selenium in plants. J. Assoc. Offic. Anal. Chem. 58:117.

Patterson, E.L., R. Milstrey and E.L.R. Stokstad. 1957. Effect of selenium in preventing exudative diathesis in chicks. Proc. Soc. Exp. Biol. Med. 95: 617.

Rotruck, J.T., A.L. Pope, H.E. Ganther, A.B. Swanson, D.G. Hafeman and W.G. Hoekstra. 1973. Selenium: biochemical role as a component of glutathione peroxidase. Science 179: 588.

Schwarz, K. and C.M. Foltz. 1957. Selenium as an integral part of Factor 3 against necrotic liver degeneration. J. Am. Chem. Soc. 79: 3292.

Schwarz, K., J.G. Bieri, G.M. Briggs and M.L. Scott. 1957. Prevention of exudative diathesis in chicks by Factor 3 and selenium. Proc. Soc. Exp. Biol. Med. 95:621.

Scott, M.L. and J.N. Thompson. 1971. Selenium content of feedstuffs and effects of dietary selenium levels upon tissue selenium in chicks and poults. Poultry Sci. 50:1742.

Scott, M.L., G. Olson, L. Krook and W.R. Brown. 1967. Selenium-responsive myopathies of myocardium and smooth muscle in the young poult. J. Nutr. 91:573.

Scott, M.L., M.C. Nesheim and R.J. Young. 1982. Nutrition of the Chicken. 3rd ed. M.L. Scott and Assoc., Ithaca, NY.

Thompson, J.N. and M.L. Scott. 1970. Impaired lipid and vitamin E absorption related to atrophy of the pancreas in selenium-deficient chicks. J. Nutr. 100:797.

COMPARISON OF ORGANIC AND INORGANIC SOURCES OF SELENIUM FOR GROWTH AND HEALTH OF CHANNEL CATFISH

R.T. LOVELL and CHINLU WANG

Department of Fisheries and Allied Aquacultures, Auburn University, Alabama, USA

Introduction

The role of selenium in nutrition was first identified as a toxic element to grazing animals in South Dakota and Wyoming causing hair loss, lameness, and hoof sloughing (Franke, 1934). Selenium was classified as an essential element in animal nutrition after it was found to prevent exudative diathesis in chickens and prevent liver necrosis in rats (Patterson *et al.*, 1957; Schwartz and Foltz, 1957). Subsequently, selenium deficiency was found to cause white muscle disease in horses (Stowe, 1967), sheep (Muth and Allaway, 1963; Kubota *et al.*, 1967) and cattle (Cawley and Bradley, 1978). Selenium deficiency also impaired the immune system in rats, chicks, horses, pigs, sheep and cattle (Combs and Combs, 1986).

Selenium is required in the diet for normal growth and physiological function of fish. Selenium deficiency caused reduced growth, low glutathione peroxidase activity and exudative diathesis in rainbow trout (Hilton *et al.*, 1980; Bell *et al.*, 1987). In channel catfish, selenium deficiency was characterized by reduced growth, poor feed efficiency and reduced glutathione peroxidase activity (Gatlin and Wilson, 1984). Dietary selenium deficiency suppressed macrophage killing ability in channel catfish (Wise *et al.*, 1993). The dietary selenium requirements as sodium selenite for normal growth were found to be 0.25 mg/kg for channel catfish and 0.38 mg/kg for rainbow trout. Dietary selenium levels of 15 mg/kg for channel catfish and 13 mg/kg for rainbow trout were toxic to the fish (Hilton *et al.*, 1980; Gatlin and Wilson, 1984).

Selenium requirement of fish consuming high fishmeal diets, such as Atlantic salmon, can be met with the selenium in the fishmeal, which contains 2 mg/kg selenium. Selenium supplementation with sodium selenite or selenomethionine did not benefit the growth and performance of Atlantic salmon fed a typical practical diet (Lorentzen *et al.*, 1994). However, selenium supplementation will be necessary in diets for those species, such as channel catfish and tilapia, which receive diets containing mainly grain and oil seed products.

Organic minerals have a higher rate of intestinal absorption, are transported intact to target tissue or organs, and have a high rate of retention in tissue as compared with inorganic sources (Ashmead and Zunio, 1992). In a selenium digestibility trial with Atlantic salmon, selenomethionine was the most digestible (91.6%), followed by sodium selenite (63.9%) and selenium in fishmeal (46.6%) (Bell and Cowey, 1989). In a digestibility study of channel catfish, Paripatananont and Lovell (1996) found that selenomethionine was more digestible than sodium selenite in a purified egg white-based diet (90.8 vs 62.8%) and in a soybean meal-based diet (88.9 vs 69.7%). Paripatananont and Lovell (1995a) indicated that dietary zinc requirement of channel catfish was reduced when zinc methionine replaced the inorganic zinc in diets of fish. Channel catfish fed with zinc methionine had stronger resistance to *Edwardsiella ictaluri* infection and better immune responses than fish fed with inorganic zinc sources (Paripatananont and Lovell, 1995b). It is possible that organic selenium in the diet will reduce the channel catfish's selenium requirement over that determined with inorganic selenium because of the higher digestibility or tissue retention of organic selenium.

Selenium concentration in soil depends on the content in parent rock. Selenium in parent rock is released into soil by the weathering process. In addition to the selenium content in parent rocks, the soil pH, oxidation-reduction condition, moisture level and degree of aeration affect the chemical forms of selenium and their availability to plants in soil. Selenium released from rock under alkaline, well aerated conditions is oxidized to selenate (SeO_4^{2-}), a more soluble and available form to plants. However, in most of the grain and soybean production areas in the United States, the selenium content in soil is low and in the less available form of selenite (SeO_3^{2-}) (Combs and Combs, 1986). Therefore, the selenium content of grains, soybean meal or other plant feedstuffs from those areas in the United States is less than 0.1 mg/kg (National Research Council, 1993). Soybean and other plant feedstuffs from most areas of the United States do not contain sufficient selenium for normal growth of livestock or fish (National Research Council, 1993).

Sodium selenite that has been used to establish selenium requirements for channel catfish is the source for selenium supplementation in commercial channel catfish feeds. There has been no investigation about the effects of organic or chelated selenium supplementation in channel catfish diets. Paripatananont and Lovell (1995a) found that the bioavailability of chelated zinc (zinc methionine) for growth in channel catfish was 352% of that of inorganic zinc (zinc sulfate) in an egg white diet and 482% of that of inorganic zinc (zinc sulfate) in a soybean meal diet. Chelated copper, iron and manganese have been found to be more available to birds and mammals than inorganic sources of these elements (Ammerman *et al.*, 1995; Littell *et al.*, 1995). There is interest in reducing the amount of selenium supplementation in fish and livestock feeds for environmental reasons. If the bioavailability of organic selenium is higher than that of an inorganic source, the use of organic sources of selenium will allow lower amounts of selenium supplementation in channel catfish feeds.

A study was conducted with channel catfish to compare the effects of sodium selenite, selenomethionine and selenoyeast (Sel-Plex 50, Alltech Inc., a high selenium strain of yeast) on weight gain, feed efficiency, tissue selenium content, glutathione peroxidase activity, and responses to challenges with the pathogenic bacterium *E. ictaluri*.

Materials and Methods

DIETS AND EXPERIMENTAL DESIGN

The basal diet presented in Table 1 was supplemented with 0, 0.02, 0.06, 0.20, or 0.40 mg of selenium/kg from sodium selenite (Sigma, St. Louis, Missouri), DL-selenomethionine (Sigma, St. Louis, Missouri), or selenoyeast (Alltech Biotechnology, Nicholasville, Kentucky). The amount of vitamin E supplemented in the diet was 25 mg/kg of diet, the minimum requirement for growth of channel catfish (Lovell *et al.*, 1984). L-arginine-HCl was added to the diet to compensate for the low arginine content of casein. The diets were extruded as semi-moist (approximately 30% moisture), 3 mm diameter pellets. The diets were stored at -18°C and thawed in a refrigerator (4°C) for 24 hours before feeding. Diets were analyzed for moisture by drying to constant weight, crude protein by the Kjeldahl method (AOAC, 1990) and selenium by a fluorometric method (AOAC, 1990). The basal diet contained 0.05 mg/kg of selenium.

Table 1. Composition of the basal diet.

Ingredient	g/kg
Casein (vitamin free)	397.2
L-arginine-HCl	2.8
Dextrin	370.0
Cellulose	80.5
Se-free mineral premix*	40.0
Vitamin premix†	14.0
Calcium carbonate	12.0
Ascorbic acid polyphosphate‡	3.5
Corn oil	30.0
Cod liver oil	30.0
Carboxymethylcellullose	20.0

* Contains (as g/kg of premix): $FeSO_4.7H_2O$, 5.0 g; $MgSO_4.7H_2O$, 132.0 g; K_2HPO_4, 240.0 g; $NaH_2PO_4.H_2O$, 231.0 g; NaCl, 45.0 g; $AlCl_3.6H_2O$, 0.15 g; $CuSO_4.5H_2O$, 0.5 g; $MnSO_4.H_2O$, 0.7 g; $CoCl_2.6H_2O$, 1.0 g; KI, 0.15 g; $ZnSO_4.7H_2O$, 3.0 g; cellulose, 341.5 g (Paripatananont and Lovell, 1995).

† Provides the following diluted in cellulose (mg/kg of diet): thiamin, 10 mg; choline, 500 mg; niacin, 150 mg; riboflavin, 20 mg; pyridoxine, 20 mg; calcium pantothenate, 200 mg; vitamin B_{12}, 0.06 mg; retinyl acetate (500,000 IU/g), 12 mg; alpha-tocopherol acetate, 25 mg; cholecalciferol (1,000,000 ICU/g), 1 mg; menadione Na-bisulfite, 80 mg; inositol, 400 mg; biotin, 1 mg.

‡ Source of ascorbic acid polyphosphate (15% ascorbic acid) was Hoffman La Roche (Basel, Switzerland).

FISH AND FEEDING PROTOCOL

Channel catfish (*Ictalurus punctatus*) fry were brought to the fish nutrition laboratory at the Auburn University Fisheries Research Unit and maintained on a vitamin refortified commercial salmon starter for 3 weeks. Subsequently, fish (average weight, 1.70 g) were stocked into 39 aquaria (40 l) at a density of 30 fish/aquarium. All fish received the casein basal diet for 2 weeks, then each test diet was randomly assigned to three aquaria. The fish were fed to satiation twice-daily for 9 weeks. Daily feed consumption was recorded for each aquarium. Fish were collectively weighed every 2 weeks. When a bacterial infection was observed in fish, a 1 hour static bath treatment with oxytetracycline (20 mg/l) was given, followed by two additional treatments at 3-day intervals.

SAMPLE COLLECTION AND ANALYSIS

At the end of the feeding trial, fish were collectively weighed and counted. Blood from five fish in each aquarium was obtained by severing the caudal peduncle and collecting directly from the caudal vein by heparinized syringes. Plasma was separated by centrifugation and analyzed for glutathione peroxidase GSH-Px activity as described by Lawrence and Burk (1976). Fresh livers from six fish per treatment (two fish in each aquarium) were individually homogenized and assayed for selenium-dependent GSH-Px activity using H_2O_2 as substrate, according to Lawrence and Burk (1976). Total protein content of plasma and liver extracts was determined as described by Lowry *et al.* (1951). Liver and muscle samples from two fish in each aquarium were collected for selenium analysis by a fluorometric method (AOAC, 1990).

BACTERIAL CHALLENGE

A strain of *E. ictaluri* (AL-95–70) isolated from an epizootic of enteric septicemia in channel catfish was obtained from the Southeastern Cooperative Fish Disease Diagnostic Laboratory at Auburn University for the challenge. The organism was recycled through live channel catfish to enhance the virulence, re-isolated from the brain of moribund fish, and inoculated into a liter of brain-heart infusion broth to prepare the cell suspension for the challenge. Twelve fish from each of three aquaria per treatment were challenged by 24 hour immersion in suspensions of $10^{5.9}$ cells/ml, which was LC_{50} for this cell, following the procedure previously described by Paripatananont and Lovell (1995b). Mortality was recorded in each aquarium for 13 days after which time there were no more mortalities. Liver and kidney from moribund fish and fish that had recently died were necropsied to confirm that *E. ictaluri* was the cause of death by the procedure described by Shotts and Plumb (1994). Antibody titers from two to 12 surviving fish from each aquarium were measured on day 14 post-challenge. Serum was separated and measured for *E. ictaluri* specific antibody titer with the serum agglutination technique described by Roberson (1990). Serum from sheep (*Ovis aries*) was tested as the negative control.

STATISTICAL ANALYSIS

Weight gain, enzyme activity, and tissue selenium data were analyzed by the General Linear Model procedure of SAS (Statistical Analysis Service, 1987) to establish the regression of fish response on increasing levels of dietary selenium. Where the regression was non-linear, the data were analyzed by the broken-line regression procedure to determine the breakpoint in the response curve which represents the optimum dietary concentration of selenium (Robbins *et al.*, 1979). Relative bioavailability of selenium from selenomethionine or Sel-Plex 50 compared to that from selenite was determined by deriving the ratio of the slopes of the linear sections of the regression lines (Littell *et al.*, 1995). The challenge data were analyzed by the SAS procedure to determine if there were differences among treatments, and specific treatment comparisons were made to determine where differences occurred. Treatment differences were considered to be significant when $P < 0.05$.

Results

Weight gain and apparent feed efficiency responded quadratically ($P<0.01$) to the graded dietary selenium levels from all three sources (Tables 2 and 3). There were significant differences in the slopes and breakpoints in the weight gain regression curves for the three selenium sources. The regression of weight gain on dietary selenium supplementation was highest for selenomethionine, followed by Sel-Plex 50 and selenite. The breakpoint in the regression curve, considered to be the minimum dietary concentration for optimal growth, was 0.28 mg selenium/kg diet for selenite, 0.09 mg/kg diet for selenomethionine and 0.11 mg/kg diet for Sel-Plex 50 (Figure 1). The ratio of the slope of the linear part of the regression curve for selenomethionine to the linear slope of the selenite curve was 3.36 and the slope ratio for Sel-Plex 50 to selenite was 2.69. Thus, the relative bioavailabilities of selenomethionine and Sel-Plex 50 for fish growth were 336 and 269%, respectively, compared to that of selenite (Table 2).

Plasma GSH-Px activity of fish increased in a linear manner ($P<0.01$) in response to dietary selenium supplementation from the three selenium sources (Table 2); there were no breakpoints (Figure 2). The slopes of the regression lines for selenomethionine and Sel-Plex 50 were greater than that for selenite. The slope ratios of selenomethionine and Sel-Plex 50 to selenite were both 1.16. Thus, the relative bioavailability of selenomethionine and Sel-Plex 50 based on plasma GSH-Px was 116% of that of selenite (Table 2). Liver GSH-Px activity increased quadratically ($P<0.01$) with graded dietary selenium from all three sources. There was a significant difference in the slope and breakpoint in the liver GSH-Px regression curves between selenite and selenomethionine or Sel-Plex 50. The breakpoint in the regression curve occurred at 0.17 mg selenium/kg diet for selenite, 0.12 mg selenium/kg diet for selenomethionine and 0.12 mg selenium/kg diet for Sel-Plex 50 (Figure 3). The ratios of the slopes of selenomethionine and Sel-Plex 50 to selenite were 1.47 and 1.49, respectively. Thus, the relative

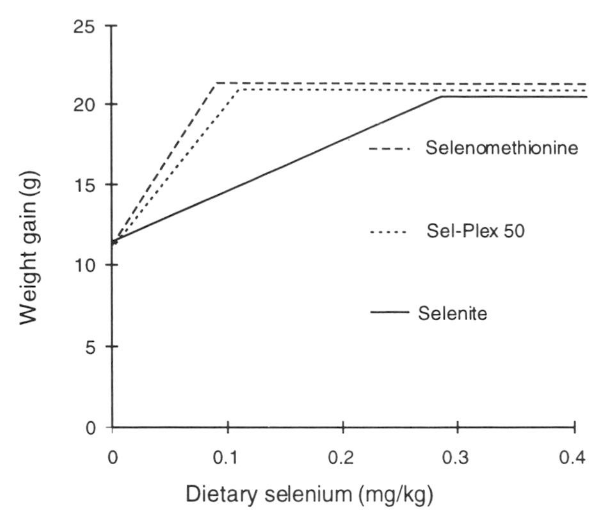

Figure 1. Regression of weight gain on supplemental dietary selenium and breakpoints in the lines for channel catfish fed a casein-based diet supplemented with sodium selenite, selenomethionine or selenoyeast (Sel-Plex 50).

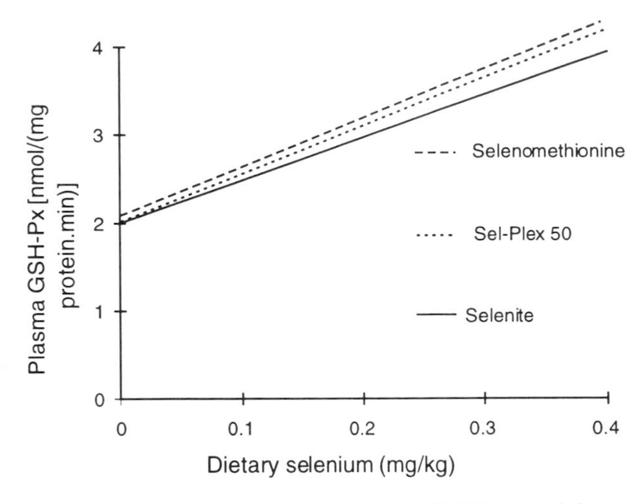

Figure 2. Regression of plasma glutathione peroxidase (GSH-Px) activity on supplemental dietary selenium and breakpoints in the lines for channel catfish fed a casein-based diet supplemented with sodium selenite, selenomethionine or selenoyeast (Sel-Plex 50).

bioavailabilities of selenomethionine and Sel-Plex 50 based on liver GSH-Px activity were 147 and 149%, respectively, of that of selenite (Table 2).

The regression of muscle selenium content on dietary selenium was quadratic for selenomethionine and Sel-Plex 50 ($P<0.01$), but linear for selenite ($P<0.01$) (Figure 4). The breakpoints were 0.10 mg selenium/kg diet for selenomethionine and 0.10 mg selenium/kg diet for Sel-Plex 50 (Table 2). Muscle selenium content of fish fed selenite was significantly lower than that of fish fed

Table 2. Regression of weight gain, plasma GSH-Px, liver GSH-Px, liver Se, and muscle Se on dietary concentratins of Se from various sources; breakpoint dietary selenium requirement; and relative bioavailability values to channel catfish.

Criterion	Selenium source	Regression			Selenium requirement[a] (mg/kg)	Relative bioavailability[b] (%)
		Equation	Significance	SEM		
Weight gain	Selenite	y=10.8 + 34.9x	Q(*P*<.01)	0.16	0.28±0.031	100
	Selenomethionine	y=10.6 + 117.0x	Q(*P*<.01)	0.11	0.09±0.007	336
	Sel-Plex 50	y=10.2 + 93.8x	Q(*P*<.01)	0.11	0.11±0.012	269
Plasma GSH-Px	Selenite	y=2.0 + 4.4x	L(*P*<.01)	0.17	–	100
	Selenomethionine	y=2.3 + 5.1x	L(*P*<.01)	0.20	–	116
	Sel-Plex 50	y=2.0 + 5.0x	L(*P*<.01)	0.15	–	116
Liver GSH-Px	Selenite	y=22.5 + 600.7x	Q(*P*<.01)	2.92	0.17±0.024	100
	Selenomethionine	y=22.5 + 883.7x	Q(*P*<.01)	2.21	0.12±0.021	147
	Sel-Plex 50	y=21.4 + 891.8x	Q(*P*<.01)	1.90	0.12±0.020	149
Liver Se	Selenite	y=0.06 + 10.3x	Q(*P*<.01)	0.078	0.18±0.045	100
	Selenomethionine	y=0.77 + 20.3x	Q(*P*<.01)	0.070	0.09±0.015	197
	Sel-Plex 50	y=0.74 + 19.0x	Q(*P*<.01)	0.075	0.11±0.025	184
Muscle Se	Selenite	y=0.16 + 0.7x	L(*P*<.01)	0.010	–	100
	Selenomethionine	y=0.16 + 3.3x	Q(*P*<.01)	0.012	0.10±0.022	478
	Sel-Plex 50	y=0.17 + 3.1x	Q(*P*<.01)	0.012	0.10±0.024	453

[a] Breakpoint in the regression line.
[b] The ratio of the slope of Selenomethionine or Sel-Plex 50 regression line to the slope of Selenite regression line.

Table 3. Weight gain by channel catfish fed a casein basal diet supplemented with 0 (control), 0.02, 0.06, 0.20 or 0.40 mg of selenium/kg from sodium selenite, selenomethionine or Sel-Plex 50.[a]

Dietary Se mg/kg	Weight gain (g) with		
	Selenite	Selenomethionine	Sel-Plex 50
0	9.43	9.43	9.43
0.02	12.39[a]	14.47[a]	13.11[a]
0.06	13.76[a]	17.69[b]	15.68[c]
0.20	17.39[a]	20.81[b]	20.65[b]
0.40	20.60[a]	21.12[a]	20.30[a]

[a] Means in the rows for each response with the same letter are not different at $P<0.05$. Means represent three replicate aquaria in each treatment.

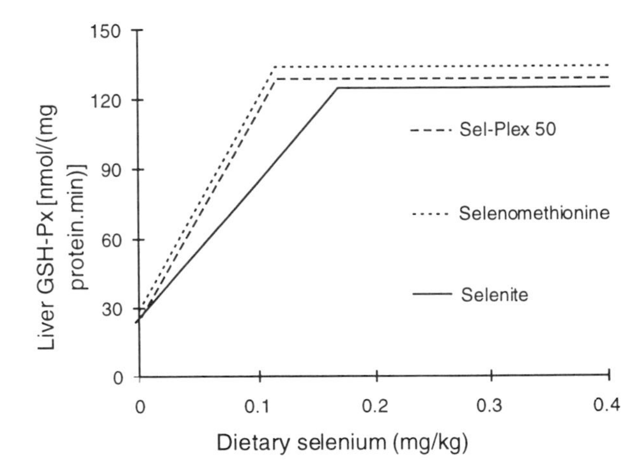

Figure 3. Regression of liver glutathione peroxidase (GSH-Px) activity on supplemental dietary selenium and breakpoints in the lines for channel catfish fed a casein-based diet supplemented with sodium selenite, selenomethionine or selenoyeast (Sel-Plex 50).

selenomethionine and Sel-Plex 50 at the maximum dietary level (0.40 mg selenium/kg diet). The slope ratio of the regression curves for selenomethionine and Sel-Plex 50 to selenite were 4.78 and 4.53 (Table 2). The relative bioavailabilities of selenomethionine and Sel-Plex 50 based on muscle selenium content were 478 and 453%, respectively, of that of selenite. Liver selenium content responded quadratically ($P<0.01$) to dietary selenium from the three sources (Figure 5). There was a significant difference in the slope and breakpoint in the liver selenium regression curves between selenite and selenomethionine or Sel-Plex 50. The breakpoints were 0.18 mg selenium/kg diet for selenite, 0.09 mg selenium/kg diet for selenomethionine and 0.11 mg selenium/kg diet for Sel-Plex 50 (Table 2). The slope ratio of the regression curve for selenomethionine and Sel-Plex 50 to selenite were 1.97 and 1.84, respectively. The relative bioavailabilities of selenomethionine and Sel-Plex

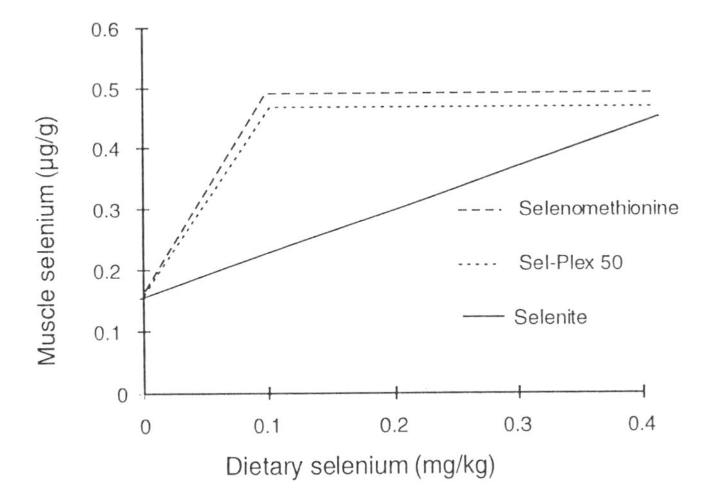

Figure 4. Regression of muscle selenium content on supplemental content on supplemental dietary selenium and breakpoints in the lines for channel catfish fed a cxasein-based diet supplemented with sodium selenite, selenomethionine or selenoyeast (Sel-Plex 50).

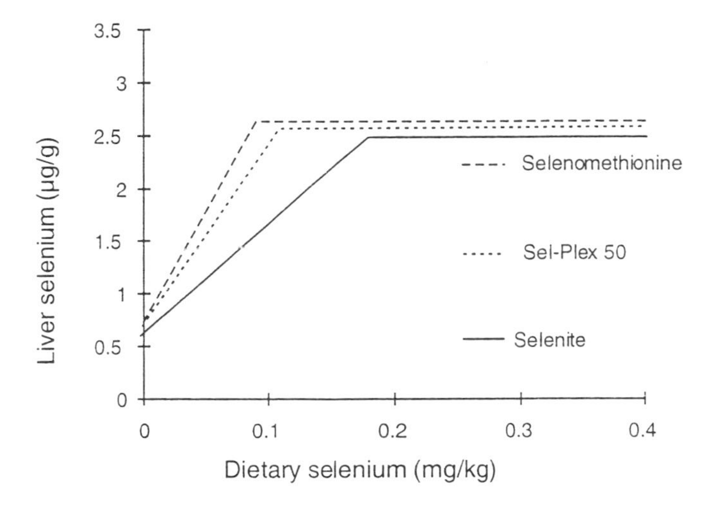

Figure 5. Regression of liver selenium content on supplemental dietary selenium and breakpoints in the lines for channel catfish fed a casein-based diet supplemented with sodium selenite, selenomethionine or selenoyeast (Sel-Plex 50).

50 for liver selenium accumulation were 197 and 184%, respectively, of that of selenite (Table 2).

Mortality among the fish challenged with *E. ictaluri* began on day 7 and ceased on day 12 post-challenge. All mortalities were confirmed by necropsy or visual examination for overt signs of *E. ictaluri* infection. Fish fed the basal diet were very sensitive to *E. ictaluri* challenge and had over 90% mortality (Table 4). Mortality did not decrease significantly with supplementation

Table 4. Percentage mortality among channel catfish fed a casein basal diet supplemented with 0 (control), 0.02, 0.06, 0.20 or 0.40 mg of selenium/kg from sodium selenite, selenomethionine or Sel-Plex 50 and subsequently challenged with *E. ictaluri*.[a]

Dietary Se (mg/kg)	Mortality (%) with		
	Selenite	Selenomethionine	Sel-Plex 50
0	91.67	91.67	91.67
0.02	86.11[a]	75.00[a]	80.55[a]
0.06	61.11[a]	52.78[b]	44.44[b]
0.20	61.11[a]	36.11[c]	47.22[b]
0.40	36.11[a]	19.45[c]	27.78[b]

[a] Means in the rows for each response with the same letter are not different at $P<0.05$. Means represent 12 fish in each of the three aquaria in each treatment.

Table 5. Antibody titer of channel catfish fed a casein basal diet supplemented with 0 (control), 0.02, 0.06, 0.20 or 0.40 mg of selenium/kg from sodium selenite, selenomethionine or Sel-Plex 50. [a]

Dietary Se (mg/kg)	Antibody titer with		
	Selenite	Selenomethionine	Sel-Plex 50
0	64	64	64
0.02	80[a]	70[a]	75[a]
0.06	149[a]	213[b]	320[c]
0.20	192[a]	352[b]	554[c]
0.40	365[a]	512[b]	717[c]

[a] Means in the rows for each response with the same letter are not different at $P<0.05$. Means represent 2 to 12 fish in each of the three aquaria in each treatment.

of 0.02 mg of selenium/kg of diet, but did with higher levels of supplementation. Minimum mortality occurred at a dietary selenium concentration of 0.20 mg/kg for fish fed selenomethionine and at 0.40 mg/kg for fish fed selenite or Sel-Plex 50. There was no difference in mortality among selenium sources at the highest selenium concentration.

Antibody titers for fish fed the basal diet and the diets containing 0.02 mg of selenium/kg were extremely low and not significantly different (Table 5). Antibody production increased with dietary selenium supplementation of 0.06 mg/kg of diet for fish fed selenomethionine and Sel-Plex 50. As dietary selenium concentration increased, antibody titers generally increased for all selenium sources. At the highest dietary selenium concentration, antibody titer was highest for Sel-Plex 50, intermediate for selenomethionine and lowest for selenite.

Discussion

The dietary selenium requirement of channel catfish for weight gain was 0.28 mg/kg from selenite, similar to the selenium requirement value of 0.25 mg/kg diet determined with selenite by Gatlin and Wilson (1984). When selenomethionine or Sel-Plex 50 were the sources, dietary selenium requirements for growth in channel catfish were markedly reduced to 0.09 and 0.11

mg/kg diet. Selenium from organic sources was 269 to 336% more available than inorganic selenium for growth. However, there was less difference among selenium sources for liver GSH-Px activity. These data indicate that channel catfish fingerlings require two to three times more selenium from selenite than from selenomethionine or Sel-Plex 50 for growth but approximately one and one-half times more selenium from the inorganic source than from the organic sources for liver GSH-Px activity.

Bioavailabilities of the two organic sources of selenium, selenomethionine and Sel-Plex 50, were fairly similar. Relative bioavailabilities of the organic sources ranged from an average of 116% for plasma GSH-Px to an average of 465% for muscle selenium content. Reasons for this variation in bioavailability among measurement criteria are not readily apparent. This indicates that the positive effect of chelation on bioavailability of selenium goes beyond absorption.

Lorentzen *et al.* (1994) fed Atlantic salmon selenium from selenite and selenomethionine and found no difference in weight gain or hepatic GSH-Px activity at the supplemental level of 1 or 2 mg/kg diet. Also, Bell and Cowey (1989) observed that plasma and hepatic GSH-Px activity were not different in Atlantic salmon fed 1 mg selenium/kg diet from selenite, selenomethionine, selenocysteine, or fish meal. In the present study, channel catfish fed selenium from the organic sources had better growth and higher GSH-Px activity than those fed selenium from the inorganic source at suboptimal levels but not at levels above the minimum dietary requirement. The dietary selenium levels fed to the Atlantic salmon were markedly above the National Research Council (1993) requirement level for salmonids and this apparently explains why the previously mentioned studies with Atlantic salmon did not show a difference between organic and inorganic selenium sources.

One reason for the higher bioavailability of organic selenium to channel catfish is enhanced absorption. Recent research at this laboratory with channel catfish (Paripatananont and Lovell, 1996) showed that net absorption of selenomethionine was higher than that of selenite in a purified egg white-based diet (90.8 vs 62.8%) and in a soybean meal-based diet (88.9 vs 69.7%). This does not explain completely why the organic selenium was more available: net absorption of selenomethionine was 140% of that of selenite while bioavailability was 336% of that of selenite. Chelated selenium might be absorbed and transported intact to target tissues and be more available in metabolic processes than inorganic selenium (Ashmead, 1992). Paripatananont and Lovell (1995b) found that bioavailability of chelated zinc for growth of channel catfish was 352% of that of zinc sulfate and subsequently found that digestibility of the organic zinc was only 140% of that of the inorganic zinc (Paripatananont and Lovell, 1996). Apparently, other factors besides net absorption are responsible for the higher bioavailability of organic compared to inorganic trace elements.

This study indicates that selenium from the organic sources has higher bioavailability for channel catfish than selenium from the inorganic source. Sel-Plex 50 seems to have nearly the same bioavailability as selenomethionine. The dietary selenium allowance of channel catfish can be reduced by at least one-half by using one of the organic selenium sources to replace conventional selenite.

The challenge data show that dietary selenium is necessary for resistance of channel catfish to *E. ictaluri* infection and that concentration and source of selenium influence immune responses. Mortality was 90% in fish fed the diet devoid of supplemental selenium and decreased to 25 to 35% when selenium was increased to meet the growth requirement. At this supplemental level, fish fed selenomethionine showed significantly lower mortality than fish fed selenite, indicating that selenomethionine is more efficacious than the inorganic source of selenium in protecting channel catfish from *E. ictaluri* infection. Sel-Plex 50 appeared to be intermediate between selenomethionine and selenite in protecting the fish against *E. ictaluri* infection.

Antibody production was responsive to dietary concentration and source of selenium. The organic sources were generally more efficacious than inorganic selenium for this immune response. Suppressed antibody production appears to be a common result of selenium deficiency among animals. According to Combs and Combs (1986), selenium plays a role in antibody production through the proliferation and protection effects of GSH-Px on B-cells. Wang (1996) found that selenium deficiency, in the presence of dietary vitamin E, seriously reduced GSH-Px activity in channel catfish, and that selenomethionine was a more potent source of selenium for GSH-Px activity than selenite. Another possible role of GSH-Px in immune response is protecting cell membranes against peroxide damage during phagocytosis. Production of hydrogen peroxide in phagocytes is necessary in the process of killing invaded microbes. However, according to Spallholz *et al.* (1990), the cell membrane of macrophages can be damaged by the overproduction or leak of peroxides into cytoplasm when there is low GSH-Px activity.

This study shows that dietary selenium is essential for optimal immune response and resistance in channel catfish infected with *E. ictaluri*, and that an organic selenium source was more efficacious than an inorganic selenium source. Both antibody production and macrophage chemotaxis were responsive to selenium concentration and source. Organic selenium was especially effective in enhancing macrophage activity, suggesting that selenium chelates may be more efficacious than inorganic selenium in this capacity.

Summary

Year-0 channel catfish fingerlings were fed casein basal diets supplemented with 0, 0.02, 0.06, 0.20, or 0.40 mg of selenium/kg from sodium selenite, selenomethionine, or selenoyeast (Sel-Plex 50) to compare the bioavailability of selenium from these various sources. Weight gain, feed efficiency, plasma, and hepatic glutathione peroxidase (GSH-Px) activity were measured at the end of 9-week feeding periods. The fish were challenged with *E. ictaluri* by 24 hour immersion in 10^6 cells/ml at 25±1°C to determine resistance to infection and antibody production.

Broken-line regression analysis showed that the minimum dietary selenium requirements for weight gain for selenite, selenomethionine, and Sel-Plex 50 were 0.28, 0.09, and 0.11 mg/kg, respectively. Broken-line analysis showed the selenium requirements for hepatic GSH-Px activity for selenite, selenomethionine, and Sel-Plex 50 were 0.17, 0.12, and 0.12 mg/kg, respectively. The relative bioavailability of selenium from the organic sources compared to that of inorganic selenium, determined from the ratio of the slope of the selenomethionine and Sel-Plex 50 regression curves to the slope of the selenite regression curve, was 336% for selenomethionine and 269% for Sel-Plex 50 for weight gain, and 147 and 149% for hepatic GSH-Px activity. Organic selenium had higher level of retention in liver and muscle than inorganic selenium.

Mortalities and antibody production subsequent to *E. ictaluri* challenge were responsive to dietary selenium concentration and source. Dietary selenium concentrations effecting minimum mortality for selenite, selenomethionine, and Sel-Plex 50, were 0.40, 0.20, and 0.40 mg/kg, respectively. Antibody production was highest for fish fed the selenium yeast, intermediate for fish fed selenomethionine, and lowest for fish fed selenite.

These results indicate that selenium from organic sources has about two times the potency of that from selenite for growth and immune response in channel catfish, therefore, dietary selenium allowance in catfish diets can be reduced appreciably by using organic sources to replace inorganic selenium.

References

Ammerman, C.B., D.H. Baker and A.J. Lewis. 1995. Bioavailability of Nutrients for Animals. Academic Press, San Diego, CA, pp. 441.

AOAC, Association of the Official Analytical Chemists. 1990. Method of Analysis, AOAC, Washington, D.C., p. 1298.

Ashmead, H.D. and H. Zunino. 1992. Factors which effect the intestinal absorption of minerals. In: H.D. Ashmead (Editor). The roles of amino acid chelates in animal nutrition. Noyes Publication, Park Ridge, NJ, pp. 21–46.

Bell, J.G. and C.B. Cowey. 1989. Digestibility and bioavailability of dietary selenium from fish meal, selenite, selenomethionine, and selenocysteine in Atlantic salmon (*Salmo salar*). Aquaculture 81:61–68.

Bell, J.G., C.B. Cowey, J.W. Adron and B.J.S. Pirie. 1987. Some effects of selenium deficiency on enzyme activities and indices of tissue peroxidation in Atlantic salmon parr (*Salmo salar*). Aquaculture 65:43–54.

Cawley, G.D. and R. Bradley. 1978. Sudden death in calves associated with acute myocardial degeneration and selenium deficiency. Vet. Rec. 103:239–240.

Combs, G.F., Jr. and S.B. Combs. 1986. The role of selenium in nutrition. Academic Press, New York, NY, p. 532.

Franke, K.W., 1934. A new toxicant occurring naturally in certain samples of plant feedstuffs. I. Results obtained in preliminary feeding trials. J. Nutr. 8:596–608.

Gatlin, M.D. and R.P. Wilson. 1984. Dietary selenium requirement of fingerling channel catfish. J. Nutr. 114:627–633.

Hilton, J.W., P.V. Hodson and S.J. Slinger. 1980. The requirement and toxicity of selenium in rainbow trout (*Salmo gairdneri*). J. Nutr. 110:2527–2535.

Kubota, J., W.A. Allaway, D.L. Carter, E.E. Cary and V.A. Lazar. 1967. Selenium in crops in the United States in relation to selenium-responsive disease of animals. Agr. Food Chem. 15:448–453.

Lawrence, R.A. and R.F. Burk, 1976. Glutathione peroxidase activity in selenium-deficient rat liver. Biochem. and Biophys. Res. Commun. 71:952–958.

Littell, R.C., A.J. Lewis and P.R. Henry. 1995. Statistic evaluation of bioavailability assay. In: C. B. Ammerman, D.H. Baker and A.J. Lewis (Editors), Bioavailability of Nutrients for Animals. Academic Press, San Diego, CA, pp. 5–35.

Lorentzen, M., A. Maage and K. Julshamn. 1994. Effects of dietary selenite or selenomethionine on tissue selenium levels of Atlantic salmon (*Salmo salar*). Aquaculture 121:359–367.

Lovell, R.T., T. Miyazaki and S. Relangnator. 1984. Requirement for alpha-tocopherol by channel catfish fed diets low polysaturated triglycerides. J. Nutr. 114:894–901.

Lowry, O.H., N.J. Rosebrough. A.L. Farr and R.J. Randall. 1951. Protein measurement with the Folin phenol reagent. J. Biol. Chem. 193:265–275.

Muth, O.H. and W.A. Allaway. 1963. The relationship of white muscle disease to the distribution of naturally occurring selenium. J. Am. Vet. Med. Assoc. 142:1379–1384.

National Research Council. 1993. Nutrient requirements of fish. National Academy Press, Washington, D.C. p. 114.

Paripatananont, T. and R.T. Lovell. 1995a. Chelated zinc reduces the dietary zinc requirement of channel catfish, *Ictalurus punctatus*. Aquaculture 133:73–82.

Paripatananont, T. and R.T. Lovell. 1995b. Responses of channel catfish fed organic and inorganic sources of zinc to *Edwardsiella ictaluri* challenge. Journal of Aquatic Animal Health 7:147–154.

Paripatananont, T. and R.T. Lovell. 1996. Comparative digestibility of chelated and inorganic trace minerals in channel catfish (*Ictalurus punctatus*) diets. Proceedings of the VII International Symposium on Fish Nutrition and Feeding, p 24.

Patterson, E.L., R. Milstrey and E.L.R. Stockstad. 1957. Effect of selenium in preventing exudative diathesis in chicks. Proc. Soc. Exp. Biol. Med. 95:617–620.

Robbins, K.R., H.W. Norton and D.H. Baker. 1979. Estimation of nutrient requirements from growth data. Nutr. 109:1710–1714.

Roberson, B.S. 1990. Bacterial agglutination. In: J.S. Stolen, T.C. Fletcher, D.P. Anderson, B.S. Roberson and W.B. van Muiswinkel, editors. Techniques in fish immunology. SOS Publications, Fair Haven, NJ. pp. 81–86.

SAS, 1987. SAS User's guide. SAS institute, Cary, NC.

Schwartz, K. and C.M. Foltz. 1957. Selenium as an integral part of Factor 3 against dietary necrotic liver degeneration. J. Amer. Chem. Soc. 79:3292–3293.

Shotts, E.B., Jr. and J.A. Plumb. 1994. Bacterial diseases of fish. VIII. Enteric septicemia. In J.C. Thoesen, editor. Suggested procedures for the detection and identification of certain finfish and shellfish pathogens, 4th edition. American Fisheries Society, Fish Health Section, Maryland.

Spallholz, J.E., L.M. Boylan and H.S. Larsen. 1990. Advances in understanding selenium's role in the immune system. Annals New York Academy of Sciences 587:123–139.

Stowe, H.D., 1967. Serum selenium and related parameters of naturally and experimentally fed horses. J. Nutr. 93:60–64.

Wang, C. 1996. Comparison of organic and inorganic sources of selenium for growth, glutathione peroxidase activity and immune responses in channel catfish. Doctoral dissertation, Auburn University, 110p.

Wise, D.J., J.R. Tomasso, D.M. Gatlin, S.C. Bai and V.S. Blazer. 1993. Effects of dietary selenium and vitamin E on red blood cell peroxidation, glutathione peroxidase activity, and macrophage superoxide anion production in channel catfish. Journal of Aquatic Animal Health 5:177–182.

NUTRITIONAL MANIPULATION OF MEAT QUALITY IN PIGS AND POULTRY

WILLIAM H. CLOSE

Close Consultancy, Wokingham, Berkshire, UK

Introduction: defining meat quality

In recent years, meat quality has assumed a greater consumer significance and public attention. There is growing awareness of the link between diet and health and this is reflected in the demand for more information and for products which are healthy and of consistently high quality. As a consequence, this has lead to the demand for meat with a high lean content. On the other hand as animals, and especially pigs and poultry, have become leaner, more complaints have been received that the eating quality of the meat has declined, with complaints of dryness, toughness and a lack of taste and flavour. Thus, many major retailers are now establishing their own specifications for the production of meat. Other concerns are about food safety and hygiene, especially the presence of micro-organisms, bacterial contamination and residues, as well as the welfare and husbandry conditions under which animals are kept. Thus, both the diets fed to the animal and the systems of animal production are being increasingly questioned. It is fortuitous that with the elucidation of the major factors influencing meat quality and hygiene new opportunities are being created for the development of a consistent, healthy, safe and attractive product which offers value for money. So, how do we define meat quality and how can we assess it?

Hofmann (1987) and Russo (1988) have broadly classified meat quality characteristics into four main categories:

- Organoleptic properties
- Technological quality
- Nutritional value
- Hygienic characteristics or food safety aspects

Organoleptic properties are the traits that influence the consumer to regularly purchase and eat meat. Technological qualities refer to the suitability of meat for further processing and are primarily determined by treatment after slaughter. Nutritional value concerns the chemical composition of the meat and its suitability for human consumption. Hygienic quality or safety implies freedom from harmful microorganisms and any residues. These can be controlled through legislation, proper feeding designs and strategies, quality management schemes on the farm and procedures in the slaughterhouse and processing plant.

Although many factors can influence meat production, this paper is only concerned with the nutritional and eating qualities of meat, particularly pork and to a limited extent poultry. It attempts to examine the possibilities for nutritional manipulation of these characteristics in the animal and to establish their likely value to both the consumer and the producer.

Components of meat quality

The main factors contributing to the eating quality of meat from pigs and poultry are tenderness, juiciness and colour. These are dependent upon several metabolic and biological phenomena within the animal or carcass and include the following:

- The content of intramuscular or marbling fat.
- Taint, especially the content of skatole, indole and testosterone.
- The type and fatty acid content of the animal's diet and hence its carcass.
- Maturation or conditioning effects, especially the breakdown of the myofibrillar structure of muscle during the post-mortem period and the activity of the calpain and calpastatin enzyme systems.
- Drip loss and the maintenance of the integrity of the cell membranes post-slaughter. This has special significance for improving shelf life and yield while reducing bacterial contamination.
- Stress, especially that during transportation and lairage.
- The potential to flavour meat.
- The effects of feeding *per se*, including fermentation in the large intestine and the role of fibrous structures in the animal's diet, especially non-starch polysaccharides (NSP).

Each of these will be considered separately to examine what opportunities exist to enhance the eating quality of meat through nutritional and biotechnological means. Although the paper is primarily concerned with pig meat, attention is also drawn to aspects relevant to poultry meat.

INTRAMUSCULAR FAT CONTENT

It is generally accepted, especially in pigs, that the effects of the fatness of meat on eating quality depends on the quantity of marbling fat. This is the lipid found in the connective tissue surrounding the muscle fibre bundles. Intramuscular fat typically contributes 0.5 to 2.5% of the loin muscle, with values at the lower end of this range being found in very fast growing, lean, modern genotypes. It has been suggested that at least 2.0% of marbling fat is required for optimal eating quality (Bejerholm and Barton-Gade, 1986; De Val *et al.*, 1988), but values considerably lower than this have been reported in several European studies. For example, Kempster *et al.* (1986) and Wood *et al.* (1986) showed that the lower the backfat thickness, the lower the percentage of marbling fat and the less the overall acceptability of pork as evaluated by a trained taste panel (Table 1). For this reason there has been considerable interest in the use of the Duroc breed in cross-breeding programmes, with its known high content of marbling fat in the final carcass. A recent large study conducted by the Meat and Livestock Commission (1992) showed that the higher the proportion of the Duroc gene in the final cross, the higher the eating quality of the meat in terms of tenderness. This increase was associated with a considerable increase in marbling fat, as well as the concentration of red oxidation muscle fibre type (Figure 1).

Table 1. The relationship between backfat thickness and eating quality of loin chops of pigs*.

	Backfat thickness (mm)	
	8	16
Marbling fat, %	0.55	0.96
Taste panel assessment[†]		
Tenderness	1.0	1.1
Juiciness	1.1	1.3
Flavour	1.5	1.7
Overall acceptability	0.7	1.0

*Kempster *et al.* (1986), Wood *et al.* (1986).
[†] Scale: –7 to +7, except juiciness, scale 0–4.

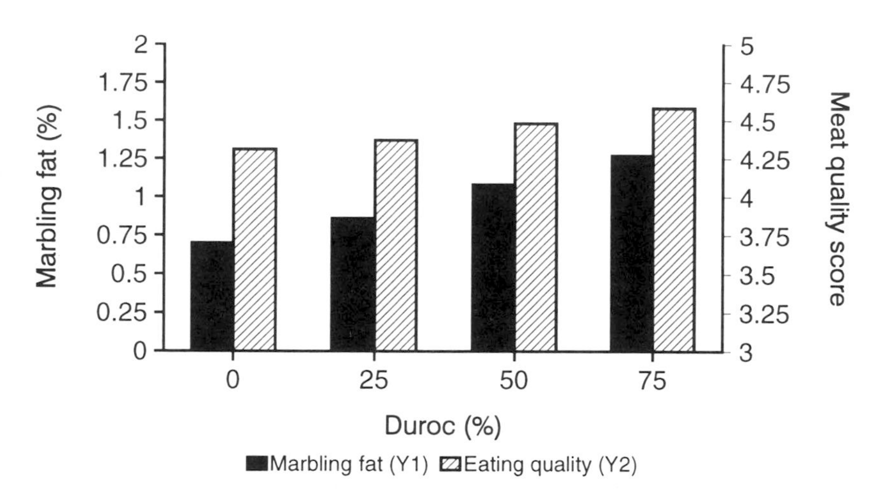

Figure 1. The relationship between marbling fat content and meat eating quality of pork containing different proportions of Duroc gene. (Meat quality is the mean assessment of tenderness, juiciness and flavour).

TAINT AND THE EFFECTS OF SKATOLE

The concentration of skatole in fatty tissue is responsible for the abnormal flavour characteristics of pork. High values lead to complaints of 'off' flavours and taste. Skatole (3-methyl indole) is a volatile compound produced in the hindgut of animals by microbial degradation of the amino acid tryptophan and which originates from dietary or endogenous proteins. Although the majority of skatole is degraded in the liver and excreted in the urine, the undegraded portion is deposited in the fat and muscle of the body. High concentrations in these tissues give rise to unpleasant smell and taste of meat, especially in entire male animals.

Several recent studies have shown the effects of different feed components and dietary elements on skatole concentration in the intestinal contents, as well as in the faeces and backfat of pigs. For example, feeding very high levels of peas increased skatole levels from 0.09 to 0.12 ppm (Madsen *et al.*, 1990) and resulted in a poorer pork flavour score and reduced overall acceptability by trained taste panellists. Yeast slurry from brewers has also been shown to elevate the content of skatole in the hindgut (Pedersen *et al.*, 1986), whereas casein reduced it (Jensen *et al.*, 1995).

The concentration of skatole in the tissue of animals can be reduced by limiting the fermentation of protein in the hindgut and by decreasing the actual quantity of undigested protein reaching the hindgut. The fermentation of these proteins and the production of microbial metabolites such as short chain fatty acids and skatole are therefore decreased. Since glycocompounds from the *Yucca schidigera* plant have been shown to bind nitrogenous compounds in the large intestine (Headon, 1992), it is not surprising that its addition to diets has been shown to reduce skatole production. In this respect, Ender *et al.*, (1993) fed pigs to various slaughter weights with and without *Yucca schidigera* extract (De-Odorase) in their diets. The addition of De-Odorase reduced the level of skatole in the backfat of entire male pigs slaughtered between 95 and 120 kg body weight and considerably improved sensory score (Table 2) Interestingly, there was also a reduction in drip loss of the meat from 5.64 to 4.60% with the inclusion of De-Odorase in the diet.

Table 2. The influences of De-Odorase on skatole content and the sensory evaluation of backfat of entire male pigs*.

Final body weight (kg)	Treatment	Skatole content (μg/g)	Sensory score†
95	Control	0.24	0.8
	De-Odorase	0.16	0.5
105	Control	0.22	0.8
	De-Odorase	0.16	0.5
120	Control	0.27	1.0
	De-Odorase	0.21	1.2

*Ender *et al.* (1993).
† Scale: 0 (no odour) to 3 (strong odour).

An alternative strategy to reduce the production of metabolites from protein fermentation is to provide easily digestible fibre, or NSP, in the diet which are preferentially metabolized by the intestinal microbes. The fermentation of the NSP sources in the large intestine produces volatile fatty acids which reduce the pH within the large intestine to a level that is well below the optimum for microbial protein function and activity. For example, the inclusion of sugar beet pulp which contains high amounts of pectin has resulted in a significant reduction of skatole concentration in the backfat of pigs and increased the overall acceptability score of the meat (Longland *et al.*, 1991; Nute *et al.*, 1994). The inclusion of oligosaccharides in the diet may also be beneficial (Terada *et al.*, 1992).

FATTY ACID COMPOSITION

In non-ruminant animals it is well established that simply changing the type and quantity of oils and fat in the diet can change the fatty acid content of fat in the carcass since many are absorbed intact from the small intestine and incorporated directly into fatty tissue (Wood *et al.*, 1994). This has relevance not only for human nutrition, but also for the further processing of the animal. On the one hand, the consumer demands fats which are predominantly unsaturated and 'healthy', whereas from a processing perspective, these have a soft and oily texture which gives rise to unacceptable fat quality, not appreciated by the meat industry. There is also the risk of the development of lipid oxidation and 'off' flavours with the increase in the unsaturated fatty acid content of carcass fat.

Cameron and Enser (1991) investigated the connection between intramuscular fatty acid content and eating quality of pork. Their results showed that in general, higher concentrations of polyunsaturated fatty acids were associated with lower values for tenderness, juiciness, flavour and overall acceptability, whereas higher concentrations of saturated and monounsaturated fatty acids resulted in higher overall score (Table 3). Similarly, there is an increased risk of rancidity and 'off' flavour when fish oils are used in high quantities in pig diets. The inclusion of fats and oils in the diet must therefore balance their main role of providing readily available sources of energy to the animal, while maintaining acceptable meat quality from both the consumer's and retailer's perspective.

Table 3. Correlations between intramuscular fatty acid composition and eating quality of pig meat*.

	Fatty acid composition									
	14.0	16.0	16.1	18.0	18.1	18.2	18.3	20.4	22.5	22.1
Tenderness	0.14	0.13	0.17	−0.04	0.19	−0.21	0.05	−0.20	−0.23	−0.17
Juiciness	0.15	0.05	0.08	0.04	0.09	−0.06	0.23	−0.20	−0.23	−0.16
Flavour	0.11	0.08	0.21	0.06	0.19	−0.19	0.10	−0.19	−0.23	−0.21
Overall acceptability	0.19	0.12	0.17	0.01	0.19	−0.20	0.15	−0.26	−0.28	−0.21

*Cameron and Enser (1991).

MATURATION AND CONDITIONING EFFECTS

It is well known that the tenderness of meat improves with conditioning and storage after slaughter. For example, Wood *et al.* (1995) demonstrated that increasing the conditioning period from 1 to 10 days at 1°C significantly improved the overall liking of pork loin joints from 3.4 to 3.9 units (scale 1–8). This improvement was considerably greater than that associated with either breed (Large White *vs* Duroc) or feed (*ad libitum vs* restricted) effects.

During conditioning, proteolytic enzymes act to break down the myofibrillar structure of muscle and thereby tenderize meat. Evidence now suggests that the myofibrillar degradation is dependent to a large extent on the activity or suppression of the calpain and calpastatin enzyme systems, respectively. Indeed, it has been suggested that 90% or more of the tenderization that occurs post-mortem is associated with the calpains (Koohmaraie, 1994). The major components of the calpain enzyme system are μ-calpain and m-calpain and these are calcium dependent (Taylor *et al.*, 1995). On the other hand, calpastatin inhibits calpain and several studies have shown that tenderness increases to a greater extent during post-mortem storage in those muscles that have higher calpain or lower calpastatin concentration than others. Thus, low calpastatin activity seems closely associated with increased post-mortem tenderization. Recently, Sensky *et al.* (1996), demonstrated that the concentrations of both μ- and m-calpain in samples of porcine *longissimus dorsi*, identified as either tough, normal or tender, remained fairly constant across the different samples whereas the activity of calpastatin was found to decrease as the tenderness of the meat increased.

Since the calpain enzyme system is calcium dependent, the injection of calcium salt solutions, such as calcium chloride, into the carcass of animals can significantly improve the eating quality of meat and reduce the toughening effects of cooking (Wulf *et al.*, 1996). The question is whether such action could be stimulated in the animal prior to slaughter by providing additional calcium or other electrolytes in the diet.

OXIDATION STABILIZATION AND DRIP LOSS

Lipids are important components of meat and enhance several desirable characteristics such as flavour, tenderness and juiciness. However, one of the major causes of deterioration of meat, even during cold storage, is lipid oxidation which ultimately results in unacceptable flavours, odours and rancidity, as well as a reduction in the concentration of polyunsaturated fatty acids, fat-soluble vitamins and pigments. There is also concern about the production of peroxides and aldehydes and the formation of 'free radicals' which produce harmful chemical products. It might be assumed that adipose tissue is the main site of lipid oxidation in meat, but the most susceptible lipids are the phospholipids present in the subcellular membranes, since they are rich in polyunsaturated fatty acids.

Lipid oxidation is therefore a major cause of deterioration in the quality of meat. It also influences the yield of saleable meat, since the disruption of the subcellular membrane destroys the integrity of the cell wall, releasing intracellular fluid. This results in considerable fluid or drip loss, a major problem and economic loss in both poultry and pork. It is therefore beneficial to reduce both the occurrence and rate of lipid oxidation.

The role of vitamin E as a protective antioxidant is well recognized and the feeding of high dietary levels to both pigs and poultry has improved meat quality by reducing the rate of lipid oxidation and by maintaining the integrity of the cell membrane post-slaughter. This resulted in the meat keeping its fresh appearance and colour for longer, as well as in reduced drip loss, allowing better presentation of both poultry (Sante and Lacourt, 1994; McKnight, 1996) and pork (Monaghan *et al.*, 1993; Cheah *et al.*, 1995). The latter have suggested that the effect of the addition of dietary vitamin E on meat characteristics is due to the inhibition of the enzyme phospholipase A2. This enzyme not only hydrolyses the phospholipids that cause 'off' flavour, but may also be partly responsible for the development of pale, soft and exudative (PSE) pork. An interesting feature of the work of McKnight (1996) with poultry was the considerable reduction in the bacterial count, as well as the improvement in taste panel assessment of the meat as the drip loss decreased (Table 4).

In addition to vitamin E (α-tocopherol), the other main element in preventing cell membrane oxidation is glutathione peroxidase. This is a selenium enzyme and a deficiency of selenium will leave membranes vulnerable to oxidation, increasing the requirement for vitamin E. The association between vitamin E and selenium is therefore well established. Glutathione peroxidase can remove peroxides from cell membranes and thus its effect is complementary to that of vitamin E. Although vitamin E and selenium cannot completely substitute for each other, the addition of Se to the diet could facilitate the effects of vitamin E, especially in relation to maintenance of cell membrane integrity and therefore drip loss.

In poultry, Edens *et al.* (1996) drew attention to the possible role of selenium in influencing quality and yield, as well as drip loss of meat. They compared both inorganic and organic forms (Sel-Plex 50) and noted a greater reduction in drip loss with the organic compared with the inorganic source. Similarly, with pigs, Muñoz *et al.* (1996) reported that Sel-Plex 50, in conjunction with other antioxidants significantly reduced drip loss in the *longissimus dorsi* muscle (Table 5). Interestingly, the longer the period of maturation post-mortem, the greater was the drip loss and the more beneficial the effects of the antioxidants.

Table 4. The influence of vitamin E supplementation on characteristics of poultry meat*.

Vitamin E (IU/kg)	Drip loss (%)	Bacterial count (breast meat)	Taste evaluation[†]
100	1.03	2.19	3.09
200	0.98	1.44	2.11
300	0.86	1.13	1.54

*McKnight (1996).
All assessments made at day 6 post-processing.
[†] Taste panel evaluation 1 = very good; 5 = very poor.

Table 5. Effect of supplementation with dietary antioxidants on drip loss for porcine longissimus dorsi muscle*.

		Time post-mortem (hours)			
		24	48	72	120
Whole muscle	Control	1.98	2.69	3.51	4.75
	Treatment†	1.57	2.25	3.00	4.13
Steaks	Control	4.70	6.74	8.75	10.74
	Treatment†	3.57	5.21	7.49	9.50

* Muñoz *et al.* (1996).

† Treatment: 0.1 ppm Se from Sel-Plex 50, 20-100 kg; 50 ppm Vitamin E, 20–100 kg; 670 ppm Vitamin C, 80–100 kg.

STRESS

In pigs, McKeith *et al.* (1995) have commented that one of the most effective ways to reduce the incidence of poor pork quality is to improve pre-slaughter management and handling and thereby reduce stress. Stress both during transportation and pre-slaughter can affect meat quality, since it can influence the rate and extent of post-mortem acidification in the muscle. If stress is induced over a long period, then muscle glycogen is depleted and dark, firm and dry (DFD) meat may result. Similarly, if the stress occurs immediately before or at slaughter, then the rate of glycolysis is increased at a time when carcass temperature is high, resulting in PSE meat. Thus, acute stress leads to PSE and chronic stress to DFD meat. Stress may also be breed-related, since there is a high correlation between genetic potential for protein deposition and the stress or halothane gene, although this is now being removed from pigs using DNA marker techniques.

Many factors therefore contribute to and are responsible for stress at the time of slaughter which can influence meat quality. However, it is known that during periods of stress there is a rapid excretion of chromium from the body. Chromium is now classified as an essential element since it actively contributes to the glucose tolerance factor which is involved in the stimulation of insulin activity. It also influences the response of the animal to stress by altering cortisol production and insulin action (Mowat, 1996). Such phenomena have been recognized in humans (Fisher, 1995). Thus the action of chromium could reduce stress; and Chang and Mowat (1992) demonstrated its beneficial effect in stressed and growing feeder calves. There were improvements in growth rate and feed efficiency as well as a decrease in serum cortisol and immunoglobulin levels in the blood. Changes in behaviour have also been observed with animals which received chromium supplementation becoming quieter and calmer. In rats, for example, enhanced insulin action with supplemental chromium prevented blood sugar-induced hypertension.

Thus organic chromium may play a role in reducing stress during transportation and in lairage, thereby minimizing the incidence of both PSE- and DFD-type meats. Chromium supplementation may therefore be particularly beneficial with abnormally active and stress-sensitive animals, as well as those which have to be transported long distances to other farms for further production or to the slaughterhouse.

EFFECTS OF FEEDING

Feeding level *per se*, i.e. the supply of nutrients to the animal, influences carcass composition since it directly affects growth rate as well as the proportion of protein and fat in the body. The higher the level of feeding, the higher the rate of lean and fat gain, especially the latter, and therefore the higher the eating quality of the meat. The Meat and Livestock Commission (1988) showed that pigs fed *ad libitum* produced more tender meat than when fed to only 80% of *ad libitum* intake. It was interesting to note that the higher levels of intramuscular fat were not solely responsible for this increase in acceptability; and Warkup and Kempster (1991) suggested that the most important factor was the higher rate of muscle deposition. It is possible that the higher rate of lean deposition, and by implication the higher rates of protein turnover in the body, may have increased the activity of the calpain enzyme system leading to increased tenderization post-slaughter. Thus, faster growth and by implication *ad libitum* feeding is conducive to producing tender meat of better quality. However, in a more recent trial, involving both Large White and Duroc pigs, Wood *et al.* (1995) found no significant differences in the eating quality of meat between *ad libitum* and restrictedly-fed (80% of *ad libitum*) pigs, despite significant differences in the overall growth rate, backfat thickness and lipid and water content of *longissimus dorsi* muscle. The effects associated with feed level are therefore by no means clear and may be related to the quality and type of raw ingredients included in the diets rather than the feeding levels *per se*.

Conclusions

The control of meat quality is becoming increasingly important from both a consumer and producer perspective. The reasons for variation in meat quality are being identified and characterized with the result that new strategies and procedures can be applied on the farm and during processing. In this respect, nutrition plays a major role and opportunities for dietary manipulation of meat quality are being developed. However, nutrition needs to be viewed in a wider context than solely the supply of essential nutrients to the animal. The results of recent experiments suggest that specific elements may be important in controlling several of the metabolic and biochemical processes in the animal which contribute to the variation in meat quality. These include vitamin E, selenium, chromium, the extract from the *Yucca schidigera* plant and the balance of electrolytes in the diet, especially calcium. The mechanisms by which these elements influence meat quality are known and their inclusion should be considered in the diets of both pigs and poultry to ensure the highest possible characteristics of the meat in terms of taste, texture and flavour, as well as health, hygiene and processing qualities.

References

Bejerholm, C. and P.A. Barton-Gade. 1986. Effect of intramuscular fat level on eating quality of pig meat. Proceedings of the 32nd European Meeting of Meat Research Workers. pp. 389–391: Ghent.

Cameron, N.D. and M.B. Enser. 1991. Fatty acid composition of lipid in longissimus dorsi muscle of Duroc and British Landrace pigs and its relationship with eating quality. Meat Science 29: 295–307.

Chang, X. and D.N. Mowat. 1992. Supplemental chromium for stressed and growing feeding calves. J. Anim. Sci. 70: 559–605.

Cheah, K.S., A.M. Cheah and D.I. Krausgrill. 1995. Effect of dietary supplementation of vitamin E on pig meat quality. Meat Science 39: 255–262.

De Val, D.L., F.K. McKeith, P.J. Bechtel, J. Novakofski, R.D. Shanks and T.R. Carr. 1988. Variation in composition and palatability traits and relationship between muscle characteristics and palatability in a random sample of pork carcasses. J. Anim. Sci. 66: 385–395.

Edens, F.W., T.A. Carter and A.E. Sefton. 1996. Influence of dietary selenium sources on post-mortem drip loss from breast meat of broilers grown on different litters. Poultry Sci. 75 (Supplement 1): 60.

Ender, K., G. Kuhn and K. Nürnberg. 1993. Reducing boar taint by yucca extract (De-Odorase). EAAP Conference, Aarhus, 16–19 August, 1993, Vol II pp. 2–16.

Fisher, J.A. 1995. The chromium program. Harper and Row, New York.

Headon, D.R. 1992. Eine Fallstudie zur Ermittlung der Glycokomponenten and Enzyme, die zur Herabsetzung der Umweltbelastung beitragen. 6. Europ. Alltech-Symposium, Würzburg, 3.3.

Hofmann, K. 1987. The concept of meat quality; definition and application. Fleischwirtschaft 17: 44–49.

Jensen, M.T., R.P. Cox and B.B. Jensen. 1995. Microbial production of skatole in the hindgut of pigs given different diets and its relation to skatole deposition in backfat. Animal Science 61: 293–304.

Kempster, A.J., A.W. Dilworth, D.G. Evans and K.D. Fisher, 1986. The effects of fat thickness and sex on pig meat quality with special reference to the problems associated with overleanness. Anim. Prod. 43: 517–533.

Koohmaraie, M. 1994. Muscle proteinases and meat aging. Meat Science 36: 93–101.

Longland. A.C., J.D. Wood, M. Enser, J.C. Carruthers and H.D. Keal. 1991. Effects of growing pigs diets containing 0, 150, 300 or 450 g molassed sugar beet feed per kg on carcass meat eating quality. Anim. Prod. 52: 559–560.

Madsen, A., R. Ysterballe, H.P. Mortensen, C. Bejerholm and P. Barton. 1990. The influence of feeds on meat quality of growing pigs. Report 673: National Institute of Animal Science, Denmark.

McKeith, F.K., M. Ellis and T.R. Carr. 1995. The Pork Quality Challenge. In Management for Quality Pork Production. pp. 1–11. Pork Industry Conference, University of Illinois.

McKnight, W.F. 1996. Effect of antioxidant vitamins on broiler meat quality. Proceedings of BASF Technical Symposium, Atlanta, Georgia, January 23, 1996. pp. 67–96.

Meat and Livestock Commission, 1988. Stotfold Pig Development Unit, First Trial Results, Milton Keynes: Meat and Livestock Commission.

Meat and Livestock Commission, 1992. Stotfold Pig Development Unit, Second Trial Results, Milton Keynes: Meat and Livestock Commission.

Monaghan, F.J., D.J. Buckley, P.A. Morrissey and P.B. Lynch. 1993. In Safety and Quality of Food for Animals. pp. 104–107. Edited by J.D. Wood and T.L.J. Lawrence. Occasional Publication No. 17: British Society of Animal Production.

Mowat, D.N. 1996. Twenty-five perceptions of trivalent chromium supplementation including effective fiber, niacin function and bloat control. In Proceedings of the 12th Animal Symposium on Biotechnology in the Feed Industry. (T.P Lyons and K.A. Jacques, eds) Nottingham University Press. Loughborough, Leics. UK. pp. 83–90.

Muñoz A., M.D. Garrido and M.V. Granados, 1996. Effect of selenium yeast and vitamins C and E on pork meat exudation. In press.

Nute, G.R., J.D. Wood, R.M. Kay and J.G. Perrott, 1994. Effects of molassed sugar beet feed on pig meat quality. Anim. Prod. 58: 471–472.

Pedersen, J.K., A.B. Mortensen, A. Madsen, H.P. Mortensen and J. Hyldegaard-Jensen. 1986. The influence of feed on boar taint in meat from pigs. Report 638: National Institute of Animal Sciences, Denmark.

Russo, V. 1988. Carcass and pork quality: industrial and consumer requirements. Proceedings of the Meeting, Pig Carcass and Meat Quality. 2–3 June 1988, Reggio Emilia. Italy. pp. 3–22.

Sante, V.S. and A. Lacourt. 1994. Tocopherol supplementation and antioxidant spraying on colour stability and lipid oxidation of turkey meat. J. Sci. Food Agric. 65: 503–508.

Sensky, P.L., T. Parr, S.N. Brown, R.G. Bardsley, P.J. Buttry and J.D. Wood. 1996. Relationship between pig meat toughness and the calpain proteolytic system. Animal Science 62: 663–664.

Taylor, R.G., G.H. Geesink, V.F. Thompson, M. Koohmaraie and D.E. Goll. 1995. Is Z-Disk degradation responsible for post-mortem tenderization. J. Anim. Sci. 73: 1351–1367.

Terada, A., H. Hara, M. Kataoka and T. Mitsuoka. 1992. Effect of lactulose on the composition and metabolic activity of the human faecal flora. Microbial Ecology in Health and Disease 5: 43–50.

Warkup, C.C. and A.J. Kempster. 1991. A possible explanation of the variation in tenderness and juiciness of pigmeat. Anim. Prod. 52: 559.

Wood J.D., R.C.D. Jones, M.A. Francombe and O.P. Whelehan. 1986. The effects of fat thickness and sex on pig meat quality with special reference to the problems associated with overleanness. Anim. Prod. 43: 535–544.

Wood, J.D., J. Wiseman and D.J.A. Cole. 1994. Control and manipulation of meat quality. In Principles of Pig Science. (D.J.A. Cole, J. Wiseman and M.A. Varley, eds). Nottingham University Press. Loughborough, Leics., UK. pp. 433–456.

Wood, J.D., S.N. Brown, F.M. Whittington, A.M. Perry and S.P. Johnson. 1995. Comparison of factors affecting the tenderness of pigmeat. Animal Science 60: 561–562.

Wulf, D.M., J.B. Morgan, J.D. Tatum and G.C. Smith. 1996. Effect of animal age, marbling score, calpastatin activity, subprimal cut, calcium injection, and degree of doneness on the palatability of steaks from Limousin steers. J. Anim. Sci. 74: 569–576.

MANIPULATING IMMUNE FUNCTION TO IMPROVE HEALTH AND FEED EFFICIENCY

MODULATION OF THE IMMUNE RESPONSE: CURRENT PERCEPTIONS AND FUTURE PROSPECTS WITH AN EXAMPLE FROM POULTRY

PAUL F. COTTER

Biology Department, Framingham State College, Framingham, Massachusetts, USA

Preface

Immunity to infectious disease results from a myriad of anatomical and physiological mechanisms designed for the purpose of maintaining internal homeostasis, or the well being of the chicken. These systems might be compared to the physics of a force acting on a stationary pendulum. Movement results, but this is gradually dampened until the device is again stationary. The encounter between a chicken and an infectious agent (bacteria, virus, etc.) is like the perturbing force that displaces the pendulum. The immune response is designed to restore equilibrium. Biological systems are infinitely more complex, so an adjustment of our model is necessary. This change could be accomplished if it is assumed that the re-equilibrated pendulum has acquired some additional mass as a result of its experience. Now a greater force than before would be required to move our device. Vaccines work in just this way.

Immune responses can vary in the amount of benefit that they provide. Too little (anergy) and the animal is overcome by infection, too great (hypersensitivity) and tissue damage results. Reactions of the latter type include the Koch phenomenon, which resulted in necrosis of skin in tuberculous guinea pigs. A Shwartzman reaction occurs at prepared skin sites following an intravenous injection of bacterial lipopolysaccharide. The mediators of these untoward reactions are macrophages and cytokines such as tumor necrosis factor, substances usually perceived as having a beneficial role in immunity (reviewed in Rook, 1990).

What does the future hold? Poultry producers have benefited from genetic gains in growth and feed conversion. Vaccines have provided the means to offset loses due to Marek's, Newcastle, and bronchitis diseases. Newer diseases such as chick anemia agent await the development of effective vaccines, while old problems like salmonella and *E. coli* are re-emerging in new forms. The goals for the future must include a greater understanding of the underlying mechanisms of immunity; and as suggested above by the hypersensitivity example, more is not necessarily better. Our efforts must be directed toward finding ways to optimize the response to infectious challenges. Figure 1 shows the nature of the problem. Three possible types of immune responses are shown. The object is to increase the area occupied by the beneficial zone while reducing those occupied by the other two. Moving the two ellipses toward one another would have this effect. Accomplishment of this task will be difficult, but progress in this direction will be made only with a more complete understanding of immunity, the development of better vaccines and new products. In the latter case dietary products that are effective as immunodulators will be especially welcome. An experimental example follows.

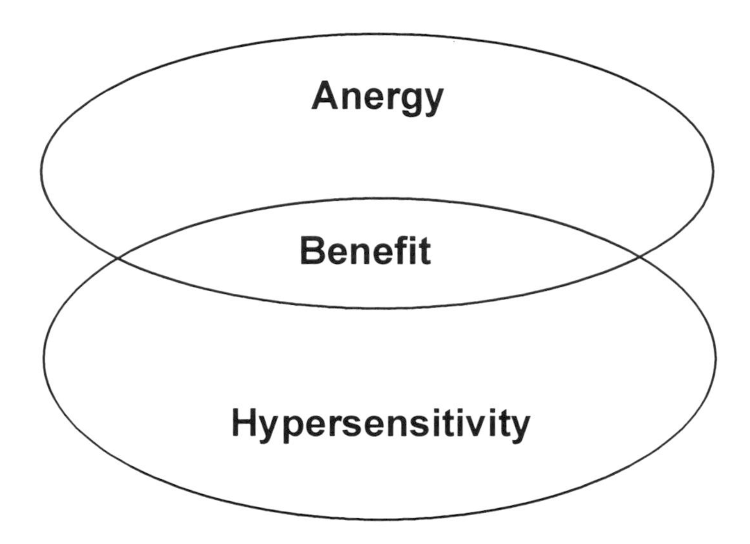

Figure 1. Three possible immune responses are represented. More area will be occupied by beneficial responses if the two ellipses are caused to come closer together.

Introduction

Poultry diets that included Bio-Mos, a mannanoligosaccharide derived from yeast cell walls, stimulated immunity in turkeys (Savage and Zakrzewska, 1996) and reduced the mortality of broilers (MacDonald, 1996). While the precise mechanisms for these beneficial effects are as yet unknown, Bio-Mos has been shown to aid in the sequestration of immunosuppressive mycotoxins sometimes found in feed and to stimulate phagocytosis. Both of these responses would presumably result in the enhancement of protective immune functions (see Killeen and Rosell, 1996, and Trenholm *et al.*, 1996) and to have a beneficial effect on growth.

Phytohemagglutinin (PHA), a lectin derived from kidney beans *(Phaseolus vulgaris),* induces an inflammatory swelling reaction when it is injected into the wattles of chickens. This is known as a delayed wattle reaction (DWR, Stadeker *et al.*, 1977) because the response develops over time and usually requires between 24 and 48 hours to reach its maximum. Thereafter it undergoes a resolution phase which may take several additional days. Thus the kinetics of this phenomenon include both the development of inflammation of the tissue at the test site, and its subsequent resolution. Additional parameters such as the frequency of responders in study populations or the degree of swelling after various treatments have also been investigated (Goto *et al.*, 1977).

The reaction is useful as an *in vivo* measure of immunocompetence in the chicken because it is easily observed and because of its convenience. Moreover, prior sensitization with PHA is not required. The PHA-DWR is known to be T-cell dependent (Goto *et al.*, 1977) and it has been used in genetic selection as a measure of immunocompetence. Because of its histological similarity to inflammatory reactions caused by PHA in other species, it is considered to be an avian example of cutaneous basophilic hypersensitivity (CBH) (McCorkle *et al.*, 1980). It is also known to be regulated in part by B-complex haplotypes (chicken MHC) and sex (Taylor *et al.*, 1987).

The present study reports that inclusion of Bio-Mos in diets fed to growing commercial strain pullets had a significant influence on the kinetics of both the secondary and tertiary PHA-DWR. Moreover, the data presented here show that the benefit provided by Bio-Mos resulted in an enhanced capacity to control the magnitude of the DWR reaction rather than to potentiate it. Thus these results demonstrate a clear example of immunomodulation (Hylands and Poulev, 1996) in the chicken, and it is suggested that the presence of PHA in the wattle tissue causes an untoward (hypersensitivity) reaction which was reduced in the Bio-Mos fed chickens.

Materials and methods

The pullets (Golden Comets, Hubbard Farms, Walpole, NH) used for this study were chosen because of their importance in commercial egg production. They were vaccinated against Newcastle and Marek's diseases at the hatchery and housed in battery cages throughout the experimental period. They were fed a commercial starter diet containing 20% protein through six weeks of age which was then replaced with a grower formula containing 15% protein for the duration of the trials (United Cooperative Farmers, Fitchburg, MA). At hatching the chicks were randomly divided into two groups, one of which was fed with diets containing Bio-Mos at 2.2 lb per ton, while the controls received similar diets without the Bio-Mos.

The PHA wattle tests were first conducted when the chicks reached six weeks of age and were repeated using the same chickens at 8 and 10 weeks. During the first trial each chick received an injection of 50 μg PHA-P (Difco Laboratories, Detroit, MI) dissolved in non-pyrogenic physiological saline. This was given in the right wattle in a volume of 0.05 ml while saline alone was used in the opposite wattle. At 8 weeks of age the test was repeated as before except that the dose of PHA was reduced to 25 μg and the opposite wattles were used for the lectin or the saline. At 10 weeks the birds were re-tested after being divided into two dosage groups one of which received 25 μg PHA, while the other received 13 μg PHA per dose. The swelling responses were determined on all occasions by measuring the wattle thickness to the nearest 0.1 mm using a digital vernier caliper. Measurements were taken just prior to the challenge and at three or four additional times post-challenge.

The overall design of the study is given in Table 1. Included are the numbers of chicks used in each trial, their ages at the time of testing, and the PHA dosages used.

Table 1. The experimental design.

	Number of chicks		
Age (weeks)	Control	Bio-Mos	PHA dose (mg)
6	12	12	50
8	12	12	25
10	6	6	25
10	6	6	13

Results

Results of the primary test, which was conducted at six weeks of age, are shown in Figure 2. The mean values for the four post-challenge swelling measurements are shown along with the pre-challenge values for both the Bio-Mos and control groups. Swelling of the test wattles was evident as early as 3 hours after the PHA challenge. This reached a maximum at 24 hours when the wattles were nearly three times their initial thickness (Table 2). Resolution of the swelling was underway by 48 hours, but it was evident that some remained for as long as 96 hours post-challenge. Although the Bio-Mos fed group had slightly more swollen wattles at 24 hours this was not significantly different from the controls.

The data for the 8-week and 10-week trials are presented using a format similar to that shown in Figure 2, except that the 96 hour swelling measurements were not taken on either of those occasions. The results of the secondary test, given at 8 weeks of age, are shown in Figure 3, while the third test results are shown in Figure 4. It was evident that the response was well developed by 3 hours. in both test groups, but the controls had noticeably larger wattles (more inflammation) than the Bio-Mos fed group. There was a clear difference at both 3 hours and again at 24 hours. All wattle swellings were reduced at the 48 hour measurement indicating the resolution phase of the reaction was well under way.

There was a noticeable change in the response kinetics during the third trial which was conducted when the pullets had reached 10 weeks of age (Figure 4). The swelling at 3 hours was much reduced in comparison to that occurring in the earlier trials. There were no significant differences between the two groups at 3 hours, although differences associated with diet were clearly evident by 24 hours. As was the case during the 8 week trial, the wattles of the Bio-Mos fed groups were considerably less swollen than were those of the controls.

Since each of these trials was conducted at different times during a period

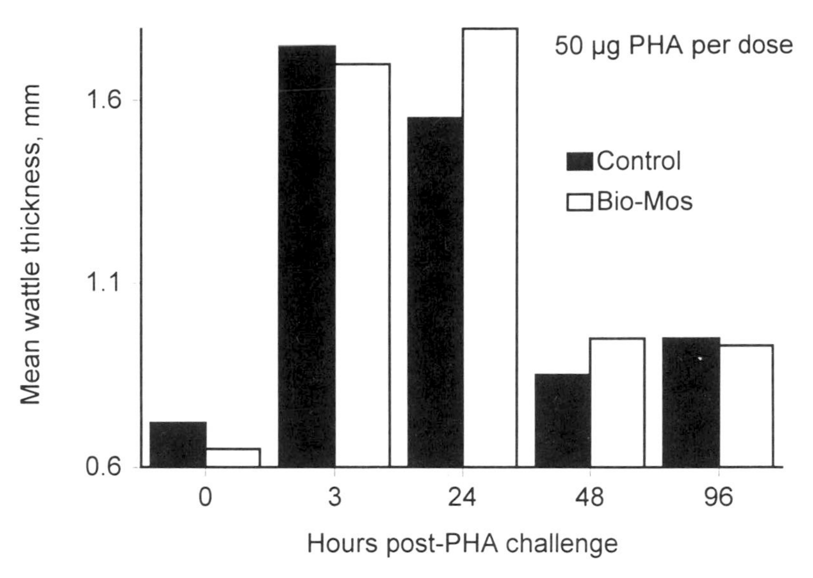

Figure 2. The kinetics of the primary PHA-DWR in 6-week-old pullets fed diets supplemented with the mannanoligosaccharide Bio-Mos at 2.2 lb per ton are compared with those of unsupplemented controls.

Table 2. Percent wattle swelling at 24 hours post-challenge.

Treatment	6 Weeks	8 Weeks	10 Weeks
Bio-Mos	283*	154	167
Control	257	218	217
Probability[†]	NS	0.005	0.01

* Percent thickness increase at 24 hours.
[†] Probability of a dietary effect. NS = non-significant.

of rapid increase in body size, some changes in the physical characteristics of the wattles themselves might be expected to have occurred. Thus a comparison of initial (pre-challenge) wattle size and the maximum response, represented by the 24 hour thickness measurements, is given in Table 2. The data represent the mean percentage increase in wattle thickness. This was calculated by dividing the 24 hour measurement for each group by its pre-challenge value. Both dosages used in the 10 week study were pooled.

Discussion

The results of this study demonstrate that including the mannanoligosaccharide Bio-Mos in diets fed to growing pullets resulted in the modulation of a standard measure of a cell-mediated immune function, the PHA wattle reaction. The group fed the Bio-Mos demonstrated significantly different response kinetics during both secondary and tertiary reactions when compared with the controls. The effect was seen mainly at 3 and 24 hours during the 8 week

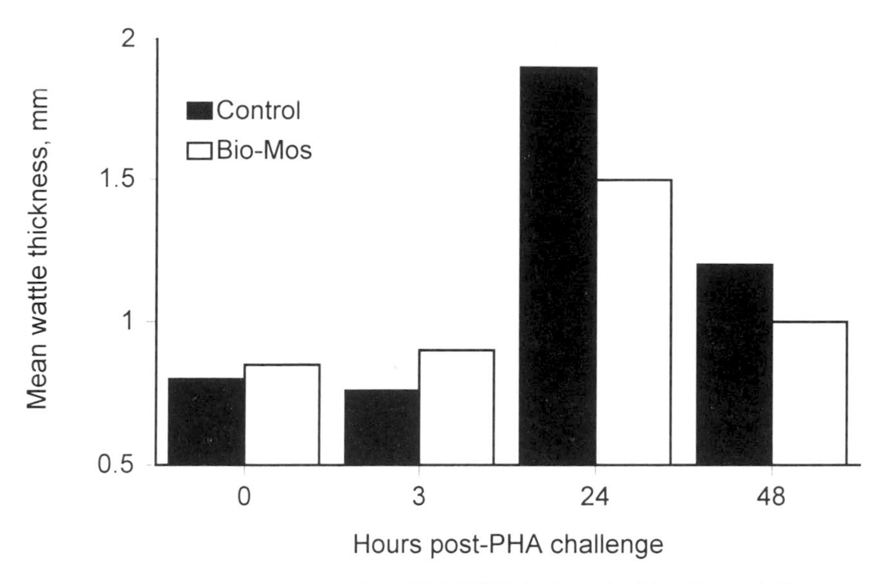

Figure 3. The kinetics of the secondary PHA-DWR in 8-week-old pullets fed diets supplemented with the mannanoligosaccharide Bio-Mos at 2.2 per ton are compared with those of unsupplemented controls.

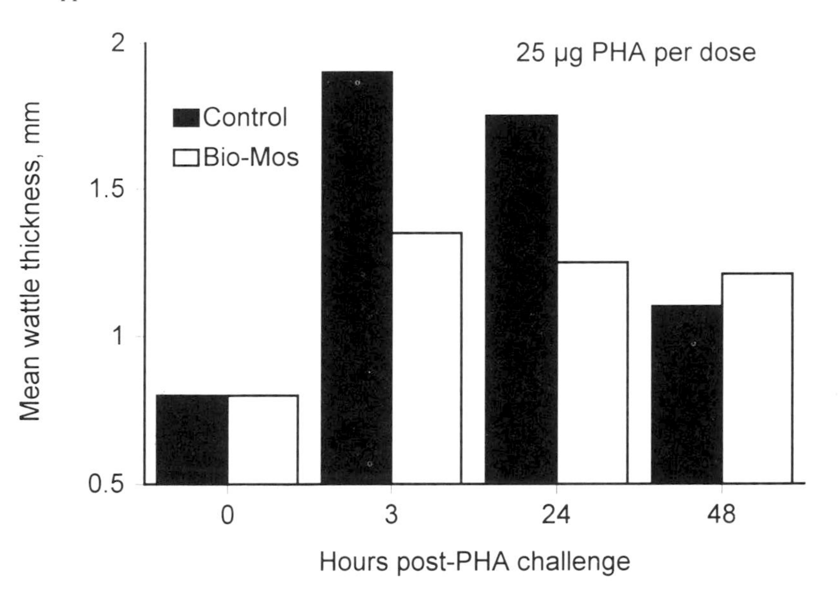

Figure 4. The kinetics of the tertiary PHA-DWR in 10-week-old pullets fed diets supplemented with the mannanoligosaccharide Bio-Mos at 2.2 lb per ton are compared with those of unsupplemented control.

trial (secondary reaction) and at 24 hours in the 10 week trial (tertiary reaction). This contrasted with the primary (6 week) reaction where no significant differences were observed.

The possibility that the wattles of the Bio-Mos fed group became more swollen than the controls between the 3 and 24 hour readings cannot be excluded since no measurements were made on either group during these times. Such a possibility is not considered likely because, after being smaller initially, it would require that they became larger at a faster rate than the controls and resolved the response more quickly as well.

Examination of the data in Table 2 suggests that the relative amount of swelling at 24 hours produced from the second and third exposure to PHA was reduced in comparison to that from the first exposure. This is likely a demonstration of immunological memory due to the initial sensitization which occurred during the first trial. Prior sensitization of broiler chickens had such an effect on the response kinetics when bovine serum albumin or human plasma proteins were used to induce wattle reactions (Cotter *et al.*, 1985; Cotter and Wing, 1987). Moreover, Edelman *et al.* (1985) reported that sensitization with PHA-P resulted in secondary responses in chickens challenged with low doses of a related material, PHA-M. Thus there can be differences in the amount of swelling that depend on a prior experience with PHA. Perhaps these observations represent the CBH equivalent of classical secondary skin graft rejections. In that example of cell-mediated immunity, second set rejections occur more rapidly than first set graft rejections.

The mechanism by which Bio-Mos exerted its modulation effect on the PHA reaction is as yet unknown. A plausible explanation may be envisioned if one assumes that the introduction of PHA into the wattle tissue causes an acute inflammatory reaction which if unchecked would result in tissue damage. By removing the insult sooner, less damage should accrue. Thus it is likely that the benefit provided by the Bio-Mos derives from its capacity to accelerate the return of tissue homeostasis.

A CBH reaction analogous to that caused here by PHA was produced by injecting wattles with an antigen made from the bacterium *Staphylococcus aureus*. The kinetics of that response were influenced by the B-complex haplotype which is the major genetic locus influencing immunity (Cotter *et al.*, 1987). These earlier observations, taken in the light of the present study, suggest a mechanism for the observed genetic differences in susceptibility to staphylococcal disease which have been reported in chickens (Cotter and Taylor, 1991; Cotter *et al.*, 1992). If the antigen responsible for the inflammatory insult to the tissues were removed more efficiently, a milder disease should result. Mechanistic support for this idea comes from *in vitro* work in which *S. aureus* stimulated macrophages were shown to release soluble factors, one of which was capable of causing resorption of cartilage (Klasing, 1994). This may explain why *S. aureus* is often associated with leg problems in poultry. Remediation of these problems could be accomplished either as a result of manipulation of the diet as was the case here, or by taking advantage of inherited (B-complex) differences. Moreover alteration of the diet for the purpose of enhancing immune response may offer an alternative in situations where genetic selection is not feasible or where effective vaccines are not available.

The example illustrated above taken together with the observations reported by Savage and Zakrzewska (1996) show evidence that the inclusion of the mannanoligosaccharide Bio-Mos added to poultry diets had salutary effects on immunity. In the latter case an increase in the amount of secretory (bile) IgA was observed in Bio-Mos fed turkeys. Presumably this represents a benefit because the gut serves as an important portal for the entry of pathogens, while IgA acts there as a barrier. In the present case the salutory effect of Bio-Mos is seen as its capacity to act as an immunological buffer reducing inflammation resulting from exposure to PHA.

Returning to the pendulum analogy used at the outset, the effect seen here would be as if there were two pendulums set in motion by the distorting force (PHA). In the control group displacement is unimpeded resulting in the maximum time to equilibration. Bio-Mos in the other group dampens this displacement of the pendulum as if it were surrounded by a liquid. At first the liquid's density is not much greater than air; but with repeated exposures density increases so as to retard the motion of the displacing force. In any event the accuracy of such a model will only be borne out if the observations made here can be extended to additional systems.

References

Cotter, P.F. and R.L. Taylor, Jr. 1991. Differential resistance to *Staphylococcus aureus* challenge in two related lines of chickens. Poultry Science 70: 1357–1361.

Cotter, P.F. and T. Wing. 1987. Kinetics of the wattle reaction to human plasma. Avian Diseases 31: 643–648.

Cotter, P.F., T. Wing and J. Swanson. 1985. The delayed and saline wattle reactions in broilers challenged with bovine serum albumin. Poultry Science 64:1293–1295.

Cotter, P.F., R.L. Taylor, Jr., T.L. Wing and W.E. Briles. 1987. Major histocompatibiltiy (B) complex-associated differences in the delayed wattle reaction to staphylococcal antigen. Poultry Science 66: 204–208.

Cotter, P.F., R.L. Taylor, Jr. and H. Abplanalp. 1992. Differential resistance to *Staphylococcus aureus* challenge in major histocompatibility (B) complex congenic lines. Poultry Science 71:1873–1878.

Edelman, A.S., P.L. Sanchez, M.E. Robinson, G.M. Hochwald and G.J. Thorbecke. 1985. Primary and secondary wattle swelling response to phytohemagglutinin as a measure of immunocompetence in chickens. Avian Diseases 30:106–111.

Killeen, G. and V. Rosell. 1996. The potential of polysaccharide supplements in diets for livestock and pets. In: Proceedings of Alltech's 12th Annual Symposium. (T.P. Lyons and K.A. Jacques, eds). Nottingham University Press, Nottingham, UK. pp. 149–158.

Klasing, K.C. 1994. Partial characterization of interleukin 1-like and tumour necrosis factor-like activities released from chicken macrophages. In: Proceedings of the Avian Immunology Group (AIRG). Reading University. International Periodical Publishers, Carfax Pub. Co., Abingdon, Oxfordshire UK, pp, 151–159.

Hylands, P.J. and A.A. Poulev. 1996. Immunostimulants maximizing the health and efficiency of animals through plant-derived biomolecules. In: Biotechnology in the Feed Industry, Proceedings of Alltech's 11th annual symposium. (T.P. Lyons and K.A. Jacques, eds). Nottingham University Press, Nottingham, UK. pp. 117–139.

MacDonald, F. 1996. Use of immunostimulants in agricultural applications. In: Biotechnology in the Feed Industry, Proceedings of Alltech's 12th Annual symposium. (T.P. Lyons and K.A. Jacques, eds). Nottingham University Press, Nottingham, UK. pp. 97–103.

Rook, G.A.W. 1990. Mechanisms of immunologically mediated tissue damage during infection. In: Recent Advances in Dermatology. Vol. 8, Churchill Livingstone Pub. pp. 193–210.

Stadeker, M.J., M Lukic, A. Dvorak and S. Leskowitz. 1977. The cutaneous basophil response to phytohemagglutinin in chickens. J. Immunol 118:1564–1568.

Savage, T.F. and E.I. Zakrzewska. 1996. The performance of male turkeys fed a starter diet containing a mannanoligiosaccharide (Bio-Mos) from day old to eight weeks of age. In: Biotechnology in the Feed Industry, Proceedings of Alltech's 12th Annual symposium. (T.P. Lyons and K.A. Jacques, eds). Nottingham University Press, Nottingham, UK. pp. 47–54.

Trenholm, H.L., L.L. Charmley and D.B Prelusky. 1996. Mycotoxin binding agents: an update on what we know. In: Biotechnology in the Feed Industry, Proceedings of Alltech's 12th Annual Symposium. (T.P. Lyons and K.A. Jacques, eds). Nottingham University Press, Nottingham, UK. pp. 327–349.

Taylor, Jr., R.L., P.F. Cotter, T.L. Wing, and W.E. Briles. 1987. Major histocompatibility (B) complex and sex effects on the phytohemagglutinin wattle response. Animal Genetics 18:343–350.

IMMUNOSUPRESSION IN POULTRY CAUSED BY AFLATOXINS AND ITS ALLEVIATION BY *SACCHAROMYCES CEREVISIAE* (YEA-SACC[1026]) AND MANNANOLIGOSACCHARIDES(MYCOSORB)

G. DEVEGOWDA, B.I.R. ARAVIND and M.G. MORTON

Department of Poultry Science, University of Agricultural Sciences, Bangalore, India

Introduction

The economic losses in the poultry industry due to aflatoxicosis, although not documented quantitatively, are well known, and may run into millions. The adverse effects of aflatoxin-contaminated diets consumed by chickens range from the undetectable to disastrous by way of reduced egg production in layers and growth depression and morbidity in broilers. Due to its effect on several organs and systems within the body accompanied by a wide variety of symptoms, it could well be described as the 'Aflatoxin Syndrome' rather than the more simple term, aflatoxicosis.

One of the major causes of worry is immunosuppression due to aflatoxin and the consequent increase in susceptibility to a wide variety of infectious diseases along with the accompanying vaccine failure and therapeutic drug failures. Immunosuppression caused by aflatoxins produces a cascading effect starting with depressed protein synthesis, lowered serum albumin and globulin levels, reduction in circulating antibody levels, impairment of the reticulo-endothelial system and reduced cell-mediated immunity, besides affecting the normal development of the thymus and bursa of Fabricius.

Several approaches have been taken in an effort to minimize the deleterious effects of aflatoxicosis. Initially, efforts were concentrated on total elimination beginning with prevention of aflatoxin formation through control of fungal growth in feedstuffs. Subsequently, attempts were made to reduce aflatoxin to a less toxigenic compound, deactivate it by both physical and chemical methods, modification of nutritional parameters and the application of biotechnological tools to reduce its adverse effects.

This paper will discuss effects of aflatoxin on the immune system and the methods adopted to minimize these effects in poultry by the use of *Saccharomyces cerevisiae* (Yea-Sacc[1026]) and mannanoligosaccharides (Mycosorb) (manufactured by Alltech, Inc.). A number of workers have been involved in research in this area, including our group at the Department of Poultry Science, University of Agricultural Sciences, Bangalore, India.

The immune system

In poultry, the bursa of Fabricius, thymus and spleen and to a lesser extent the cecal tonsils, Payer's patches and bone marrow constitute the organs contributing to humoral and cellular immunity which are interdependent. The T-cells which originate from the thymus control the cell-mediated immune response including helper and suppressor influences on antibody formation, delayed hypersensitivity, macrophage activation and cytotoxic responses. The B-cells which originate from the bursa of Fabricius and bone marrow controls the humoral immune system involving the production of antibodies or immunoglobulins, mainly IgM and IgA.

Aflatoxin and the immune system

Reduction in the relative size of the bursa of Fabricius accompanied by mortality due to aflatoxin has been observed by Thaxton *et al.* (1974) and Kubena *et al.* (1990). According to Reddy *et al.* (1982), the minimum effective dose level of aflatoxin needed to bring about a significant reduction in the relative weight of the bursa of Fabricius was 1.25 ppm. Thaxton *et al.* (1974) reported the atrophy of thymus and bursa of Fabricius on feeding aflatoxin-contaminated diets to chickens. Devegowda *et al.* (1994) observed a significant reduction in the size of the bursa of Fabricius in broilers fed diets containing 500 ppb aflatoxin and in ducklings fed 100–200 ppb aflatoxin (Tables 1 and 2). Virdi *et al.* (1989) observed thymic aplasia and reduced weight of the bursa of Fabricius up to the extent of 38% in chickens fed aflatoxin-contaminated diets. A serial decrease in the size of the bursa of Fabricius was observed in broilers fed graded levels of aflatoxin (Manegar *et al.*, 1996) (Table 3).

Tung *et al.* (1975) reported a reduction in total serum proteins and globulins by feeding 1.25–2.5 ppm dietary aflatoxin. Total serum proteins and albumin decreased in birds fed aflatoxin at 0.5 ppm (Umesh and Devegowda, 1990) and 1.0 ppm (Ghosh *et al.*, 1990). Devegowda *et al.* (1994) observed a decrease in serum proteins in broilers fed 500 ppb and in ducklings fed 200 ppb of dietary aflatoxin (Tables 1 and 2). Ghosh *et al.* (1990) reported that 1.0 ppm of dietary aflatoxin significantly decreased T-lymphocyte counts resulting in

Table 1. Effect of Yea-Sacc[1026] on mortality, total plasma protein, relative weight of bursa of Fabricius and HI titer level against Newcastle disease in broilers fed diets containing aflatoxin.*

	Aflatoxin (ppb)		
	0	500	500
Yea-Sacc[1026] %	0.0	0.0	0.1
Mortality, %	3.87[a]	38.40[c]	3.30[a]
Bursa weight, g/100g BW	0.32[a]	.20[b]	0.29[a]
Total protein, g/dl	2.68[a]	2.12[b]	2.56[a]
NCD HI titer, log 2 value	2.46[a]	1.74[b]	2.62[a]

[ab] $P<0.05$.

* Devegowda *et al.* (1994).

Table 2. Effect of Yea-Sacc[1026] on weight of bursa of Fabricius, total plasma proteins and albumin in ducklings fed diets containing aflatoxin.*

Parameters	100 ppb aflatoxin			200 ppb aflatoxin	
	Control	0% Yea-Sacc	1% Yea-Sacc	0% Yea-Sacc	1% Yea-Sacc
Bursal weight, g/100g BW	0.25[c]	0.19[b]	0.24[c]	0.11[a]	0.23[c]
Total protein, g/dl	1.94[d]	1.11[b]	1.92[d]	0.81[a]	1.81[c]
Albumin, g/dl	0.49[c]	0.37[b]	0.51[d]	0.29[a]	0.47[c]

[ab] $P<0.05$.
* Devegowda *et al.* (1994).

suppression of cell-mediated immunity. Cell-mediated responses appear to be affected by low levels of aflatoxin whereas high levels affect immunoglobulin production and antibody responses (Corrier, 1991).

The presence of low levels of aflatoxin in feed may not manifest itself as a disease but plays a major role in depression of vaccination immunity and may lead to the occurrence of the disease in properly vaccinated flocks. Edds *et al.* (1973) demonstrated that chickens fed a diet containing 200 ppb aflatoxin were more susceptible to challenge inoculation with Marek's disease virus as compared with controls. Decrease in humoral response measured by antibody titers against Newcastle disease in birds was observed by Boulton *et al.* (1981), Umesh and Devegowda (1990) and Devegowda *et al.* (1994) due to the incorporation of aflatoxin at various levels in broiler diets (Table 1). Dose-dependent reductions in antibody titer levels for Newcastle disease and infectious bursal disease were observed on supplementing aflatoxin to the diet of broilers (Manegar *et al.*, 1996) (Table 3).

Susceptibility to bacterial and protozoan diseases is also enhanced in poultry fed aflatoxin-contaminated diets. Chickens receiving aflatoxin-contaminated diets showed higher susceptibility to duodenal coccidiosis (Southern *et al.*, 1984) and cecal coccidiosis (Edds *et al.*, 1973). Similar results were observed in turkey poults for cecal coccidiosis (Witlock *et al.*, 1982). In addition, aflatoxin decreased the serum concentration of antibiotics used to combat bacterial infections (Hamilton, 1987).

Table 3. Effect of graded levels of aflatoxin on weight of bursa of Fabricius, titers against Newcastle disease and infectious bursal disease in broilers.*

Aflatoxin (ppb)	Bursa weight (%)	NCD titers (ELISA)	IBD titers, (ELISA)
0	0.103 ± 0.005[a]	5341 ± 49.12[a]	4318 ± 139.54[a]
50	0.103 ± 0.005[a]	4806 ± 141.69[b]	3085 ± 312.75[b]
100	0.090 ± 0.009[ab]	3943 ± 88.84[c]	2600 ± 140.05[b]
200	0.060 ± 0.008[bc]	2920 ± 103.16[d]	2332 ± 99.09[c]
400	0.066 ± 0.005[bc]	2006 ± 106.41[c]	2006 ± 87.34[cd]
600	0.056 ± 0.007[c]	300 ± 245.24[f]	1644 ± 60.19[d]

[ab] $P<0.05$.
* Manegar and Devegowda (1996).

The role of Yea-Sacc[1026] viable yeast culture in modifying the immune response during aflatoxicosis

With the advent of using yeast cultures as growth promoters in poultry diets, several beneficial effects have been recorded. The first report by McDaniel (1991) observed that the supplementation of yeast culture to diets of broiler breeders improved hatchability. A significant improvement in body weight gain and feed efficiency was observed in a biological trial in broilers fed varying levels of energy and protein (Taklimi *et al.*, 1993). In another biological trial, Stanley *et al.* (1993) observed that supplementation of 0.1% *S. cerevisiae* yeast culture to diets of broilers improved body weights significantly. On the addition of 5 ppm aflatoxin to the diet, body weights were depressed to a larger extent, although a better performance was observed in the group receiving 0.05% *S. cerevisiae* yeast culture which was further enhanced by supplementation of 0.1% of the additive (Figure 1).

In a controlled study, Devegowda *et al.* (1995) examined the effects of two levels of Yea-Sacc[1026] (0.1 and 0.2%) added to broiler diets containing two levels of aflatoxin (0.50 and 1.0 ppm). A significant improvement in weight gain, organ weights and serum proteins and enzymes accompanied by lowered mortality was recorded in the groups supplemented with Yea-Sacc[1026] when

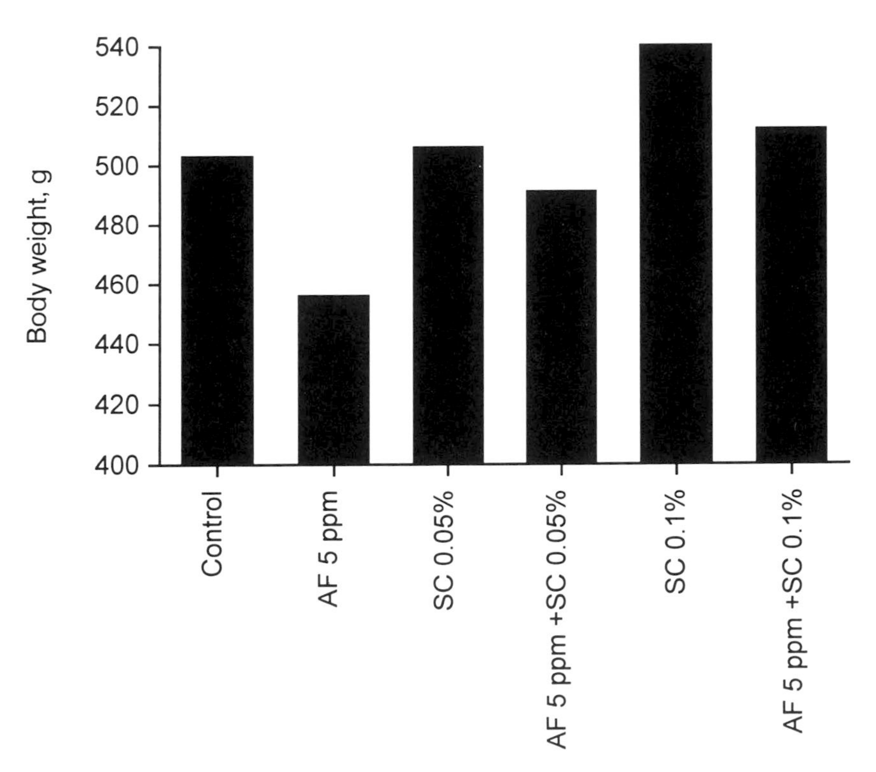

Figure 1. Effect of aflatoxin (AF) and *Saccharomyces cerevisiae* (SC) cultures on body weights of broilers (Stanley *et al.*, 1993).

compared to the groups fed only aflatoxin-contaminated diets. In a significant finding, the hemagglutination inhibition (HI) titer which was reduced considerably in the birds fed only aflatoxin-contaminated diets was reversed on the supplementation of Yea-Sacc^{1026} to the diets. This finding clearly indicated that Yea-Sacc^{1026} plays a vital role in enhancing the response of the immune system which supports the claim of Shalitin (1989) who observed that some strains of *S. cerevisiae* enhanced antibody production.

To clearly establish the fact that Yea-Sacc^{1026} had a direct influence on aflatoxins, an *in vitro* study was conducted to determine whether aflatoxin was degraded by Yea-Sacc^{1026} using two levels of aflatoxin (250 and 500 ppb) and three levels of Yea-Sacc^{1026} (0.0, 0.5 and 1.0 g) at three incubation times (48, 72 and 96 hours). Although the actual mode of action could not be established, i.e. whether the toxin was degraded or bound, in a dose-dependent response a significant amount of aflatoxin was removed ranging from 56% in 48 hours to 90% in 96 hours incubation period using 250 ppb aflatoxin and from 57% in 48 hours to 92% in 96 hours incubation period using 500 ppb aflatoxin (Devegowda *et al.*, 1996) (Table 4).

An exhaustive study to further establish the immunopotentiating properties of Yea-Sacc^{1026} in the presence of aflatoxin in poultry diets was conducted combining the methodology adopted in the earlier works including two biological trials in broiler chickens, one biological trial in ducklings and an *in vitro* study (Devegowda *et al.*, 1994). In the two biological trials in broilers, total serum protein and HI titer against Newcastle disease were significantly reduced in groups fed diets containing aflatoxin. Supplementation with Yea-Sacc^{1026} counteracted the adverse effects of aflatoxin for both parameters. The relative weights of the bursa of Fabricius, which were decreased in the aflatoxin supplemented groups, were significantly increased in groups supplemented with Yea-Sacc^{1026} comparable to control (Table 1).

In the biological trial conducted in ducklings, the adverse effects of aflatoxin included high mortality (up to 37%), lower concentrations of total serum proteins and albumin and decreased relative weight of the bursa of Fabricius. This trend was reversed upon the supplementation of Yea-Sacc^{1026} to the

Table 4. Percent degradation of aflatoxin B_1 at two levels (250 ppb and 500 ppb) by Yea-Sacc^{1026} at three levels (0.0, 0.5 and 1.0 g) at different incubation periods.*

Treatments		Degradation (%) after incubation period of		
Aflatoxin B_1 (ppb)	Yea-Sacc ^{1026} (g)	48 hrs	72 hrs	96 hrs
250	0.0	10.83±2.03^a	7.31±2.03^a	7.31±4.07^a
250	0.5	47.20±0.00^c	62.45±4.07^b	82.40±0.00^b
250	1.0	56.00±3.52^d	64.80±0.00^b	90.03±2.03^{cd}
500	0.0	7.31±4.07^a	4.96±0.00^a	7.31±4.07^a
500	0.5	39.57±2.03^b	72.43±2.03^c	85.92±3.17^{bc}
500	1.0	57.76±3.05^d	73.60±5.28^c	91.79±1.02^d

^{ab} $P<0.05$.

* Devegowda *et al.* (1995).

aflatoxin-contaminated diets with mortality reduced to 6% which was comparable to the control group (Table 2).

In the *in vitro* experiment, regardless of aflatoxin level, the pattern of aflatoxin binding was uniform. There was a gradual increase in the amount of aflatoxin bound as the incubation period was increased (50% binding at 48 hours and 88% binding at 96 hours). Aflatoxin binding was not significantly affected by the two levels of Yea-Sacc[1026] used for 72 hours of incubation, whereas for 48 and 96 hours, the percentage of aflatoxin bound increased with higher levels of Yea-Sacc[1026].

These results clearly indicated the beneficial effects of viable yeast culture when supplemented to aflatoxin-contaminated diets in poultry. Besides the positive effects seen on body weight, feed efficiency and mortality, the most significant contribution is in the ability to modify immune response, which was reflected in the improvement in size of the bursa of Fabricius and increased levels of serum protein and albumin, thereby enhancing the levels of circulating immunoglobulins. HI titers for Newcastle disease, which were reduced significantly during aflatoxicosis, were increased on the supplementation of viable yeast culture.

The role of Mycosorb in modifying the immune response during aflatoxicosis

Carbohydrates are more commonly known for their nutritive value in providing energy besides forming an integral part of structural elements such as nucleic acids, glycolipids and glycoproteins. However, biotechnologists are discovering the role carbohydrates play in various other biological functions, particularly in enhancing the immune response and the ability to block colonization and to bind various pathogens (Parekh, 1993).

Three major oligosaccharides have been identified which play a role in improving animal production – mannanoligosaccharides, fructooligosaccharides and galactooligosacchrides, the latter two with limited success.

Mannanoligosaccharides are derived from the yeast cell wall which exhibits a high degree of antigenicity mainly due to its mannan and glucan components. The antigenicity varies with different strains of yeast due to the different linkages and degree of phosphorylation in the mannan structure. The role of the mannan-glucan-protein structure of the yeast cell wall and its ability to stimulate the immune response was established years ago in mice and rats where it increased the bactericidal activity of the sera (Blattberg, 1956). Mannose sugars in the mannanoligosaccharides influence the immune system by stimulating the secretion of mannose-binding protein from the liver which binds to the capsule of invading bacteria and triggers the complement fixation system. Newman (1995) also established the adsorption property of mannanoligosaccharide to several strains of *Clostridia* and *Salmonella* in an *in vitro* study. In another *in vitro* study, Lyons (1994) established that a commercial mannanoligosaccharide (Mycosorb) stimulated immune response by increasing phagocytic activity of macrophages incubated with the peripheral blood of Wistar rats. The researchers also observed the response of broiler chicks challenged with wild strains of *Salmonella* and found that birds provided Mycosorb at the rate of 1 g per kg in the diet were able to withstand the challenge. Similarly, enhanced phagocytosis was observed in African catfish supplemented mannanoligosaccharide in the diet (McDonald, 1995).

The ability of mannanoligosaccharides to bind to several pathogens in the gastrointestinal tract is based on an ability to bind to specific receptor sites on the bacterial cell wall, thereby preventing colonization. Newman (1994) observed strains of enteric pathogens which exhibited this agglutinated mannan property to include several strains of *E. coli*, *Salmonella* and *Clostridia*.

In another trial, Sisak (1994) observed an 18% reduction in *Salmonella* colonization compared to 76% colonization in the control in birds supplemented with mannanoligosaccharides at the rate of 1 kg per ton of feed. He also established that this oligosaccharide enhanced the immune response by increasing phagocytic activity measured by luminol-enhanced chemiluminescence. Commercial turkey flocks supplemented with mannanoligosaccharide at 1 kg per ton of feed for 3 weeks followed by 0.5 kg per ton through 17 weeks had a better overall performance, decreased incidence of enteric diseases and higher titers to hemorrhagic enteritis and Newcastle disease as compared to controls (Olsen, 1995).

Trenholm *et al.* (1994) conducted an *in vitro* experiment in which Mycosorb was found to bind zearalenone up to 80% at pH 4 which was marginally higher than that at pH 3. In another *in vitro* experiment (Devegowda *et al.*, 1996) aflatoxin was prepared in aqueous solution in two concentrations (250 and 500 ppb) at two pH levels (4.5 and 6.8) to simulate *in situ* gastrointestinal tract varying pH levels. Mycosorb was added at two levels (500 and 1000 mg) and incubated for 2 hours at 37°C to determine the amount of aflatoxin binding. Regardless of the aflatoxin concentration, the binding ability of Mycosorb to toxin was more or less the same at pH 4.5 with minor variations (Figure 2). However, at pH 6.8, Mycosorb bound the toxin at higher

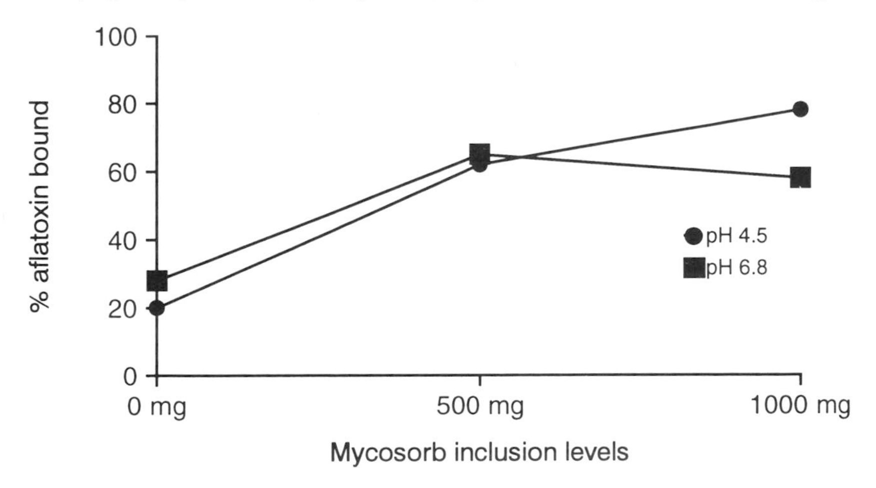

Figure 2. Effect of Mycosorb level on percentage of aflatoxin (250 ppb) bound (Devegowda *et al.*, 1996).

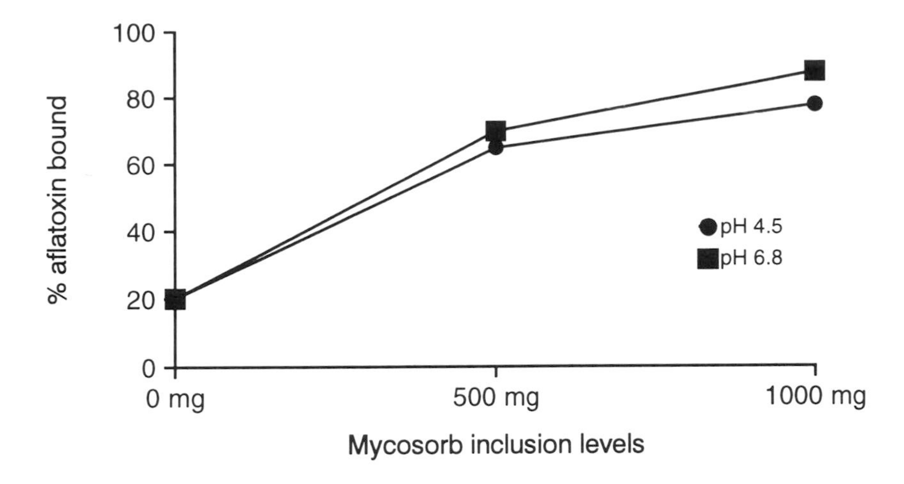

Figure 3. Effect of Mycosorb level on percentage of aflatoxin (500 ppb) bound (Devegowda *et al.*, 1996).

levels. Further, there was a gradual increase in the amount of toxin bound to Mycosorb at higher levels of inclusion (Figure 3).

Mahesh and Devegowda (1996a,b) examined the aflatoxin binding ability of Mycosorb and Novasil (a hydrated sodium calcium aluminosilicate compound) in contaminated poultry feeds and in liquid media. At the highest levels of inclusion, both products bound aflatoxin at about 80% in poultry feed (Table 5) whereas at lower levels of inclusion Mycosorb bound aflatoxin

Table 5. Comparative ability of Novasil and Mycosorb to bind aflatoxin in contaminated poultry feed *in vitro*.

	Novasil (%)			Mycosorb (%)		
Aflatoxin (ppb)	0.1	0.2	0.4	0.125	0.025	0.05
50	8*	26	54	33	58	83
100	14	47	78	48	58	69
200	25	65	78	51	62	79

* Values are percent aflatoxin bound.

Table 6. Comparative ability of Novasil and Mycosorb to bind aflatoxin in liquid media.*

	Novasil (%)			Mycosorb (%)		
Aflatoxin (ppb)	100	200	400	12.5	25	50
50	78†	81	82	76	82	84
100	80	78	85	78	82	85
200	88	82	88	79	86	87

* Mahesh and Devegowda (1996).
† Values are percent aflatoxin bound.

to a greater extent than Novasil. In the liquid media, both products, at all three inclusion levels, bound more than 70% aflatoxin (Table 6).

Conclusion

Mycotoxins continue to pose a major threat to poultry and livestock production and at low levels remain undetected although causing major economic losses by way of lowered productivity and immunosuppression. Methods adopted to counteract the effect of aflatoxin have varied from attempts at total elimination to means of minimizing deleterious effects. Application of biotechnological tools has widened the scope for keeping this toxin in check. The use of viable yeast cultures in poultry diets has given promising results and research has taken one step further by identifying the yeast cell wall as the active component which aids in counteracting aflatoxicosis. Mannanoligosaccharides provide new insights into counteracting several pathogens and toxins besides their major impact on modifying the immune response.

References

Blattberg, B. 1956. Proc. Soc. Exp. Biol. Med. 92:745.

Boulton, S.L., J.W. Dick and B.L. Hughes. 1981. Effect of dietary aflatoxin and ammonia inactivated aflatoxin on Newcastle disease antibody titres in layer breeders. Avian Disease. 26:1–4.

Corrier, D.E. 1991. Mycotoxins: mechanisms of immunosuppression. Vet. Immun. Immunopath. 30:73–87.

Devegowda, G., B.I.R. Aravind, K. Rajendra, M.G. Morton, A. Baburathna and C. Sudarshan. 1994. A biological approach to counteract aflatoxicosis in broiler chickens and ducklings by the use of *Saccharomyces cerevisiae* cultures added to feed. In: Proc. Alltech's 10th Annual Symposium on Biotechnology in the Feed Industry (T.P. Lyons and K.A. Jacques, eds) Nottingham University Press, Loughborough, Leics., UK. pp. 235–245.

Devegowda, G., B.I.R. Aravind, M.G. Morton and K. Rajendra. 1995. A biotechnological approach to counteract aflatoxicosis in broiler chickens and ducklings by the use of *Saccharomyces cerevisiae*. In: Proc. Feed Ingredients Asia '95, Singapore. 161–171.

Devegowda, G., B.I.R. Aravind and M.G. Morton. 1996. *Saccharomyces cerevisiae* and mannanoligosaccharides to counteract aflatoxicosis in broilers. In: Proc. of Australian Poultry Science Symposium, Sydney, Australia. 8:103–106.

Edds, G.T., K.P.C. Nair and C.F. Simpson. 1973. Effect of aflatoxin B_1 on resistance in poultry against caecal coccidiosis and Marek's disease. Am. J. Vet. Res. 34:819–826.

Ghosh, R.C., H.V.S. Chauhan and S. Roy. 1990. Immunosuppression in broilers under experimental aflatoxicosis. Brit. Vet. J. 146:457–462.

Hamilton, P.B. (1987). Why the animal industry worries about mycotoxins. In: Proc. Recent Developments in the Study of Mycotoxins. Kaiser Chemicals, Illinois, USA.

Kubena, L.F., R.B. Harvey, T.D. Phillips, D.E. Corrier and W.E. Huff. 1990. Diminution of aflatoxicosis in growing chickens by the dietary addition of hydrated sodium calcium alumiunosilicate. Poult. Sci. 69:727–735.

Lyons, T.P. 1994. Biotechnology in the feed industry: 1994 and beyond. In: Proc. Alltech's 10th Annual Symposium on Biotechnology in the Feed Industry (T.P. Lyons and K.A. Jacques, eds) Nottingham University Press, Loughborough, Leics., UK. pp.1–48.

Mahesh, B.K. and G. Devegowda. 1996a. Ability of aflatoxin binders to bind aflatoxin in liquid media – an *in vitro* study. In: Proc. XX World's Poultry Congress, New Delhi, India. 4:295–296.

Mahesh, B.K. and G. Devegowda. 1996b. Ability of aflatoxin binders to bind aflatoxin in contaminated poultry feeds – an *in vitro* study. In:Proc. XX World's Poultry Congress, New Delhi, India. 4:296.

Manegar, A.G., B. Umakantha and G. Devegowda. 1996. Studies on tolerance limits of aflatoxin. Indian J. Poult. Sci. (in press).

McDaniel, G. 1991. Effect of Yea-Sacc[1026] on reproductive performance of broiler breeder males and females. In: Proc. Alltech's 7th Annual Symposium on Biotechnology in the Feed Industry. Alltech Technical Publications, Nicholasville, Kentucky, USA.

McDonald, F. 1995. Use of immunostimulants in agricultural applications. In: Proc. Alltech's 9th Annual Asia-Pacific Lecture Tour. 9:61–68.

Newman, K.E. 1994. Mannan-oligosaccharides: Natural polymers with significant impact on the gastrointestinal microflora and the immune system. In: Proc. Alltech's 10th Annual Symposium on Biotechnology in the Feed Industry (T.P. Lyons and K.A. Jacques, eds) Nottingham University Press, Loughborough, Leics., UK. pp. 167–174.

Newman, K.E. 1995. Mannanoligosaccharides and animal nutrition. In: Proc. Alltech's 9th Annual Asia-Pacific Lecture Tour. 9:55–60.

Olsen, R. 1995. Mannanoligosaccharides: Experience in commercial turkey production. In: Proc. Alltech's 11th Annual Symposium on Biotechnology in the Feed Industry (T.P. Lyons and K.A. Jacques, eds) Nottingham University Press, Loughborough, Leics., UK. pp. 389–392.

Parekh, R. 1993. Carbohydrate engineering in modern drug discovery. In: The Biotechnology Report. Campden Publishing Ltd, London. 135.

Reddy, A.R., V.R. Reddy, P.V. Rao and B. Yadagiri. 1982. Effect of experimentally induced aflatoxicosis on the performance of commercial broiler chickens. Ind. J. Anim. Sci. 52:405–410.

Shalitin, C. 1989. Yeast as source of oncoproteins. Yeast. (Special issue) Vol. 5.

Sisak, F. 1994. Stimulation of phagocytosis as assessed by luminol-enhanced chemiluminescence and response to salmonella challenge of poultry fed diets containing mannanoligosaccharides. Poster presented at the 10th Annual Symposium on Biotechnology in the Feed Industry. Alltech Inc., Nicholasville, Kentucky, USA.

Southern, L.L., D.H. Baker and D.D. Schmeisser. 1984. *Eimeria acervulina* infection during aflatoxicosis in the chick. Nut. Rap. Int. 29:35–44.

Stanley, V.G., R. Ojo, S. Woldensenbet and D.H. Hutchinson. 1993. The use of *Saccharomyces cerevisiae* to suppress the effects of aflatoxicosis in broiler chicks. Poult. Sci. 72:1867–1872.

Taklimi, S.M., B.I.R. Aravind, C.V. Gowdh and G. Devegowda. 1993. Effect of live yeast culture (Yea-Sacc[1026]) on performance of broilers fed varying energy and protein levels. In: Proc. 6th Anim. Nutr. Res. Worker's Conf. Bhubaneswar, India.

Thaxton, J.P., H.T. Tung and P.B. Hamilton. 1974. Immunosuppression in chickens by aflatoxin. Poult. Sci. 53:721–725.

Trenholm, L., B. Stewart, L. Underhill and D. Prelusky. 1994. Ability of Grain Gard to bind zearelenone and vomitoxin *in vitro*. Poster presented at the 10th Annual Symposium on Biotechnology in the Feed Industry. Alltech Inc., Nicholasville, Kentucky, USA.

Tung, H.T., R.D. Wyatt, P. Thaxton and P.B. Hamilton. 1975. Concentration of serum proteins during aflatoxicosis. Toxicol. Appl. Pharmacol. 34:320–326.

Umesh, D. and G. Devegowda. 1990. Reducing adverse effects of aflatoxin through nutrition in broiler chickens. In: Proc. 13th Annual Poul. Sci. Conf. and Symp., Bombay, India.

Virdi, A.S., R. Tiwari, M. Saxena, V. Khanna, G. Singh, S.S. Saini and D.V. Vadhera. 1989. Effect of aflatoxin on the immune system of the chick. J. Appl. Toxicol. 9:271–275.

Witlock, B.R., R.D. Wyatt and W.I. Anderson. 1982. Relationship between *Eimeria adenoides* infection and aflatoxicosis in turkey poults. Poult. Sci. 61:1293–1297.

HERBS AND SPICES: THEIR ROLE IN MODERN LIVESTOCK PRODUCTION

KYLE NEWMAN

North American Biosciences Center, Alltech Inc., Nicholasville, Kentucky, USA

Animal behaviorists have observed that many animals instinctively seek out specific plants when they are sick or for other unknown reasons. Anyone who has ever owned and watched a dog or cat knows that grazing on grass usually means they have a stomach ache. By the same token, the documented effect of catnip on cats goes back to at least 1754 when the British horticulturist Philip Miller wrote an exasperated description of cats rolling in his catnip patch until it was absolutely flat. His solution was to sow the catnip rather than set it and to place thorn hedges around the garden to keep out the cats.

The basis for use of these compounds in many cases is scientifically sound. Chamomile and catnip are both known for their calming effects on smooth muscles. For this reason, they are recognized in human use for their abilities to calm an overactive stomach or as a mild sleeping aid. Cis-trans-nepetalactone is the major component of the volatile oil in these herbs and has a chemical structure similar to the valepotriates (found in the herb valerian) which are proven sedatives. Since chamomile is not a prohibited substance, it is commonly used in equine events to calm nerves without affecting performance. More and more of these herbal compounds are being scientifically investigated to remove speculation about their use. In the past 10 years, the western world has been learning what many Asians and Native Americans have known for centuries, namely that plant extracts and spices can play a significant role in health and nutrition.

Antimicrobial activities of herbs and spices

There are a number of scientific studies concentrating on the bactericidal and bacteriostatic aspects of various plants and plant extracts. Osborn (1943) found more than 60 genera of plants that demonstrated inhibitory properties toward the growth of either *Escherichia coli* or *Staphylococcus aureus* or both.

Garlic has been called the wonder drug of the herbal world. The ancient Egyptians actually worshipped garlic and fed it to their slaves to keep them strong and healthy. The Romans ate garlic for strength before battle since it was the herb of Mars, the god of war. More recently, it has been reputed to keep evil and vampires away. In Europe, it is the main ingredient of Four Thieves Vinegar, a legendary remedy that has been sold in France since early in the 18th century. According to legend, four criminals were recruited to bury the dead from a plague in Marseilles. These criminals never became ill, reportedly because they drank a mixture of crushed garlic and wine vinegar (Kowalchik and Hylton, 1987).

Certain compounds have been isolated from garlic which may help account for some of the 'magical' properties this herb seems to have. When the raw garlic clove is damaged, alliin is converted to sulfenate, pyruvate and ammonia by alliinase. Sulfenate leads to allicin (also known as diallyl thiosulfinate), which has been identified by Cavallito *et al.* (1944) as an antimicrobial compound. Another sulfur-containing compound (ajoene) has been isolated from garlic and has also been shown to exhibit broad spectrum antimicrobial properties (Table 1; Naganawa *et al.*, 1996).

In addition to the inhibitory compounds found in garlic, a number of other plants or plant compounds have been found to be inhibitory to microorganisms. Onions contain smaller amounts of the active ingredients identified in garlic and may be used for some of the same applications. As with garlic, the palatability of the material remains a concern.

Cranberry juice has been used for a number of years as an aid in combating urinary tract infections. It was thought that the benzoic acid content of cranberries was primarily responsible for this activity, although studies have found that cranberry may also prevent attachment of potential pathogens to the host. Use of cranberry extract for this application in animal feeds has not been documented, but it is thought that the tart nature of cranberry makes direct use without extraction impractical.

Table 1. Minimum inhibitory concentrations of ajoene, a garlic-derived compound, on various microorganisms.

Microorganism	Minimum inhibitory concentration (μg/ml)
Bacillus cereus	4
B. subtilis	4
Staphylococcus aureus	16
Micrococcus luteus	136
Mycobacterium smegmatis	4
M. phlei	14
Lactobacillus plantarum	19
Streptococcus sp.	56
Escherichia coli	116
Klebsiella pneumoniae	152
Pseudomonas aeruginosa	>500
Xanthomonas maltophilia	118
Candida albicans	13
Hanseniaspora valbyensis	11
Pichia anomala	11
Saccharomyces cerevisiae	12

Protozoal infections may also be prevented or slowed by herbal formulations. For years, malaria treatment in humans has been through the use of an extract from the bark of the chinchona tree (quinine). Now, quinine resistance has emerged in Southeast Asia, but a new extract from the wormwood shrub has proven an effective replacement. One problem is the fact that drug companies are slow in this development since they cannot apply for patents on a drug that is already used in traditional Chinese medicine and therefore profits cannot be guaranteed. Another setback is that the crude plant has been shown to contain a central nervous system depressant. One of the active ingredients of wormwood is thujone, which in large amounts can cause convulsions and has properties similar to tetrahydrocannabinol (THC, the active agent in marijuana). Black walnut extract is also reported to be effective for parasitic and fungal infections, although, again, proper documentation is lacking.

Yucca

On several occasions at these symposia benefits from extracts of Mojave yucca (*Yucca schidigera*) have been reported (Headon *et al.*, 1991; Cole and Tuck, 1995; Ender *et al.*, 1996). For the most part, these studies have revolved around the odor-reducing capacity of yucca extract and associated improvements in performance and animal health. Other reports have shown significant reductions in ascites mortality in broilers receiving yucca extract (De-Odorase; Menocal, 1995). In human nutrition, one report has shown that oral administration of yucca was safe and effective in treating various arthritic conditions (Bingham *et al.*, 1975). This trial, combined with anecdotal evidence, has led to the use of yucca-based products in pet and horse diets.

Microalgae

Algae are eucaryotic photosynthetic organisms which are distinguished from plants by their lack of tissue differentiation. Most traditional algal classifications include blue-green algae, but these organisms are actually prokaryotes and have been classified as cyanobacteria. The most commonly encountered of the cyanobacteria are *Spirulina* spp. For convenience purposes many investigators have lumped true algae (such as *Chorella* spp., *Scenedesmus* spp. and *Dunaliella* spp.) and cyanobacteria into the general term *microalgae*. Studies on microalgae as a feedstuff and as a source of certain chemicals are widespread and have demonstrated positive effects on health and nutrition.

Early studies with microalgae focused on their potential as a source of protein and vitamins. Spirulina contains 60–70% protein and is rich in vitamins such as B12 and ß -carotene and minerals (iron), and is one of the few sources of γ-linolenic acid (Belay *et al.*, 1993). Gamma linolenic acid (GLA) is an essential fatty acid and a prostaglandin precursor. The prostaglandin PGE1 is involved in many physiological functions including regulation of blood pressure and cholesterol synthesis. Because of this interaction, the therapeutic effects of Spirulina have been examined. In 1983, Devi and Venkataraman first reported reductions in serum cholesterol in rats. Subsequent studies in both animals and humans have confirmed these results (Kato *et al.*, 1984; Nakaya *et al.*, 1988). In the body, linolenic acid is converted to GLA by the enzyme δ-6-desaturase but this enzyme can be inhibited by saturated fats and alcohol. Spirulina has a high content of GLA (1–1.5% dry weight basis) and therefore does not require δ-6-desaturase. Together with mother's milk and oil of evening primrose, Spirulina represents the vast majority of GLA found in nature.

Other studies have focused on growth parameters, immune function, pigmentation and alterations in the intestinal microflora. In broilers, effects on growth performance are mixed. Ross and Dominy (1990) found significant growth depression when diets containing 10 or 20% Spirulina were fed (Table 2). This is in contrast to Yoshida and Hoshii (1980) who reported no adverse growth effects even at 20% Spirulina. The quality of the microalgae was implicated as the cause for the discrepancy between these studies.

Another interesting component of Spirulina and other microalgae is a high xanthophyll content. Avila and Cuca (1974) reported that diets containing Spirulina produced significantly darker yolks than diets containing the same xanthophyll concentration from marigold meal. Lower concentrations (less than 10%) of Spirulina have been shown to have a beneficial impact on egg yolk coloration with 1.0% Spirulina providing optimum pigmentation in a diet that was otherwise free of xanthophylls (Anderson *et al.*, 1991; Table 3). However, in both performance and egg yolk color tests, the method of manufacture and drying of the Spirulina may have a dramatic impact on the results observed. Freeze-drying, used by Anderson and coworkers, yielded xanthophyll contents of 5.8 g/kg compared to 0.67 g/kg xanthophylls for oven-dried material (125°C for 4 hours).

Other benefits have also been associated with microalgal dietary supplementation. Qureshi *et al.* (personal communication) found that leghorn chicks supplemented with 10,000 ppm (10 kg/tonne) *Spirulina platensis* had improved immune function compared to unsupplemented birds. Performance parameters were not affected by treatment.

In pigs, reports have demonstrated that the bitter taste led to a decrease in palatability if Spirulina level exceeded 10% of the total diet (Hintz *et*

Table 2. Effects of graded amounts of dehydrated Spirulina on performance of 3-week old cockerels.

Spirulina content (%)	Weight gain (g)	Feed:gain ratio
0	173[a]	1.99[a]
5	167[ab]	2.03[a]
10	164[b]	2.06[a]
15	165[ab]	1.97[a]
20	146[c]	2.04[a]

[a,b,c] Values in the same column with no common superscript are significantly different ($P<0.05$).
Each diet was fed to three groups of eight birds per group (Ross and Dominy, 1990).

Table 3. Average Roche egg yolk color scores of eggs from Japanese quail fed graded levels of freeze-dried Spirulina*.

Spirulina content (%)	Roche color score†
0	1.6[a]
0.25	3.5[b]
0.5	6.1[c]
1.0	8.3[d]
2.0	10.5[e]
4.0	12.8[f]

* Spirulina contained 5787 mg/kg xanthophylls.
† Optimal yolk score was determined to be between 8 and 9.
[a-f] Means in a column with no common superscript are significantly different ($P<0.05$).

al., 1966; Hintz and Heitman, 1967). In addition, early studies showed that while algae are high in protein, the cell wall is difficult to degrade and therefore only 54% digestible by the pig (Hintz *et al*., 1966). Improvements in processing provide for a more available protein source but recent trials have not been conducted.

Antihistamines and antioxidants

Herbs with antihistamine properties also have found their way into human and animal use. In the gastrointestinal tract, histamines are known to adversely affect digestion. Reishi mushroom, marjoram and willow have been directly or indirectly found to exhibit antihistamine properties. Herbal combinations with this activity have been shown to enhance cellulolytic bacterial populations, propionate concentrations and milk yields in buffaloes (Randhawa and Randhawa, 1996). Herbal formulations to improve animal production are demonstrating an efficacy in all species tested. Spices such as fennel and anise have been used successfully to stimulate intake in pigs and cattle. Additionally, since the middle ages anise tea has been used by nursing mothers to increase milk production, improve digestion and decrease flatulence. Anecdotal evidence in dairy cattle has indicated increases in milk production. Both anise and fennel oil contain 50–60% anethol and for this reason are sometimes used interchangeably.

Other spices such as rosemary have also turned up in natural foods and feeds. Preservatives such as BHA, BHT and ethoxyquin are being replaced in some cases by vitamin E and oil of rosemary. The extract of rosemary is a powerful antioxidant, although the strong odor and bitter taste have limited its use. Stephen Chang and Chi-tang Ho at Rutgers University have patented a process that produces a bland extract of rosemary which has been shown to have better stability than BHT and BHA at higher temperatures.

Herbal combinations

A specific Chinese herbal combination was recently examined for its effects on pig performance (Bourne, personal communication). The trial involved pigs initially weighing 8 kg during a 25-day period with treated pigs (5 kg herbal mixture/tonne) showing better gain, feed conversion and cost per kg of liveweight gain (Table 4). In a separate trial with pigs weaned at 21 days, the use of a formula called Chinese Herb 112 led to a 12% improvement in gain, and a 9.5% increase in intake (Table 5).

Table 4. Efficacy of a Chinese herbal formula on the post-weaning performance of pigs.

	Control	Treatment
Average initial weight, kg	7.9	8.1
Average final weight, kg	18.4	19.4
Average gain/pig, kg	10.5	11.3
Daily gain, g	421.0	452.0
Feed conversion	1.12	1.05

Table 5. Effect of Chinese Herb 112 on production parameters in weanling pigs.

	Control	Chinese Herb 112	S.E.D.	*P* value
Daily gain, g	418.6	470.3	15.2	0.004
Feed intake, g/day	650.7	713.3	20.3	0.019
Feed conversion	1.55	1.55	0.06	0.77

Conclusions

Public awareness of benefits from the use of herbs and spices is on the rise. With the demand for herbs increasing, a conscious effort to understand appropriate uses must be made. The herbal area is not without a dark side, however, and a few cautions should be exercised. Many herbs are extremely dangerous and should not be used. The similarities between comfrey and foxglove have caused more than one brush with death. A couple in Washington heard about the benefits of comfrey on arthritis and went out and picked what they thought was that plant. Instead they picked foxglove which is a source of the heart drug digitalis. Ephedra (called Ma huang in China) is another herb which can be quite dangerous but has been used as a 'health food' cold remedy. Heart palpitations, shortness of breath and other problems have moved it up the list of compounds receiving FDA scrutiny.

Since many of the drugs that are now in use were originally isolated from flora of one sort or another (penicillin from a bread mold, digitalis from foxglove, and ephedrine from ephedra), it makes sense that there are many unknown properties that can be of benefit to health and growth in animal and human use. Experience and familiarity with these herbs can eliminate any of the obvious problems that may be encountered from their use. As with any new areas of science, the mystique of the herbs lies in the unknown. There is a seemingly infinite range of possibilities that have yet to be explored with herbs and spices and the information that is becoming available will provide more tools for tomorrow's nutritionist.

References

Anderson, D.W., C.S. Tang and E. Ross. 1991. the xanthopylls of spirulina and their effect on egg yolk pigmentation. Poult. Sci. 70:115–119.

Avila, E.G. and E.M. Cuca. 1974. Effecto de la alga *Spirulina geitleri* sobre la pigmentation de la yema de huevo. Tec. ECU. Mex. 30:30–34.

Belay, A., O. Yoshimichi, K. Miyakawa and H. Shimamatsu. 1993. Current knowledge on potential health benefits of Spirulina. J. Appl. Phycol. 5:235–241.

Bingham, R., B.A. Bellew and J.G. Bellew. 1975. Yucca plant saponin in the management of arthritis. J. Appl. Nutr. 27:45–51.

Cavallito, C.J., J.S. Buck and C.M. Suter. 1944. Allicin, the antibacterial principle of *Allium sativum*. I. Isolation, physical properties and antibacterial action. J. Am. Chem. Soc. 66:1950–1951.

Cole, D.J.A. and K. Tuck. 1995. Using yucca to improve pig performance while reducing ammonia. In: Biotechnology in the Feed Industry. (T.P. Lyons and K.A. Jacques, eds.) Nottingham University Press, Loughborough, Leics. UK.Vol. 11. pp. 421–426.

Devi, M.A. and L.V. Venkataraman. 1983. Hypocholesteremic effect of blue-green algae *Spirulina platensis* in albino rats. Nutr. Rep. Int. 28:519–530.

Ender, K., G. Kuhn and K. Nurnberg. 1996. Reducing skatole in pig meat with dietary *Yucca schidigera* extract. In: Biotechnology in the Feed Industry. (T.P. Lyons and K.A. Jacques, eds) Nottingham University Press, Loughborough, Leics. UK. Vol. 12. pp. 275–280.

Headon, D.R., K.Buggle, A. Nelson and G. Killeen. 1991. Glycofractions of the yucca plant and their role in ammonia control. In: Biotechnology in the Feed Industry. (T.P. Lyons, ed). Alltech Technical Publications, Vol. 7 pp. 95–108.

Hintz, H.F., H. Heitman, Jr., W.C. Weir, D.T. Torell and J.H. Meyer. 1966. Nutritive value of algae grown on sewage. J. Anim. Sci. 25:675–681.

Hintz, H.F. and H. Heitman, Jr. 1967. Sewage-grown algae as a protein supplement for swine. Anim. Prod. 9:135–140.

Kato, T., K. Takemoto, H. Katayama and Y. Kuwabara. 1984. Effects of Spirulina (*Spirulina platensis*) on dietary hypercholesteremia in rats. J. Jap. Soc. Nutr. Food Sci. 37:323–332.

Kowalchik, C. and W.H. Hylton (eds). 1987. Rodale's Illustrated Encyclopedia of Herbs. Rodale, Press Emmaus, PA. pp. 215–217.

Menocal, J.A. 1995. Some considerations to reduce ascites syndrome in broilers. In: Biotechnology in the Feed Industry. (T.P. Lyons and K.A. Jacques, eds). Nottingham University Press, Loughborough, Leics. UK. Vol. 11. pp. 427–443.

Naganawa, R., N. Iwata, K. Ishikawa, H. Fukuda, T. Fujino and A. Suzuki.

1996. Inhibition of microbial growth by Ajoene, a sulfur containing compound derived from garlic. Appl. Environ. Microbiol. 62:4238–4242.

Nakaya, N., Y. Honma and Y. Goto. 1988. Cholesterol lowering effect of Spirulina. Nutr. Rep. Int. 37:1329–1337.

Osborn, E.M. 1943. On the occurance of antibacterial substances in green plants. Brit. J. Exp. Path. 25:227–231.

Randhawa, C.S. and S.S. Randhawa. 1996. Herbal remedies stimulate rumen function. Feed Mix. 4:31–34.

Ross, E. and W. Dominy. 1990. The nutritional value of dehydrated, blue-green algae (*Spirulina plantensis*) for poultry. Poult. Sci. 69:794–800.

Yoshida, M. and H. Hoshii. 1980. Nutritive value of Spirulina, green algae, for poultry feed. Jpn. Poult. Sci.17:27–30.

ORGANIC CHROMIUM

THE EUROPEAN PERSPECTIVE ON ORGANIC CHROMIUM IN ANIMAL NUTRITION

ARCHIMEDE MORDENTI,[1] ANDREA PIVA[1] and GIANFRANCO PIVA[2]

[1]*University of Bologna, Bologna, Italy*
[2]*Catholic University of Milano, Piacenza, Italy*

Chromium: essential trace element

To consider a mineral element as essential in human and animal nutrition one should demonstrate that purified diets, which are totally free of or deficient in that particular element, are capable of inducing a deficiency syndrome and that regression of symptoms is observed as soon as the diet is properly supplemented with the missing or deficient element.

Up until 1950 the total number of mineral elements regarded as essential was 13 and among those available in very small amounts (trace elements or micronutrients), only iron, copper, iodine, manganese and cobalt were recognized as such. In 1953 molybdenum was added to the list, in 1957 selenium and in 1959 chromium (McDonald *et al.*, 1992).

Schwartz and Mertz first demonstrated in 1957 that yeast contained a substance capable of increasing glucose uptake and improving the efficiency of insulin utilization when it was compromised. This substance was called GTF (glucose tolerance factor). Subsequently the same investigators identified chromium as the GTF active ingredient. Apart from chromium, the complex called GTF also includes nicotinic acid, glycine, glutamic acid and cysteine. Chromium was found to occur in appreciable amounts in beer yeast as well as in many other types of food for human and animal consumption, such as liver, kidney, pepper, spinach, unrefined sugar, whole oat meal, etc. The chromium content of all refined products such as sugar, cereals, bread, etc. is always markedly lower than that of the raw materials since the 'refining' process removes up to 80% of the naturally occurring chromium (Burton, 1995).

The fact that chromium is an essential element was first demonstrated in rats by Schwartz and Mertz (1959) and then confirmed in man in 1977 (Jeejeebhoy *et al.*, 1977) with studies on a woman who had been on parenteral nutrition for five years. The patient became diabetic with marked glucose intolerance, weight loss and altered nerve pulse conduction. Insulin treatment was not effective and it was only after adding 250 μg of chromium in the form of chloride that the patient was relieved of diabetic symptoms and required no more insulin treatment.

Several investigations have shown that chromium interacts with glucose and (or) lipid metabolism in cats, monkeys, guinea pigs, rabbits, squirrels, poultry, turkeys, pigs, calves and humans (Anderson, 1988).

GTF, of which chromium is an essential component, is indispensable for normal sugar, protein and lipid metabolism. Furthermore, chromium favours the interactions between insulin and its specific receptors located in target organs such as muscles and fat tissue (Moorandian and Morley, 1987). This explains the enhanced trophic effect of GTF on insulin. When insulin binds to its specific receptor, cellular uptake of circulating glucose and amino acids is facilitated. The sugar is subsequently used as a source of energy which, under hormonal effects (GH, IGF-1) favours protein synthesis and, consequently, muscular development. The symptoms caused in rats by chromium deficiency are summarized in Table 1.

Chromium requirements (recommended daily dose) for adult humans ranges, according to American guidelines, between 50 and 200 μg. Considering that in modern nutrition we often resort to refined products, it is estimated that 90% of American adults consume chromium-deficient diets (supplying only 40–60% of the requirement) with average daily intake of about 25 μg (Anderson, 1987; Anderson *et al.*, 1992). Deficiencies may result even if 33 μg/day of chromium as suggested by the World Health Organization (WHO, 1996) are consumed.

Chromium requirements, on the other hand, would seem to increase in humans and in animals as a consequence of factors generally described as 'stressors' such as fatigue, trauma, pregnancy and several forms of nutritional (high glucose diets), metabolic, physical, environmental and also emotional stress (Burton, 1995). During stress there is increased cortisol secretion which acts as an insulin antagonist by both increasing plasma glucose concentration and reducing glucose utilization in peripheral tissues (especially muscle and fat). The increase in blood glucose stimulates mobilization of chromium reserves which are then irreversibly excreted through the urine (Borel *et al.*, 1984; Mertz, 1992). All factors which can induce stress increase chromium excretion via the urine. This explains the fact that typically all factors which favour the preservation of high blood glucose levels or, more generally, high

Table 1. Chromium deficiency symptoms in rats*.

Inability to metabolize carbohydrates normally
Decreased sensitivity of peripheral tissues to insulin
Impaired protein metabolism
Reduced growth rate
Reduced longevity[†]
Elevated serum cholesterol
Increased aortic plaques
Corneal lesions
Reduced sperm count and fertility

* Lindemann (1996a).

[†] 1000 ppb of chromium as picolinate increases the longevity in rats by 25% (Evans and Meyer, 1994).

insulin, create predisposing conditions for the onset of chromium deficiency (Anderson *et al.*, 1990).

Chromium oxidation states and toxicity

The form of chromium used as a supplement is very important. Anderson *et al.* (1993) added nine different chromium forms to a rat diet and found that for some forms, including chromium chloride, there is very little tissue uptake. This variability changes greatly depending on the species of chromium assumed while the total chromium content in the diet provides no information as to the availability of chromium. Chromium is theoretically available in all oxidation states from −2 to +6, but is most frequently found in 0; +2; +3 and +6 (Ducros, 1992) of which the first (metal state) is biologically inert.

Bivalent chromium (Cr^{2+}). On exposure to air, Cr^{2+} oxidizes and changes into Cr^{3+}. For this reason, in biological systems chromium is not available in the bivalent form.

Hexavalent chromium (Cr^{6+}). In this state, chromium is bound to oxygen in the form of chromates (CrO_4^{2-}) or bichromates ($Cr_2O_7^{2-}$) with strong oxidizing properties. Hexavalent chromium easily crosses biological membranes and once inside the cell reacts with protein components and nucleic acids and undergoes reduction to Cr^{3+}. The reaction with genetic material is the basis of the carcinogenic properties of hexavalent chromium.

Trivalent chromium (Cr^{3+}). Trivalent chromium is the most stable oxidation state in which chromium is found in living organisms. It does not easily cross cell membranes (Mertz, 1992) and is very unreactive. This accounts for the substantial bioactive difference from hexavalent chromium. Furthermore, the latter is easily and rapidly reduced to Cr^{3+}. For this reason the hexavalent form is not available in tissues and biological products. In addition, the trivalent form shows a strong tendency to form coordination compounds with poor reaction capacity. This shows that in living organisms chromium preferably plays structural roles rather than performing active functions of the enzymatic type (Mertz, 1992).

In the analysis of chromium, special attention has been paid to the comparison between Cr^{3+} and Cr^{6+}. While hexavalent chromium has long been thought to have toxic properties, the trivalent form is regarded by many investigators as harmless and even beneficial if used in adequate amounts (WHO, 1996). Both forms of toxicity (acute and chronic) are therefore typically ascribed to hexavalent chromium which is found to be irritating, carcinogenic, antigenic and corrosive as it is absorbed at the intestinal and pulmonary level and in some circumstances also by the skin. On the other hand, based on several experimental findings it is now commonly accepted that trivalent chromium has very low toxicity (it is virtually non-toxic) especially if administered in the form of organic compounds. More specifically, chromium provided as chromium yeast would be by no means toxic (Flodin, 1988).

The useful/toxic dose ratio for Cr^{3+} is of the order of 1:10,000. From this perspective, chromium is therefore a highly safe trace element (Mowat *et al.*, 1995). In fact, Cr^{3+} toxicity is lower than all other listed minerals such as copper, iodine, zinc, manganese, and especially, selenium (Lindemann, 1996a). It would in any case be advisable to assess also the potential risks for humans and animals following inhalation or contact.

Chromium absorption and excretion

Chromium may be available in the diet as an inorganic compound or in the form of organic complexes. Elementary chromium, available following release or contamination, is virtually unabsorbed and has no nutritional value (Ducros, 1992). Hexavalent chromium reaches man and animals mainly through inhalation or industrial contamination.

At the intestinal level, inorganic Cr^{6+} is absorbed much more easily than Cr^{3+} which is only scarcely utilizable. In fact in humans, and most likely also in animals, the absorbed percentage of Cr^{3+} is always poor and shows an inverse relationship to the amount consumed ranging from 2% in case of a 2 μg daily intake to 0.5% if the daily amount is 40 μg (Flodin, 1988). The average amount excreted via the urine, if the daily Cr^{3+} intake is 38 μg, is in the order of 0.3 μg/day.

Trivalent chromium tends to accumulate not only in external epidermal tissues (hair, etc.) but also in liver, kidney, skeleton, spleen, lung and large intestine. Accumulation in other tissues, especially muscle, would be very limited or virtually absent (Wallach, 1985).

In any case, 98% of the inorganic chromium consumed by man with food would not be absorbed and would thus be eliminated with the faeces (Offenbaucher *et al.*, 1986). On the contrary, organic chromium bioavailability seems to be much higher, of the order of 25–30% (Mowat, 1994). The presence of amino acids, ascorbic acid, high level of sugars, oxalate and aspirin in the diet enhances chromium absorption while phytate and anti-acids seem to reduce chromium in blood and tissues (Hunt and Stoecker, 1996).

Chromium excretion, mainly via the urinary tract, may increase 10- to 300-fold in stress situations or if dietary sugar content is increased. This also becomes nutritionally important because in such conditions it is necessary to increase the trace element concentration in the diet.

Effects of chromium on animal performance

Experimental evidence of response to chromium has been reported in a broad range of scientific literature. In practical breeding conditions very often environmental, nutritional, social and metabolic *stresses* (early weaning, intensive farming, trauma, transport, advancing age, excess energy and protein, thermal fluctuations, high production, pregnancy, etc.) depress natural defence mechanisms and favour the onset of the so-called induced diseases which are the expression of metabolic or infectious disorders (immunodepression). Several studies recently carried out in Canadahave showed that the addition of organic chromium to the diet significantly improved the growth of stressed calves and reduced the incidence of respiratory diseases and the need for antibiotic treatment (Mowat, 1994). An important finding was that the addition of chromium was ineffective in calves treated with antibiotics (Chang and Mowat, 1992). In dairy cows, improved reproductive efficiency was also observed (Mowat *et al.*, 1995), together with reduced incidence of ketosis, improved body condition score, higher milk yield and better immune response (Burton *et al.*, 1993).

When fed to slaughter animals, chromium may change carcass composition. Experiments carried out by Page *et al.* (1993) showed that in pigs there is increased muscular mass and reduced backfat thickness (Table 2). Mooney and Cromwell (1995) have recently found added organic chromium in diets for growing and fattening pigs increased the percentage of carcass muscle and at the same time reduced carcass fat. In heavy pigs, chromium yeast was found to be effective in improving feed efficiency, average daily gain and lean cuts (Savoini *et al.*, 1996). Organic chromium had analogous results when fed to rabbits, reducing fat deposition in the carcass and increasing feed intake (Tucci and D'Imperio, 1996). In humans, especially in athletes, administration of chromium picolinate seems to increase the development of skeletal muscles (Evans and Pouchink, 1993), thus confirming the trophic effect of chromium in subjects undergoing considerable physical fatigue and stress.

Positive findings have also been reported in relation to pig reproductive efficiency with chromium supplemented at a rate of 200 ppb for six to seven months prior to the beginning of the reproductive evaluations (Table 3). In a large study conducted in Australia, involving several hundred sows, chromium supplementation reduced abortion, natural sow mortality, sows returning to oestrus and open sows at the time of farrowing (Campbell, 1996). Evans and Mayer (1994) also observed positive effects on longevity in rats.

Table 2. Carcass response in swine supplemented with chromium picolinate*.

	Chromium (ppb)		Response	
	0	200	Absolute	Relative (%)
Experiment 1[†]				
Loin muscle area, cm^2	34.9	37.2	2.3	6.6
Tenth rib backfat depth, cm	2.83	2.44	0.39	13.8
Experiment 2				
Loin muscle area, cm^2[a]	34.0	39.9	5.9	17.4
Tenth rib backfat depth, cm[a]	3.15	2.63	0.52	16.5
Experiment 3				
Loin muscle area, cm^2[b]	31.5	38.4	6.9	21.9
Tenth rib backfat depth, cm[b]	3.07	2.39	0.68	22.1

* As cited by Lindemann (1996b).

† Adapted from Page *et al.* (1993); initial weight of the pigs was 37.8, 30.5 and 22.4 kg, respectively for experiments 1, 2, 3.

a Means differ (P<0.05).

b Means differ (P<0.01).

Table 3. Effects of organic chromium on litter size in swine*.

	Organic chromium, ppb	
	0	200
Trial 1		
Litter size, n		
Total born	9.6	11.8
Born live (see below)	8.9	11.2
Day 21	8.2	10.3
Litter weight, kg		
Total born	13.8	17.0
Born live	12.9	16.3
Day 21	46.5	54.6
Trial 2		
Litter size n°		
Total born	10.7	11.3
Born live (see below)	9.6	10.5
Day 21	9.0	9.5
Litter weight, kg		
Total born	16.4	16.8
Born live	15.4	16.1
Day 21	50.4	54.7
Born live by parity		
Parity 1	9.2	9.6
Parity 2	9.7	10.5
Parity 3	10.4	12.4
Combined	9.6	10.5

[a] Means differ ($P<0.05$).
[b] Means differ ($P<0.10$).
* Adapted from Lindemann *et al.* (1995) and Lindemann (1996b).

It has been suggested that chromium, especially if fed in the form of organic compounds, is beneficial in diets fed to race horses (Pagan *et al.*, 1995).

Because of its effects on carcass composition, nitrogen retention (Page *et al.*, 1993; Wang, 1995) and sow productivity (Lindemann, 1996b) organic chromium can significantly reduce pollution through nitrogen savings. In summary, dietary chromium intake may prove to be beneficial in those cases where marginal deficiencies are likely to develop (Mowat, 1994), in other words when the following conditions are encountered:

- high performing animals;
- high-energy but also high-protein diets;
- working animals (racehorses, etc.);
- early weaning;
- stressed animals (due to transportation, etc.);
- unfavourable environmental conditions (high temperatures).

More generally, stressed animals are likely to benefit from chromium supplementation. A comprehensive overview of these issues has recently appeared (Lindemann, 1996b).

Chromium in food and feedstuffs

The amount of chromium (almost always trivalent) contained in human food and animal feedstuffs is highly variable and depends on the following factors:

- nature and origin of nutritional sources;
- technological processing that the food or feedstuff has undergone;
- type of specific environmental contamination;
- composition of the soil (in the case of plants) and fermentation substrate on which microorganisms (more specifically, yeasts) are grown and produced.

All the above factors reduce reliability of dietary chromium content calculations made on the basis of tables in reference books for either humans or animals. In spite of this, certain indications can be provided to gain some idea of actual chromium intake with different types of foods or feedstuffs. However it should be stressed that only a very small proportion of dietary chromium (typically below 25%) is bioavailable and thus utilizable by man or animals.

CHROMIUM IN HUMAN FOODS

Chromium in human diets should not exceed the threshold value of 500 µg/kg (0.5 ppm). Chromium content of cereals and potatoes has been measured at 0.1 ppm though in some corn samples values exceeding 20 ppm were found. Apart from a few exceptions, data from various sources agree that cereal seeds contain little chromium via uptake from soil in comparison with forages (Adrian, 1991).

Yeasts are rich in chromium but have an·extremely variable concentration of this trace element depending on the nature of the fermentation substrate. They are, in any case, one of the richest nutritional sources of bioavailable chromium since values of 1 ppm of dry matter are often exceeded (i.e. 10 times the content found in cereals), occasionally reaching even 5 ppm.

Meat and its derivatives have a relatively constant chromium content (variations due to its presence in the feedstuff are only minor) while in the organs (kidney, liver, heart, lung, spleen, etc.) there is strong variability partly due to the different availability of this element in the diet. In the present conditions one may expect that slaughter products meant for human consumption have a low and relatively constant chromium content, on average about 0.15 ppm.

The chromium content of milk tends to be lower during lactation. It is more abundant in colostrum and then decreases over time. This is also true for human milk. Chromium in milk is mainly bound to its fat components. For this reason 'skimmed' products have less chromium than whole milk, butter or cheese (Adrian, 1991). The average chromium content in milk is of the order of 0.015 ppm. This low concentration is due to the fact that the mammary gland performs an effective filter function thus limiting the transfer of the element from the blood to the mammary secretions, i.e. the milk. Nevertheless the chromium content in the human milk is about 670 μg/litre (Ballarini, 1996).

In the edible portion of the egg chromium content ranges between 0.005 and 0.02 ppm, but in the yolk it reaches concentrations from two to 20 times higher. This confirms that in laying hens, as is the case for dairy cows, chromium transfer is mainly a function of the lipid fraction.

In fish chromium concentrations are extremely variable depending on the possibility that environmental (industrial) pollutants accumulate in marine organisms.

In any case, based on recent investigations human diets tend to be chromium deficient (Anderson *et al.*, 1992), especially if one considers certain stress conditions and special diets. Over the years the reported concentrations of chromium in food, serum and urine show values lower by three orders of magnitude (Hunt and Stoecker, 1996). The lower values reflect improvements in analytical instruments and greater attention to possible chromium contamination in sample collection and analysis and to the use of more purified diets.

CHROMIUM IN ANIMAL FEEDSTUFFS

As far as feedstuffs for animal nutrition (forage and feedstuffs) are concerned, the variability in chromium content is much higher both in relation to raw material origin (it is well known that chromium-rich by-products are quite often used) and to soil chromium content in the soil where forage is grown. Thus, based on the findings obtained by Italian investigators the chromium content of forages would range from trace amounts to levels of as much as 5 or 10 ppm on a dry matter basis (100 times the average chromium content in cereals!) (Table 4).

Also Bonomi *et al.* (1997) found that chromium content in hay samples in a 70-farm survey varied (Table 5) and did not always correspond to the theoretical needs of dairy cattle. The same authors observed more severe sanitary problems in farms with lower level of chromium in forages. These findings are consistent with the high content of some hay meal products available in the marketplace.

Chromium content of by-products from the fish industry are on average markedly higher than in the organisms that they originate from. Fish meals

Table 4. Chromium content of forages from Umbria and Marche*.

	Cr (mg/kg dry matter)
Corn silage	1.4
Grass hay	4.8
Oat hay	1.2
Alfalfa hay, first cut	1.7
Alfalfa hay, second cut	5.0
Sainfoin hay	9.5

* Begliomini *et al.* (1979).

Table 5. Chromium content of the Parma district hays*.

	Location		
	Plain	Hill	Mountain
Number of samples	30	20	20
Cr, mg/kg DM±S.D.	0.72±0.12	0.68±0.19	0.75±0.14

Bonomi *et al.* (1997).

are obtained not only from the whole carcass but also from offal (head, bones, intestines, etc.) where chromium is contained in much higher amounts.

Among the raw materials that can be used for animal nutrition special mention should be made of the lysates obtained from the by-products of leather processing. Average chromium content is close to 50 ppm, but without preliminary treatment (Montoneri *et al.*, 1994) can reach levels of 4% (40,000 ppm).

The chromium content of mineral feedstuffs is also highly variable. In phosphates, for example, this micronutrient can range from approximately 60 to 500 ppm, with average values around 200 ppm and minimum and maximum values of 65 and 538 ppm, respectively (Sullivan *et al.*, 1994) (Table 6). The bioavailability of chromium originating from phosphates is very limited. Chromium content of zeolites is also high (50–100 ppm) though biological availability is limited.

This trace element is also found in molasses at quite high levels of approximately 2 ppm. This could also account for the high chromium content of some yeasts grown on this substrate.

Under normal conditions, the chromium content of raw materials used may vary from 0.1 mg/kg in cereals to 200 mg/kg and more in phosphates. As a consequence, complete or complementary feedstuffs may show markedly different chromium levels depending on the type of raw materials used. It is most likely that total chromium levels are lower than:

- 1 mg/kg in diets which do not contain protein hydrolysates, phosphates, hay meals or yeasts;
- 5 mg/kg in diets containing phosphates and (or) yeasts and (or) hay meals but less than 10% ash;
- 10 mg/kg in diets containing phosphates and (or) yeasts, protein hydrolysates, hay meals but between 10 and 20% ash;
- 25 mg/kg in low-dose premixes where chromium in individual raw materials may be available even at very high concentrations.

The abundant scientific literature on trace elements shows that Cr^{3+} absorption is very low and, as previously reported, reaches values of only 0.5–3.0% of the intake if administered in the inorganic form. Chromium available in organic complexes (chromium yeasts or bound to compounds like proteinates, orotate, picolinate, nicotinate or again available in feedstuffs of animal origin) would seem to be more readily absorbed. In any case absorption would be inversely proportional to the amount consumed (Flodin, 1988).

As specified above, chromium tends to build up in the external epidermal adnexa (hair, bristles, etc.) and also in kidney, liver, skeleton, spleen and lung;

Table 6. Levels of chromium in nine US mono-di-calcium phosphates used in animal nutrition*.

Sample	Cr (mg/kg)
1	82
2	194
3	74
4	85
5	538
6	65
7	493
8	200
9	96
Mean	203
S.E.M.	61

* Sullivan *et al.* (1994).

on the contrary, accumulation in muscles and, therefore, meat seems to be very limited or indeed nonexistent (Wallach, 1985). It seems most reasonable to suggest that the chromium-rich diets used in animal breeding do not involve any risks for consumers.

Chromium requirements

In a large well-documented and thorough study carried out by WHO (1996) and in a recent review of Hunt and Stoecker (1996), chromium in trivalent form was regarded as an essential nutrient. It was also suggested that man's specific chromium requirements range between 33 and 200 µg/day with a recommended daily allowance of about 75 µg/day.

Being an essential trace element, chromium would seem to behave like other micronutrients performing different activities depending on the amount fed. At low doses it is thought to play a useful biological role (meeting specific requirements), at higher rates it might have some pharmacological effects while at very high concentrations it could be toxic (Venchikov, 1974). Compared to many other approved trace elements (especially selenium), the toxic levels of chromium are much higher than the useful ones thus demonstrating that Cr^{3+} toxicity is really very low.

This information suggests that in the case of chromium one cannot exclude pharmacological effects, although this phenomenon should be related to dosage and be conceptually extended to all trace elements (iron, zinc, cobalt, copper, iodine, manganese, molybdenum and selenium) which are regularly used as additives in animal feedstuffs.

Regulatory aspects

Hexavalent chromium is taken into consideration due to its toxicity and as such Cr^{6+} contamination of feed must be below 500 µg/kg feed. In contrast,

Cr^{3+} has a very low toxicity and it is used in the organic form to compensate chromium deficient diet in human nutrition. More often Cr^{3+} is included in athlete diets or in diet for people subjected to stress. The wide consumption of refined processed products (sugar, bread, etc. with low chromium content due to processing) enhances the need for chromium supplementation in order to reach recommended daily allowances for adults of 70–75 µg of Cr^{3+}.

In Europe, mineral supplements for human consumption are available over-the-counter and are designed to satisfy RDAs. To market supplements containing chromium for humans (including chromium yeast) the present regulation requires the producer to submit to authorities the label when the product is put on the market (Ballarini, 1996).

Today a broad and constantly increasing utilization of organic chromium in animal nutrition occurs in numerous non-European and European (Switzerland) countries with regulations similar to those for human nutrition. In the European Union (EU), on the other hand, and particularly in Italy, any supplement to farm animal diets of additives not specifically authorized ('positive list') is forbidden. Chromium supplements for animal feed are therefore not allowed. Whenever the farmer or the feed manufacturer may suspect a deficiency of chromium, the only option available is to use authorized feedstuffs naturally rich in chromium.

Trivalent chromium:

- is a trace element for man and animals;
- is poorly absorbable and non-toxic (safety limit 1:10,000);
- is often deficient in western diet rich in refined and purified foods;
- does not accumulate in animal products (meat, milk and eggs);
- contributes to limit the environmental impact of animal breeding.

It is not clear why this nutrient even in Europe should not be included among feed additives for animals as well as it is for man.

Conclusions

Chromium is an essential element in animal nutrition as it is in human nutrition. Requirements are variable depending on individual (genetic, age, sex, physiology, production, health status, etc.) and environmental aspect (stress, diet, raw material quality, etc.), that influence digestion and metabolism. Natural sources of Cr^{3+} are extremely variable due to type, source and processing of raw materials before being fed to animals. As a consequence, deficiency may arise in different species and stages of production and become evident showing:

- higher stress sensitivity;
- immune system depression;
- decrease in reproductive efficiency;
- decrease in productive efficiency (growth, milk, egg production, feed efficiency);
- decreased production quality (lower lean cuts);
- increased pollution load due to nitrogen.

The low chromium toxicity is well documented and indeed lower than almost all the other trace elements already authorized. There is no reason why the EU should not authorize its use in animal production as already happens in other non-EU countries.

It may be hypothesized that a very limited increase in chromium content of animal products is likely to occur. The scientific literature agrees that following the widespread use of refined foods and hyperglycaemic diets in the United States and western countries, approximately 90% of the population is exposed to possible chromium deficiencies. The type of dietary chromium is the factor which determines its utilization efficiency. The toxicity of trivalent chromium, the form commonly available in feedstuffs, is very low and does not imply any risk for animal or human health or for the environment.

References

Adrian, J. 1991. Le chrome des produit alimentaires: origines et teneurs. Science des Aliments 11:417–464.

Anderson, R.A. 1987. Chromium. In: Trace elements in human and animal nutrition. (Mertz W., Ed.) (5th Ed.) 225–244. Academic Press, Inc. New York, NY.

Anderson, R.A. 1988. Chromium. In: Trace Minerals in foods. (K. Smith, Ed.). Marcel Dekker Inc., New York, NY. pp. 231–247.

Anderson, R.A., N.A. Bryden, M.M. Polansky and S. Reiser. 1990. Urinary chromium excretion and insulinogenic properties of carbohydrates. Am. J. Clin. Nutr. 51, 864–868.

Anderson, R.A., N.A. Bryden and M.M. Polansky. 1992. Dietary chromium intake. Biol. Trace Element Res. 32:117–121.

Anderson, R. A., N.A. Bryden and M.M. Polansky. 1993. Form of chromium affects tissue chromium concentration. FASEB J. 7: A204.

Ballarini, G. 1996. Il cromo nell'alimentazione animale. Quaderni ASSALZOO 64/96, 5–42.

Begliomini A., F. Camilli, M. Morcellini and A. Morozzi. 1979. Composizione chimica e valore nutritivo delle principali foraggere coltivate in Umbria e grado di contaminazione delle stesse da pesticidi e metalli pesanti. Zoot. Nutr. Anim. 5:495–500.

Bonomi A., E.M. Zambini, B.A. Bonomi and G. Guareschi. 1997. Ricerche ed osservazioni sul contenuto di cromo nei foraggi prodotti nell'agro parmense. In press.

Borel J.S., T.C. Majerus, M.M. Polansky, P.B. Moser and R.A. Anderson. 1984. Chromium excretion of trauma patients. Biol. Trace Element Res. 6:317–326.

Burton, J.L. 1995. Supplemental chromium; its benefits to the bovine immune system. Animal Feed Sci. Technol. 53:117–133.

Burton, J.L., B.A. Mallard and D.N. Mowat. 1993. Effects of supplemental chromium on immune response of periparturient and early lactation dairy cows. J. Anim. Sci. 71:1532–1539.

Campbell, R.G. 1996. The effects of chromium picolinate on the fertility and fecundity of sows under commercial conditions. Proc. of the 16th Annual Prince Feed Ingredient Conference. Quincy, Ill.

Chang, X. and D.N. Mowat. 1992. Supplemental chromium for stressed and growing feeder calves. J. Anim. Sci. 70:559–565.

Ducros, V. 1992. Chromium metabolism. Biol. Trace Element Res. 32:65–77.

Evans, G.W. and L.K. Mayer. 1994. Life span is increased in rats supplemented with a chromium-pyridine 2 carboxylate complex. Adv. Sci. Res. 1 (1):19–22

Evans, G.W. and D.J. Pouchink. 1993. Composition and biological activity of chromium-pipidine-carbonilate complexes. J. Inorg. Biochem. 49:117–187.

Flodin, N.W. 1988. Pharmacology of micronutrients, 14 Chromium, (N.W. Flodin, Ed.) 20:247–254.

Hunt, C.D. and B.J. Stoecker. 1996. Deliberations and evaluations of the approaches, endpoints and paradigms for boron, chromium and fluoride dietary recommendations. J. Nutrition. – Supplement – R.D.A. Workshop, 2441S-2451S.

Jeejebhoy, K.N., R.C. Chu, E.B. Marliss, G.R. Greenberg and A. Bruce-Robertson. 1977. Chromium deficiency, glucose intolerance and neuropathy reversed by chromium supplementation in a patient receiving long-term total parenteral nutrition, Am. J. Clin. Nutr. 30:531–538.

Lindemann, M.D. 1996a. Novel use of biotechnology in farm animal nutrition. In European and Southern African Lecture Tour, Alltech Inc., pp. 35–53.

Lindemann, M.D. 1996b. Supplemental chromium may provide benefits but cost must be weighed. Feedstuffs Dec.23, 68 (52), 14–17.

Lindemann, M.D., C.D. Wood, A.F. Harper, E.T. Kornegay and R.A. Anderson. 1995. Dietary chromium picolinate additions improve gain/feed and carcass characteristics in growing finishing pigs and increase litter size in reproducing sows. J. Anim. Sci. 73:457–465.

McDonald, P., R.A. Edwards and J.F.D. Greenhalg. 1992. Nutrizione Animale. Tecniche Nuove, Milano, Italy.

Mertz, W. 1992. Chromium: history and nutritional importance. Biol. Trace Element Res. 32:3–8.

Montoneri, E., G. Rizzi, A. Rizzi, A. Mordenti, A. Bauli, M. Riolfatti and L. Pellegini. 1994. Hydrolysis of tanner wastes to protein meal for animal feedstuffs: a process and product evaluation. J. Chem. Tech. Biotech. 59:91–99.

Mooney, K.W. and G.L. Cromwell. 1995. Effects of dietary chromium picolinate supplementation on growth, carcass characteristics and accretion rates of carcass tissues in growing-finishing swine. J. Anim. Sci. 73:3351–3357.

Moorandian, A.D and J.E. Morley. 1987. Micronutrient status in diabetes mellitus. J. Clin. Nutr. 45:877–885.

Mowat, D. N. 1994. Cromo organico, nuovo supplemento minerale per gli animali stressati. In: 8th European Lecture Tour, Alltech Inc., pp. 45–56.

Mowat, D.N., A. Subijatno and W.Z. Yang. 1995. Chromium deficiency in first parity cows. In: Proc. 11th Alltech Annual Symposium. (Eds. T.P. Lyons, K. Jacques), Nottingham University Press, Loughborough, U.K.

Offenbacher, E.G., H. Spencer, H.J. Dowling and F.X. Pi-Sunier. 1986. Metabolic chromium balances in men. Am. J. Clin. Nutr. 44:77–82.

Pagan, J.D., S.G. Jackson and S.E. Duran. 1995. The effect of chromium supplementation on metabolic response to exercise in thoroughbred horses. In: Proc. 11th Alltech Annual Symposium. (Eds. T.P. Lyons, K. Jacques), Nottingham University Press, Loughborough, UK.

Page, T.G., L.L. Southern, T.L. Ward and D.L. Thompson, Jr. 1993. Effect of chromium picolinate on growth and serum and carcass traits of growing-finishing pigs. J. Anim. Sci. 71:656–662.

Savoini G., C. Sgoifo Rossi, D. Cevolani, F. Polidori and V. Dell'Orto. 1996. Utilizzazione di lievito coltivato su substrato contenente cromo nella dieta del suino pesante. Riv. Suinicoltura 37 (4):145–149.

Sullivan T.W., J.H. Douglas and N.J. Gonzales. 1994. Level of various elements of concern in feed phosphates of domestic and foreign origin. J. Poultry Sci. 73:520–528.

Shwartz, K. and W. Mertz. 1957. A glucose tolerance factor and its differentiation from factor 3. Arch. Biochem. Biophys. 72:515–518.

Shwartz, K. and W. Mertz. 1959. Chromium (III) and the glucose tolerance factor. Arch. Biochem. Biophys. 85:292–295.

Tucci, F. and V. D'Imperio. 1996. Il ruolo del lievito di cromo nella vita del coniglio. Large Anim. Rew. 2 (4):79–81.

Venchikov, A. 1974. Zones of display of biological and pharmacological action of trace elements. In: Trace elements metabolism in animals, II, (Eds. W.G. Hoekstra, H.E. Gantherl and W. Mertz.). University Park Press, Baltimore, U.S.A.

Wallach, S. 1985. Clinical and biochemical aspects of chromium deficiency. J. Am. Coll. Nutr. 4:107–120.

Wang, Z. 1995. Influence of supplemental chromium picolinate on nitrogen balance, dry matter digestibility and leanness in growing-finishing pigs. M.S. thesis. Virginia Polytechnic and State University.

World Health Organization, Geneva. 1996. Gli oligoelementi nella nutrizione e nella salute dell'uomo. Schede informative 3–4/96, 116–120.

EFFECTS OF SUPPLEMENTAL TRIVALENT CHROMIUM ON HORMONE AND IMMUNE RESPONSES OF CATTLE

BONNIE A. MALLARD and PETER BORGS

Department of Pathobiology, Ontario Veterinary College, University of Guelph, Guelph, Ontario, Canada

Introduction

For more than three decades, chromium III (Cr) has been recognized in humans as an essential micronutrient, required for the normal metabolism of carbohydrates, proteins and lipids (Mertz, 1969; Schwartz and Mertz, 1959). Cr was first reported as the active component of glucose tolerance factor (GTF) in 1959, and as such is required to potentiate the action of insulin in controlling the rate of glucose entry into cells (Schwartz and Mertz, 1959). The exact structure of GTF remains elusive, but it is thought to be composed of nicotinic acid-Cr-nicotinic acid molecules with glutamic acid, glycine, and cysteine ligands (Lyons, 1995). The exact mechanism(s) by which GTF-Cr potentiates insulin activity is also still unknown but may involve pre- and (or) post-receptor insulin binding events (Kahn, 1978). Experiments conducted by Mooradian and Morley suggested that GTF-Cr may act by enhancing the binding of insulin to its receptor, although this has not been confirmed (Mooradian and Morley, 1987). More recently, in domestic animals dietary Cr has been shown to assume an equally important role, not only in glucose metabolism, but as an immunomodulatory agent affecting health and performance, particularly during periods of intensive production or performance (Burton *et al.*, 1994a, 1996; Wright *et al.*, 1995; Chang *et al.*, 1995; Wenk, 1995; Pagan *et al.*, 1995; Subiyatno *et al.*, 1996; Ward *et al.*, 1993; Hossain, 1995). As such, the successful development of Cr as a commercial feed supplement should provide livestock managers with a new tool to aid production and defend against stress-mediated infections which threaten performance and health.

Hormonal responses and chromium supplementation

Numerous studies with humans (Anderson *et al.*, 1991), pigs (Wenk, 1995), horses (Pagan *et al.*, 1995) and cattle (Subiyatno *et al.*, 1996) have confirmed the ability of supplemental Cr to alter glucose tolerance and insulin resistance. However, some confusion in understanding the action of Cr has arisen because of various definitions given to insulin resistance and its components. According to Kahn (1978), insulin resistance is a generic term which describes the state in which normal concentrations of insulin produce a less than normal biological response. Others (Berson and Yalow, 1970) have defined insulin resistance as the condition in which greater than normal amounts of insulin are required to produce a normal response, but this definition limits usefulness in that the state of resistance is defined by greater than normal amounts of insulin and implies that a normal response can actually be achieved, which is not always the case (Kahn, 1978). This distinction is important because biological responses to insulin, or any hormone, are composed of two separate components. The first is responsiveness which is determined by the maximum response that can be achieved at very high concentrations of insulin (indicated as the response plateau in Figure 1A). The second component is sensitivity to insulin and is usually characterized as the insulin concentration required to produce half the maximum response (i.e. the higher the concentration of insulin required to reach the half-maximal response, the lower the sensitivity, Figure 1B). Furthermore, the shape of the dose-response curve which measures biological activity against insulin concentration can also convey information about the different forms of insulin resistance. For instance, increased responsiveness alone would be the case in which the response plateau is higher, but the half-maximal response remains unaltered. Conversely, sensitivity may be increased or decreased in situations where the maximum response remains unaltered. Moreover, a combined change in insulin responsiveness and sensitivity would be reflected in simultaneous alterations to both the maximum response plateau and the half-maximal response (Figure 1C). The effector mechanisms which lead to altered insulin responsiveness and (or) sensitivity can vary and may relate to pre- or post-receptor events including the number of insulin receptors, the effective hormone concentration, the affinity of insulin for its receptor, or intracellular signalling events. Thus far, the precise manner in which Cr alters insulin resistance has not been clearly elucidated.

In order to more fully understand the form(s) of insulin resistance associated with supplementing Cr to periparturient dairy cattle we attempted to obtain precise insulin dose-response curves using a euglycemic/hyperinsulinemic clamp test. This test consisted of a continuous intravenous (iv) infusion of bovine insulin starting at 0.4 mU/kg/min for 45 min and increasing to 0.7, 1.5, 3.0 and 7.0 mU/kg/min in successive 45 min intervals. Concurrently, a continuous iv infusion of glucose (50% w/v solution) was initiated and adjusted every 5 min, in accordance with the measured blood glucose concentration, so as to maintain basal (pre-insulin infusion) blood glucose concentration or euglycemia. When euglycemia is achieved and maintained at each new level of insulin infusion, the system is said to be clamped and the corresponding glucose infusion rate equates to the metabolic rate of glucose utilization by the peripheral tissues at that dose of insulin (Sano *et al.*, 1991, 1993). Preliminary results from this experiment suggest a resistance to insulin in non-supplemented cows one week postpartum (Figure 2). A significant increase in insulin responsiveness with only a slight change in sensitivity was achieved

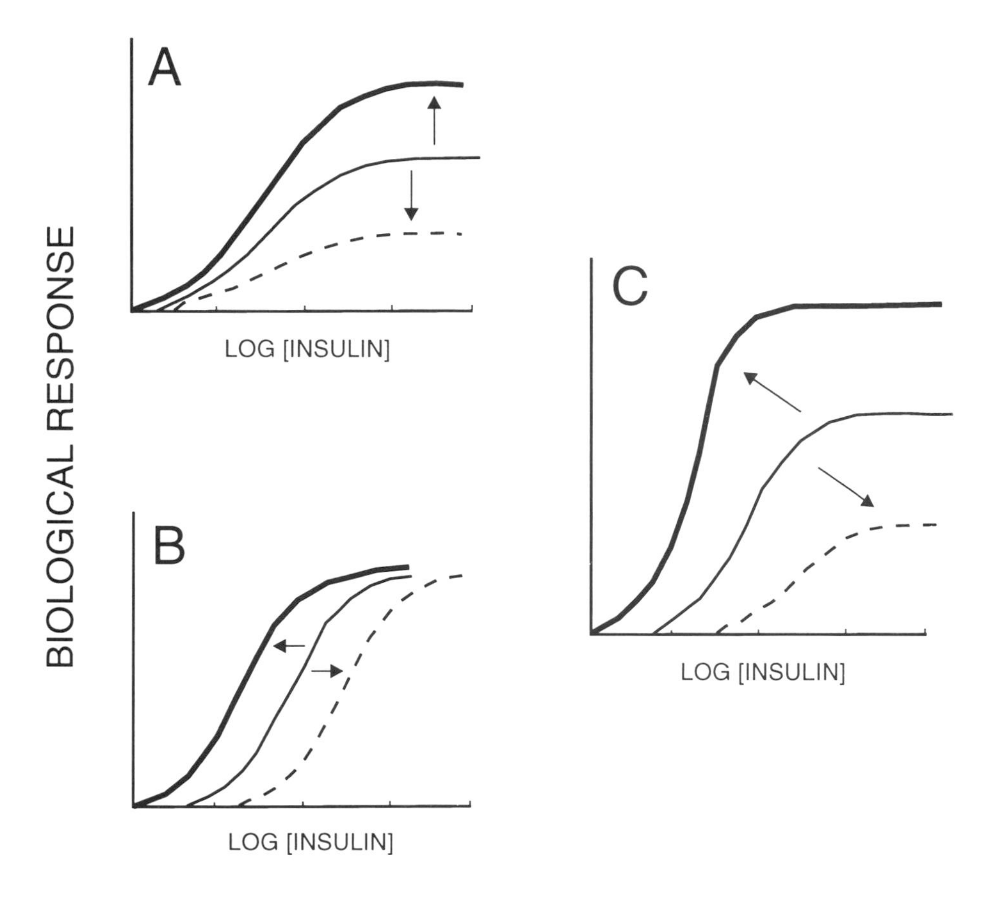

Figure 1. Types of resistance to insulin action. A) responsiveness, B) sensitivity, and C) combined responsiveness and sensitivity. Increased (dashed curve) and decreased (thickened curve) insulin resistance are indicated in comparison to the normal response curve. (Adapted from Kahn, 1978).

by supplementing Cr, as a high Cr-containing yeast (Biochrome, or Co-Factor III; Alltech Inc.), at 0.2 and 0.4 ppm, but this effect was lost at 0.6 ppm. This pattern of response to insulin over varying incremental amounts of dietary Cr (0, 0.2, 0.4 and 0.6 ppm) is not unusual since biological responses, and in particular the responses to trace elements, often increase and then plateau or decline with increasing stimulation within their biological zone of action (Venchikov, 1974). These results lend credence to the idea that during late pregnancy and early lactation, when cows experience the increased physical and metabolic stresses of calving and require increased glucose metabolism for milk production, Cr is deficient, leading to insulin resistance (Subiyatno *et al*., 1996). Pregnancy of humans and laboratory animals has previously been associated with Cr deficiency (Anderson, 1988). This is likely caused

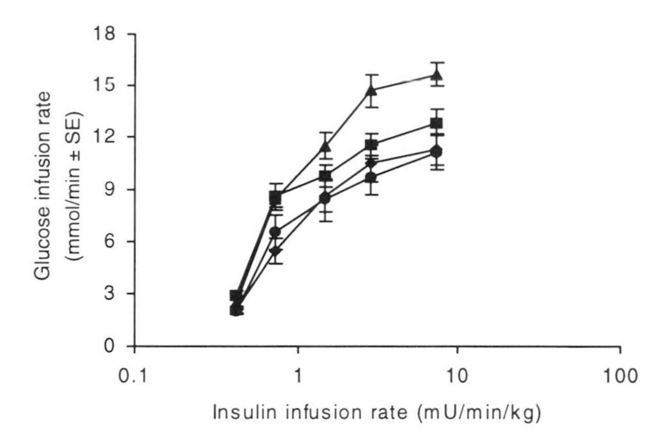

Figure 2. The effect of organic Cr from Biochrome on post-partum insulin resistance in primiparous dairy cows as assessed using the euglycemic/hyperinsulinemic clamp test. Dietary supplementation was at 0.0 (●), 0.2 (■), 0.4 (▲) and 0.6 (◆) ppm Cr.

by the increased urinary excretion and placental transport of Cr which occurs during pregnancy (Wallach, 1985). An increased insulin resistance of peripheral tissues during pregnancy has also been reported in pigs, as well as cattle (Vernon, 1988). The increased resistance of peripheral tissues to insulin may be a mechanism which helps ensure adequate glucose transport to the fetus and mammary gland, although disproportionate insulin resistance can be detrimental leading to hyperglycemia. Chromium supplementation reduces the probability of this occurrence. In addition, a recent study which examined the effects of insulin plus glucose on nutrient and hormone profiles of peripartum dairy cows suggested that low concentrations of post-partum insulin limit the ability of IGF-1 to respond to recombinant growth hormone (GH) (Leonard and Block, 1997). Early lactation is typically characterized by uncoupling of GH and IGF-1, low plasma insulin, low dry matter intake, high milk yield, and high mobilization of nutrients from body tissues. Although recombinant GH is able to increase milk yield over the entire lactation, its effects are more variable in early lactation. This seems to relate to the lower circulating concentrations of insulin post-partum, and its subsequent effects on the GH-insulin-IGF axis (Leonard and Block, 1997). Treatments such as Cr which decrease insulin resistance could increase milk yield by increasing the IGF-1 response to GH, and therefore enhance the regulation of nutrient partitioning so critical to milk production during early lactation.

Whether the noted increase in insulin responsiveness with Cr supplementation is mediated through pre- or post-receptor events is not clear. Causes of insulin resistance at the pre-receptor stage would include factors which reduce plasma insulin concentrations. This may involve decreased insulin production, increased insulin degradation, and (or) insulin binding to non-insulin receptor proteins such as anti-insulin antibodies (Kahn, 1978). Causes of insulin resistance at the receptor level may relate to changes in receptor number or affinity. Finally, causes of insulin resistance at the post-receptor level would involve extra- or intracellular signalling events which alter the efficiency of insulin action (Kahn, 1978). These possibilities are difficult to disentangle because of their complexity. However, work from our laboratory using the euglycemic clamp test in dairy cattle may shed some light on the situation since insulin is experimentally regulated and held constant for a period of time while changes in glucose utilization are monitored. This would tend to eliminate those changes in insulin-resistance involving decreased insulin concentrations seen at the pre-receptor level. In addition, because insulin operates on the 'spare' receptor principle in most tissues, a considerable decrease in receptor number would be necessary in order to cause a notable change in insulin responsiveness (i.e. maximal responses should be attainable unless insulin receptors are decreased by 90% or more) (Kahn, 1978). This means that in a spare receptor system there is no mechanism by which a lone decrease in insulin receptor numbers can result in a decrease in insulin responsiveness independent of any changes in sensitivity. Therefore, in our studies, the reduction of insulin resistance at 0.4 ppm Cr, primarily in terms of responsiveness, is not likely to result solely, if at all, from changes in receptor numbers. Other ongoing studies in our laboratory involve examining the number of insulin receptors and their affinity on bovine lymphocytes using an RIA. Preliminary results suggest no changes in the number of receptors or their affinity in cows receiving supplemental Cr. Collectively, these experiments may suggest that Cr has effects distal to insulin receptor binding, such as cell signalling events, at least in bovine lymphocytes. This may also make sense when considering the increased RNA synthesis noted in certain cell types exposed to Cr (Okada et al., 1989).

Plasma cortisol concentrations have also often been notably lower in individuals receiving dietary Cr (Chang and Mowat, 1992; Mowat *et al.*, 1993; Burton *et al.*, 1993; Pagan *et al.*, 1995). This may relate to cortisol's antagonistic response to insulin, particularly during stress-related events when cortisol increases drive insulin concentrations down. This may further contribute to the increased insulin resistance observed in post-partum dairy cows. Studies by Mowat *et al.* (1995) and Subiyatno *et al.* (1996) showed that post-partum dairy cows receiving supplemental Cr responded to glucose infusion by producing greater amounts of insulin compared to non-supplemented cows. If Cr supplementation operates not only to augment insulin receptor-related events or cell signalling events, but also is able to increase insulin release this may account for the decreases in cortisol observed in Cr supplemented animals. Cortisol has been defined as one of the classic stress hormones and is noted for its adverse effects on the immune system. This may be one of the pathways in which Cr is able to increase immune function.

Immune responses and chromium supplementation

The immune system is divided into specific and non-specific aspects of host defense. The non-specific or innate components of host defense include anatomic and physiologic barriers, phagocytic cell functions, and mediators of inflammatory responses. Specific or acquired immune responses are mediated by T and B lymphocytes and, unlike innate immunity, exhibit the properties of memory, diversity and the ability to specifically distinguish self from non-self. Innate and acquired aspects of immunity do not, however, occur independently. For example, phagocytic cells such as macrophages are important in antigen presentation and in activating specific immune responses. In addition, soluble factors such as cytokines produced during lymphocyte activation can enhance phagocytosis. Compounds such as Cr which are able to alter host defense, are referred to as immunomodulators, and may act at the level of innate or acquired host defense.

Accumulating evidence indicates that various forms of stress and distress induce neuro-endocrine alterations which disturb immune system homeostasis, leading to immunosuppression and increased disease susceptibility (Griffin, 1989). Disease incidence in dairy and beef cattle is highest during periods of maximum stress, including calving, weaning, peak lactation, and transport. It has been postulated that resistance to diseases, such as pneumonia and mastitis, may relate to an animal's ability to mount and maintain a beneficial immune response during high stress periods. Current evidence suggests a beneficial effect of feeding supplemental Cr on innate and specific aspects of host defense, disease incidence, and production traits in cattle.

Research at the University of Guelph was the first to show the beneficial effects of supplemental Cr on production traits and immune responsiveness in both beef and dairy cattle (Chang and Mowat, 1992; Burton *et al.*, 1994b). Specifically, significant increases in response to vaccination were noted in transport-stressed beef steers housed at the University of Guelph Beef Research Station. Fifty-five newly weaned calves were divided into two groups, 28 that received a supplemental Cr chelate (0.5 ppm/day) and 27 that did not. Each calf was then vaccinated with a commercial live attenuated IBR/PI3 vaccine according to the manufacturer's specifications. In order to determine serum antibody titers in response to vaccination, blood was taken from each animal prior to and at days 14 and 28 post-vaccination. Results from viral neutralization (IBR) and hemagglutination inhibition (PI3) assays showed that 35 of the 55 calves did not respond to either the IBR or PI3 components of the vaccine (Burton *et al.*, 1994b). This may be due to the reduced immunocompetence associated with stress. However, of the 20 calves that did respond to IBR, 75% were from the Cr supplemented group. In addition, least squares analysis of the IBR antibody titers revealed that Cr supplementation significantly ($P<0.05$) increased the magnitude of the antibody response by day 28 post-vaccination. No differences were seen in response to the PI3 component of the vaccine (Burton *et al.*, 1994b). A subsequent study in beef cattle showed that Cr supplementation reduced morbidity and increased antibody response to vaccination against bovine viral diarrhea, but not other components of a shipping fever vaccine (4K Vaccine, Langford Inc.) (Chang *et al.*, 1996a). The reasons for increased antibody responses to some but not other vaccine components may relate to antigen type and amount. This type of phenomenon is not an uncommon characteristic of immunomodulatory compounds or adjuvants.

The effects of dietary Cr on immunity were also investigated in dairy cows during the physical and metabolic stress periods associated with late pregnancy, calving and peak lactation (Burton *et al.*, 1994a). Ten periparturient cows housed at the University of Guelph Dairy Research Station were supplemented with chelated Cr (0.5 ppm/day) from six weeks pre-partum through 16 weeks postpartum, and 10 cows were unsupplemented controls. To assess humoral immune responses, all cows were immunized with ovalbumin antigen (OVA, subcutaneous) and human erythrocyte antigens (HRBC, intravenous) over a four-week interval. Lymphocyte activity was assessed *in vitro* using OVA antigen and the mitogen concanavalin A to stimulate peripheral blood mononuclear cells to proliferate. Cows receiving the supplemental Cr had higher serum antibody responses to OVA and HRBC compared to controls (Burton *et al.*, 1994a). *In vitro* lymphocyte proliferative responses to the mitogen ConA, but not OVA, were higher in supplemented cows. An inverse relationship may exist between OVA antibody response and OVA-induced lymphocyte proliferation. In addition, primiparous cows receiving supplemental Cr exhibited an 11% increase in actual milk yield (Mowat *et al.*, 1995). These results confirmed the preliminary observations that dietary Cr can alter specific immune responses and production parameters of stressed cattle.

Two subsequent studies were designed to determine if supplemental Cr was also able to alter aspects of non-specific immunity (Wright *et al.*, 1995; Chang *et al.*, 1996b). Specifically, the acute phase response, as indicated by serum haptoglobin and total haemolytic complement, was measured in 72 cross-bred steer calves. Results showed that serum haptoglobin, but not haemolytic complement, was lower in Cr supplemented calves (Wright *et al.*, 1995). The reduced haptoglobin concentrations were particularly evident on seven days after shipping and vaccination when morbidity was highest, and suggests reduced morbidity in Cr supplemented calves. Studies by Chang *et al.* (1996b) evaluated the phagocytic function of blood neutrophils in 40 peripartum dairy cows fed with (0.5 ppm, Cr chelate) or without Cr, but no significant differences were observed between treatment groups. Cumulatively, these observations suggest that supplemental Cr tends to preferentially modify aspects of the specific immune response, rather than innate defense mechanisms.

In an attempt to more precisely determine how Cr increases antibody responses and lymphocyte proliferation, a series of experiments were conducted to evaluate cytokine response and hormone interactions in cultures of bovine lymphocytes. Experiments in which bovine lymphocytes were cultured alone or with the addition of insulin, cortisol, growth hormone, or IGF-1, and then with the further addition of Cr, revealed that insulin, cortisol and Cr may interact to increase *in vitro* lymphocyte proliferative responses (Chang *et al.*, 1996b). Insulin receptors are known to be present on lymphocytes and insulin has been previously reported to increase *in vitro* bovine lymphocyte proliferation (Burton *et al.*, 1993). These data further substantiated the hypothesis that Cr enhances lymphocyte activity by altering insulin receptor or post-receptor events.

Subsequent studies evaluated the ability of Cr to affect the capacity of bovine peripheral blood lymphocytes to produce activating cytokines in response to T-cell mitogens. Results from these experiments showed lower amounts of the three cytokines, IL-2, IFN-γ and TNF-α, in the culture supernatants from dairy cows receiving supplemental Cr, particularly around peak lactation (Burton *et al.*, 1996). Cytokine profiles regulate a myriad of innate and acquired host defense mechanisms, including the types of antibody and cell-mediated immune responses that can be generated. For instance, higher concentrations of IFN-γ and IL-2 are associated with so-called T-helper 1 cells, and tend to support cell-mediated immune responses and IgG2a antibody production in mice and humans. Conversely, higher concentrations of IL-4, IL-5, and IL-10 are associated with T-helper 2 cells, and are known to support strong IgA, IgE and IgG1 antibody responses (Mosmann and Sad, 1996). The appropriate choice of either the T-helper 1 or T-helper 2 response is an important prerequisite for an effective immune response.

Compounds such as Cr which may influence cytokine production and direct the course of an immune response are of great practical value, finding application in vaccine formulations and therapeutic modalities in both human and veterinary medicine. Current studies in our laboratory are seeking to further characterize the immunomodulating properties of Cr, particularly its noted ability to modify bovine cytokine production.

References

Anderson, R.A. 1988. Chromium. In: Trace Minerals in Foods. (Ed. K.T. Smith). Marcel Dekker Inc., New York, NY, p. 231.

Anderson, R.A., M.M. Polansky, N.A. Bryden and J.J. Canary. 1991. Supplemental chromium effects on glucose, insulin, glucagon and urinary chromium losses in subjects consuming controlled low-chromium diets. Am. J. Clin. Nutr. 54:909.

Berson, S.A. and R.S. Yalow. 1970. Insulin antagonists and insulin resistance. In: Diabetes Mellitus: Theory and Practice. (Eds M. Ellenberg and H. Rifkin). McGraw-Hill, New York, p. 388.

Burton, J.L., T.H. Elsasser, T.S. Rumsey, J.K. Lunney and B.W. McBride. 1993. Immune responses of growing beef steers treated with estrogen/progesterone implants or insulin injections. Dom. Anim. Endocrinol. 10:31.

Burton, J.L., B.A. Mallard and D.N. Mowat. 1994a. Effects of supplemental chromium on immune responses of periparturient and early lactation dairy cows. J. Anim. Sci. 71:1532.

Burton, J.L., B.A. Mallard and D.N. Mowat. 1994b. Effects of supplemental chromium on antibody response of newly weaned feedlot calves

to immunizations with infectious bovine rhinotracheitis and parainfluenza 3 virus. Can. J. Vet. Res. 58:148.
Burton, J.L., B.J. Nonnecke, P.L. Dubeski. T.H. Elsasser and B.A. Mallard. 1996. Effects of supplemental chromium on production of cytokines by mitogen-stimulated bovine peripheral blood mononuclear cells. J. Dairy Sci. 79:2237.
Chang, G.X. and D.N. Mowat. 1992. Supplemental chromium for stressed and growing feeder calves. J. Anim. Sci. 70:559.
Chang, G.X., B.A. Mallard, D.N. Mowat and G.F. Gallo. 1996a. Effects of supplemental chromium on antibody responses of newly arrived feeder calves to vaccines and ovalbumin. Can. J. Vet. Res. 60:140.
Chang, G.X., B.A. Mallard and D.N. Mowat. 1996b. Effects of chromium on health status, blood neutrophil and *in vitro* lymphocyte blastogenesis of dairy cows. Vet. Immunol. Immunopath. 52:37.
Griffin, J.F.T. 1989. Stress and Immunity: a Unifying Concept. Vet. Immunol. Immunopath. 20:263.
Hossain, S. 1995. Effect of chromium yeast on performance and carcass quality of broilers. Alltech's Eleventh Annual Symposium on Biotechnology in the Feed Industry, Poster Presentation.
Kahn, C.R. 1978. Insulin resistance, insulin insensitivity, and insulin unresponsiveness: a necessary distinction. Metabolism 27(12):1893.
Leonard, M. and E. Block. 1997. Effects on nutrient and hormonal profile of long-term infusions of glucose or insulin plus glucose in cows treated with recombinant bovine somatotrophin before peak milk yield. J. Dairy Sci. 80:127.
Lyons, T.P. 1995. Biotechnology in the Feed Industry. Proc. of Alltech's Eleventh Annual Symposium. (Eds: T.P. Lyons and K.A. Jacques), Nottingham University Press, Loughborough, Leics. UK. p. 1.
Mertz, W. 1969. Chromium occurrence and function in biological systems. Physiol. Rev. 49:163.
Mooradian, A.D. and J.E. Morley. 1987. Micronutrient status in diabetes mellitus. Am. J. Clin. Nutr. 45:877.
Mosmann, T.R. and S. Sad. 1996. The expanding universe of T-cell subsets: Th1, Th2 and more. Immunol. Today 17:138.
Mowat, D.N., G.X. Chang and W.Z. Yang. 1993. Chelated chromium for stressed feeder calves. Can. J. Anim. Sci. 73:49.
Mowat, D.N., A. Subiyatno and W.Z. Yang. 1995. Chromium deficiency in first parity cows. In: Proc. of Alltech's Eleventh Annual Symposium on Biotechnology in the Feed Industry. (Eds: T.P. Lyons and K.A. Jacques), Nottingham University Press, Loughborough, Leics. UK. p. 309.
Okada, S., H. Tsukada and M. Tezuku. 1989. Effect of chromium (III) on nucleolar RNA synthesis. Biol. Trace. Elem. Res. 21:35.
Pagan, J.D., S.G. Jackson and S.E. Duren. 1995. The effects of chromium supplementation on metabolic response to exercise in thoroughbred horses. In: Proc. of Alltech's Eleventh Annual Symposium on Biotechnology in the Feed Industry. (Eds: T.P. Lyons and K.A. Jacques), Nottingham University Press, Loughborough, Leics. UK. p. 249.

Sano, H., M. Nakai, T. Kondo and Y. Terashima. 1991. Insulin responsiveness to glucose and tissue responsiveness to insulin in lactating, pregnant, and nonpregnant, nonlactating beef cows. J. Anim. Sci. 69:1122.

Sano, H., S. Narahara, T. Kondo, A. Takahashi and Y. Terashima. 1993. Insulin responsiveness to glucose and tissue responsiveness to insulin during lactation in dairy cows. Dom. Anim. Endocrinol. 10:191.

Schwartz, K. and W. Mertz. 1959. Chromium (III) and the glucose tolerance factor. Arch. Biochem. Biophys. 85:292.

Subiyatno, A., D.N. Mowat and W.Z. Yang. 1996. Metabolite and hormonal responses to glucose or propionate infusions in periparturient cows supplemented with chromium. J. Dairy Sci. 79:1436.

Venchikov, A.I. 1974. Zones of display of biological and pharmacotoxicological action of trace elements. In: Trace Element Metabolism in Animals – 2. Eds: W.G. Hoekstra, J.W. Suttie, H.E. Ganther and W. Mertz. University Park Press, Baltimore, p. 295.

Vernon, R.G. 1988. The partition of nutrients during the lactation cycle. In: Nutrition and Lactation in the Dairy Cow. (P.C. Garnsworthy, Ed) Butterworths, Toronto, Ontario, pp. 32.

Wallach, S. 1985. Clinical and biochemical aspects of chromium deficiency. J. Am. Coll. Nutr. 4:107.

Ward, T.L., L.L. Southern and S.L. Boleman. 1993. Effect of dietary chromium picolinate on growth, nitrogen balance and body composition of growing broiler chicks. Poult. Sci. 72 (Suppl. 1):37.

Wenk, C. 1995. Organic chromium in growing pigs: observations following a year of use and research in Switzerland. In: Proc. of Alltech's Eleventh Annual Symposium on Biotechnology in the Feed Industry. (Eds: T.P. Lyons and K.A. Jacques), Nottingham University Press, Loughborough, Leics. UK, p. 301.

Wright, A.J., B.A. Mallard and D.N. Mowat. 1995. The influence of supplemental chromium and vaccines on the acute phase response of newly arrived feeder calves. Can. J. Vet. Res. 59:311.

ORGANIC CHROMIUM: POTENTIAL APPLICATIONS IN PET FOOD FORMULATIONS

R.G. SHIELDS, JR

Heinz Pet Products, Newport, Kentucky, USA

Summary

The pet food industry is extremely large and consolidated worldwide, although national brands are formidable in each individual country. Pet food products can be distinguished based on both form and purpose, being based on consumer need, animal need, or both. Yeast and yeast culture inclusion in pet foods has a long-standing history. Validation for this practice continues to increase as evidence of new functional benefits continue to emerge. The fact that yeast can contain an organic source of chromium with bioactivity is one such attribute. This nutrient has a myriad of benefits for humans and animals based on direct or indirect effects on insulin. Various food, animal and environmental factors can influence the requirement for and therefore response to supplemental chromium. The pet food industry makes extensive use of meat by-products which are thought to contain high levels of chromium. Vegetable proteins are still used in substantial quantities, however, and the bioavailability of chromium from natural sources for pets is unknown. Although natural levels of chromium in pet diets may exceed that present in animal feeds, the potential applications requiring higher supplemental levels are much broader as well. Once regulatory hurdles are cleared, organic chromium could potentially have a bright future in pet food formulations.

Pet food industry

The global sales of pet foods are now slightly under 23 billion dollars, of which slightly over one-third each are represented by North America and Western Europe (Figure 1). Asia and Latin America represent the most promising areas for near term expansion. Slightly under half of this volume is represented by the top five companies worldwide. In the United States, growth of dry complete pet foods and treats/snacks has continued at the expense of canned and semi-moist main meal products. The trends for dry and canned products have been toward larger and smaller sizes, respectively, as canned products are used more as a supplement rather than as the total diet. The latter form represents a much greater portion of the total for cats than for dogs. The non-grocery outlets are growing at the expense of grocery outlets. These product size, form and distribution trends are by no means universal and tend to reflect the original and current mix of pet food manufacturers placing emphasis on a geographical region.

The pet food industry is designed to service approximately 98 million cats and 92 million dogs worldwide. This is likely a conservative estimate since results of some surveys would suggest that the United States and Western Europe alone could account for this size population. In the United States, the current population is estimated at approximately 66 million cats and 55 million dogs.

With the size and maturity of the pet food industry, it should be no surprise that there are numerous types of products in the marketplace. Since pets are considered by many owners to be members of the family, there is also the need to consider a lifetime feeding plan.

Early pet foods were distinguished primarily by form rather than formula. Baked biscuits entered the marketplace first in the latter half of the 19th century, followed by canned products in the early 1920s. Dry extruded pet foods followed in the 1950s and semi-moist in the 1960s. A soft-dry category consisting of a blend of dry and semi-moist pieces emerged with the introduction of Kibbles 'N Bits® in the early 1980s. Innovation of product form within each of these product categories continues to emerge today.

Pet food formulation took a dramatic turn in 1963 with the introduction of Purina® Puppy Chow® . This was the first time needs of specific life stages were considered rather than development of all purpose diets. This concept came to fruition in the 1970s when Cycle® (now a part of the Heinz Pet Products portfolio) and Hill's® Science Diet® began marketing products for geriatric pets. Formulation using a life stage concept is now universal throughout the pet food industry. The specific nutrient targets used by various companies are markedly different, however. Additionally, the distinction among puppy,

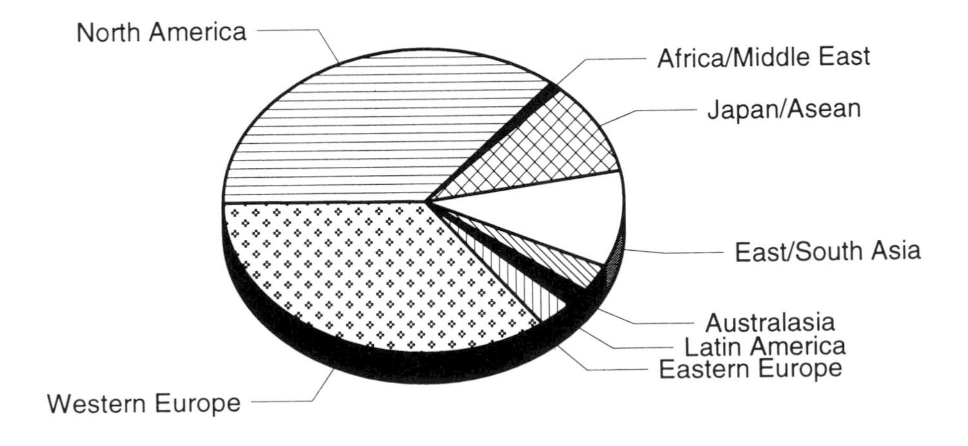

Figure 1. Global sales of pet foods.

adult and geriatric formulas is much greater in some product lines than in others.

The introduction of low calorie diets for overweight or less active pets began with the introduction of the Cycle® and Hill's® Science Diet® lines listed above and also by Purina® in the 1970s. This started the proliferation of other special purpose diets through the next several decades which continues today. Today this class of diets includes performance formulas, diets designed to improve digestibility and (or) reduce stool volume or for the performance animal. Following consumer trends, categories of natural diets and limited antigen diets have also found a strong following in consumer segments. During the last year, diets for large breed puppies have been marketed by several companies as lower nutrient density counterparts to performance puppy diets.

There is another smaller but equally important segment of pet food diets, namely medical foods. This pet food diet type started in the United States with the pioneering efforts of Dr Mark L. Morris, Sr, who developed a diet for a dog exhibiting clinical signs of kidney disease. These products are typically distributed only by veterinarians and they fall into one of three groups. The first are designed for internal organ dysfunction which may result from nutrition or environmental stress, disease challenge, or from structural or metabolic defects of the pet or genetic origin. Included would be diets designed for liver, heart, thyroid, kidney or pancreatic malfunction. These diets are typically differentiated in source of calories and in macromineral balance.

A second group would include diets for integument and gastrointestinal dysfunction. I have separated these from other organ-specific diets because adverse food reactions and atopy are typically addressed by a combination of nutrient and ingredient adjustments. Avoidance of particular ingredients, use of unusual food ingredients and inclusion of probiotics, pre-biotics and various fiber sources are typical of these diets.

Many but not all of the medical foods listed above are suitable for long term feeding as pets with these conditions seldom completely recover from their disorders. A third category of medical foods may represent critical care nutrition products. These products are designed for short term feeding situations such as recovery from surgery, inappetence, orphaned puppies, obesity and dissolution of specific uroliths. In the EU, medicinal feeds would be classified as those containing antimicrobials or antiparasitic agents and thus the above description would be categorized as dietetic feedstuffs (Wolter, 1995).

Based on the above discussion, it should be readily apparent that the concept of functional foods has been used in pet food marketing efforts for some time. Only regulatory hurdles limit the rate at which further proliferation of this concept develops in the exciting pet food industry.

Up to this point I have only discussed main meal products. Treats and supplements are also extremely important segments of the pet food industry. The true penetration of supplements in pet food diets is difficult to judge because of the widespread use of human supplements which have wider regulatory latitude than that present in the pet food industry. Numerous nutraceuticals are purchased in this manner.

Yeast in pet foods

Yeasts and yeast cultures have a long standing tradition of use in animal feeds and pet foods. Before the widespread availability and use of synthetic vitamins, yeast was added to pet food formulations as a natural source of B vitamins. With the current growth potential of the natural foods market, perhaps they will find use for this purpose once again.

Various yeast products continue to be widely used in the pet food industry as palatability enhancers. This enhancement is thought to result from the high levels of glutamate in yeast or the presence of 5′ nucleotides. These agents can mask bitterness in other natural food ingredients.

Yeast was often included in diets for livestock and pets as a source of unidentified growth factors in the 1950s. Some of these were later identified as the numerous B vitamins indicated above. During the following two decades, most research focused on yeast as a source of single cell protein (Lyons, 1990). More recently, the beneficial effects of yeast and yeast culture on improvement of gastrointestinal health have been examined. Live yeast can be considered a source of microbial culture which is fairly stable to acid and heat processing. Yeast cultures often contain supplemental enzymes which may assist in protein, carbohydrate and fat digestion. Phytase, an enzyme which can improve the utilization of plant phosphorus is also present. Finally, the beta glucans present in yeast cell walls may have beneficial effects both as pre-biotics and for non-specific improvement of the pet's immune system. Any attribute which can improve gastrointestinal tract health and immunity is extremely valuable in pet diets where, unlike livestock feeds, antimicrobial agents are not used.

Yeast is also an excellent source of selenium and chromium, two trace minerals which may have positive effects on pet health. Acceptance of the efficacy of both trace minerals has been hampered because of the overemphasis on their toxicity rather than their essentiality. The use of yeast and yeast cultures as a source of functional food ingredients has resulted in no small part from the continued research and marketing efforts of Alltech.

Organic chromium: functional food?

Several nutrients are being used as supplements in pet diets in the United States which have not been recognized as essential by the Association of American Feed Control Officials (AAFCO). Many of these nutrients as well as other compounds found naturally in the body are, however, being sold as human or pet supplements in this country or in complete pet foods in other countries. Examples of these include carnitine, glutamine, glycine, antioxidant vitamins, omega-3 fatty acids, gamma linolenic acid, glucosamine, chondroitin sulfate, coenzyme Q and chromium. Although many of these may not be considered to be essential dietary components for all pets, they may play an essential role in certain segments of the pet population and certainly all are essential from a metabolic standpoint. Combs recently presented (1994) an excellent paper which challenged criteria for defining nutrient essentiality and the 'requirement' for the nutrient in question. We must all change our paradigm regarding this issue as attention shifts from prevention of deficiency to maintenance of optimum health.

Chromium represents an interesting nutrient which, like selenium, had a focus more on toxicity than essentiality. The level of selenium allowed in pet foods and animal feeds in the United States has been severely limited because of concerns regarding toxicity and, while most premium pet foods are fortified with this nutrient, exports to countries such as Japan have been hampered because of toxicity concerns. Chromium, like most transition elements, is considered a toxic mineral (especially in hexavalent form) but is metabolically essential none the less. Attention shifted in the late 1950s when Schwarz and Mertz (1959) found that chromium from yeast improved glucose tolerance in rats.

There have been several excellent reviews relating to the essential role of chromium in human nutrition (Mertz, 1993) as well as a general papers on organic chromium presented at this conference last year (Lindemann, 1996; Mowat, 1996). Based on the numerous positive results in swine studies, a letter of no objection was obtained from the Food and Drug Administration's Center for Veterinary Medicine allowing the inclusion of chromium picolinate at levels not to exceed 200 ppm. Neither chromium yeast nor chromium picolinate has been cleared for use in pet foods in the United States and chromium has not been recognized as a required nutrient for either dogs or cats (AAFCO, 1996).

The most well recognized role of chromium is as a component of glucose tolerance factor. This compound, identified originally from brewer's yeast, has not been fully identified but is thought to be a complex containing glutathione, nicotinic acid and possibly glutathione (Toepfer *et al.*, 1977). Chromium may also be part of a high molecular weight protein containing five or six atoms of chromium per molecule which can bind nucleolar chromatin and stimulate RNA synthesis (Okada *et al.*, 1983, 1984; Ohba *et al.*, 1986; Mertz, 1993). Chromium appears to improve insulin internalization into muscle cells followed by uptake of glucose and amino acids (Roginski and Mertz, 1969; Evans and Bowman, 1992). A deficiency of chromium results in reduced sensitivity of peripheral tissues to insulin, resulting from a reduction in number of receptors, their affinity for insulin, or both (Anderson, 1986). The net result is hyperglycemia and glucosuria. Supplementation of chromium picolinate results in increased glucose disappearance and reduced glucose half-life in the blood (Amoikin *et al.*, 1995), resulting in improved glucose tolerance and reduced circulating insulin and glucagon levels (Anderson *et al.*, 1991a). It has also resulted in increased serum growth hormone, but not IGF-1 levels (Evock-Clover *et al.*, 1993; Page *et al.*, 1993).

Effects on insulin and therefore glucose and amino acid utilization would be expected to affect a wide variety of organ systems. Effects on growth rate and (or) efficiency, reproduction of both males and females, cardiovascular and ocular health and longevity have been noted (Lindemann, 1996). Response during the period of rapid growth has been mixed and likely relates to the dietary protein fed, amount of chromium in the basal diet, length of exposure to the high chromium diet, genetic potential of the animals, and feeding management (Page *et al.*, 1993; Mooney and Cromwell, 1995; Wang *et al.*, 1995; Ward *et al.*, 1995; Wenk, 1995). Effects on leanness of swine have also been mixed (Page *et al.*, 1993; Boleman *et al.*, 1995; Mooney and Cromwell, 1995). This variability has also been observed in trials relating to fat supplementation, where magnitude of response relates to all of the above factors as well as food intake. Effects on insulin or growth hormone could certainly influence these parameters.

During the reproductive cycle, chromium supplementation has resulted in increased litter size at birth and weaning with improvements increasing as parity increased and exposure to dietary chromium lengthened (Lindemann *et al.*, 1995a,b). The effect on litter size and weight at weaning was apparently not achieved at the expense of sow body stores. The litter effects likely result from the effect of insulin on both ovulation rate and early embryo survival and the magnitude may be influenced by heat stress on the sow (Trout, 1995; Mowat, 1996). These effects may be manifested directly through insulin or indirectly through luteinizing hormone or progesterone (Trout, 1995). In a separate study conducted in Australia (Purser, 1995), farrowing rate rather than litter size was affected. As with most nutrients, the magnitude of the response likely is influenced by the production rates achieved by animals on the control diet. While we typically focus efforts on reproductive efficiency in female animals, Anderson and Polansky (1981) found that sperm count was reduced by 50% and fertility 25% in rats fed diets containing less than 100 ppb chromium.

Chromium supplementation would certainly have potential in populations at high risk for many disorders including heart disease, renal disease, diabetes, or hyperlipidemia conditions. Since these conditions are over-represented in geriatric patients and since tissue chromium (Anderson, 1987) and glucose tolerance and insulin efficiency (Harris, 1990) tend to decrease with aging, routine supplementation to geriatric diets may be appropriate. In some preliminary studies with rats, longevity was increased 25% when chromium was supplemented in the diet at 1000 ppb (Evans and Meyer, 1992, 1994). It is possible that this effect results from prevention of the many disorders typical in the aging process. Marginal chromium deficiency can increase risk for diabetes (Mertz, 1969) and coronary heart disease (Schroeder, 1968). Deficiency can result in increased serum cholesterol (LDL), triglycerides, aortic lipids and plaque (Schroeder and Balassa, 1965; Riales and Albrink, 1981; Mossop, 1983). These conditions as well as high blood glucose can be reversed rapidly in patients with high cholesterol or non-insulin dependent diabetes (Evans, 1989; Press *et al.*, 1990).

Chromium supplementation has also improved immunocompetence in stressed cattle (Chang and Mowat, 1992; Burton *et al.*, 1993; Kegley and Spears, 1995) and pigs (van Heugten and Spears, 1994). Chromium has in fact also potentiated response to vaccination in feeder calves (Wright *et al.*, 1994). These effects could be manifested in part by the reduction in plasma cortisol noted in some studies where chromium was supplemented (Jacques and Stewart, 1993; Pagan *et al.*, 1995; Mowat, 1996). Supplementation also reduced plasma lactate concentrations in exercising horses (Pagan *et al.*, 1995), which may have beneficial effects on exercise performance.

With all these beneficial effects, it is surprising that chromium has not been generally accepted as an essential nutrient. As mentioned earlier, part of the reluctance to categorize it as an essential nutrient relates to the overemphasis on its role as a toxic nutrient. Additionally, as with any nutrient, defining a requirement has been hindered by the lack of a specific test to evaluate chromium status. Responses in several studies, while statistically significant are not clinically impressive (Mertz, 1993). Perhaps expectations regarding response have been disappointing because of overly zealous marketing efforts through the human supplement channel. Chromium concentration in the blood is unfortunately not in equilibrium with that of the body tissues. Urinary chromium may reflect recent chromium intake but is influenced by the inherent glucose tolerance of the animal. Hair chromium may be of some diagnostic value when considering groups of animals since it is reduced in diabetes and during pregnancy (Canfield, 1979).

Another factor limiting consensus on essentiality relates to the inconsistency of response. This has resulted from dietary, species, and animal chromium status differences. This should not be unexpected since it applies to considerations in supplementation of most nutrients. This variability of response may result from a combination of animal factors, food factors and environmental factors which may influence response to this nutrient. There also appear to be species differences with respect to potency of different chromium sources (Lindemann, 1996), and there have been recent indications which have led some researchers to believe that chromium picolinate may be carcinogenic which was not found to be the case with other chromium sources. Much obviously remains to be learned regarding this nutrient.

Factors influencing chromium requirements

Animal factors which may influence chromium requirements include life stage or activity level or metabolic abnormalities resulting from infection, aging or inborn errors of metabolism. Dose response curves to maximize different response criteria have unfortunately not been conducted. It is conceivable that geriatric animals could have higher requirements than normal adults since progressive impairment of glucose tolerance is observed with aging in humans (Harris, 1990; Anonymous, 1991). Additionally, research results have suggested that urinary chromium excretion increases with the intensity of exercise (Anderson *et al.*, 1991b) and may be three times higher in diabetics than normal adults (Morris *et al.*, 1992), so requirements for these two situations may be higher. Current minimum recommendations for diabetic humans are 200 micrograms/day, compared to approximately 50 for healthy adults (Anderson *et al.*, 1991a). Similar high levels have had beneficial effects on blood lipids (Evans, 1989; Mertz, 1993) and may therefore be of some benefits in patients with hyperlipidemia.

It would be expected that requirements of chromium for reproduction would differ from that of growth or maintenance, as happens for many nutrients.

Chromium content of human tissues declines with age and during pregnancy and illness (Anderson, 1987). Results of studies conducted by Lindemann and coworkers (1995a,b) suggest that longer exposure to dietary chromium resulted in enhanced differences relative to control diets. It is possible that shorter exposure to higher levels would also provide this benefit, provided that urinary chromium excretion did not fully compensate for enhanced intake.

The prevalence of obesity in both humans and pets is significant and 'diets' are common in both conditions. It is probable that the dietary chromium concentrations necessary to minimize loss of lean body mass during weight loss would be greater than those required in conventional adult diets. First of all, insulin resistance is more common in obese than fit individuals (Olefsky and Kolterman, 1981), and this is apparently restored upon weight loss (Franssila-Kallunki *et al.*, 1992). Secondly, when weight loss is too rapid in cats, hepatic lipidosis can occur (Biourge *et al.*, 1994; Jackson *et al.*, 1994). Finally, during weight reduction food intake is typically restricted below that of healthy animals, requiring increased dietary concentrations to even maintain similar daily consumption.

In addition to the animal factors mentioned above, various food factors are equally important. Dietary zinc, iron and phytate are all important because they adversely affect chromium status. The interaction of chromium with phytate and zinc likely occurs at the level of the intestine while that with iron likely occurs post-absorption where iron and chromium are both transported with transferrin in the blood (Jacques and Stewart, 1993). Dietary protein is also likely important since responses have been greater in diets limiting in protein (Lindemann, 1996) and urinary chromium is increased in protein-calorie malnutrition (Canfield, 1979). Since many geriatric diets are more limiting in protein, this is perhaps another reason for supplementation in this age group. Another nutrient which may require consideration is niacin. This nutrient is thought to represent part of the glucose tolerance factor complex and some organic chromium sources contain both chromium and niacin. It is possible that dietary niacin affects the absorption or biopotency of even inorganic chromium compounds since chromium may complex with niacin as a ligand with improved absorption potential (Mowat, 1996). Form of chromium is likely important but has not yet been evaluated in pet food formulations. In some species absorption of organic forms can be 10-fold higher than inorganic forms.

The ingredients used in the basal formulations are also likely important with regard to dietary chromium needs. Vegetable material contains lower levels of chromium than animal protein products, particularly by-products including liver (Canfield, 1979; Mertz, 1993; Page *et al.*, 1993). Further processing decreases the chromium content of grains. Yeast is also a rich natural source of chromium. Because pet foods contain high levels of meat by-products and yeast is supplemented in many dry products, the need for supplemental chromium in the diet may be less than for other species. On the other hand, Lindemann (1996) stated that chromium may be most beneficial in diets containing high levels of sugar. Inclusion of simple sugars is required as part of the stability system in semi-moist complete pet foods and treats to control water activity. Similar to chromium supplement source, the endogenous dietary chromium may differ widely in bioavailability among ingredients. For example, although liver contains high levels of copper, that of pork origin is almost completely unavailable to chicks unless it is autoclaved or hydrolyzed while that in beef, lamb and poultry liver is highly available relative to copper sulfate (Aoyagi *et al.*, 1993, 1995). Results of this research are presented in Table 1.

Table 1. Relative bioavailability of copper from various ingredients compared to copper sulfate for chicks

Ingredient	% Bioavailability
$CuSO_4{\cdot}5H_2O$	100
Liver source	
Pork	0
Autoclaved	32
Enzyme digested	63
Acid treated	46
Rat	21
Beef	82
Sheep	113
Chicken	125
Turkey	83
Poultry by-Product ML	92/42
Beef & bone ML	4
Pork & bone ML	53
Porcine plasma	99
Soybean meal	38
Corn gluten meal	48
Soybean mill run	47
Peanut hulls	44

Source: Aoyagi *et al.* (1993, 1995).

Environmental factors also require consideration. Chromium may have beneficial effects in stress situations as it appears to reduce serum cortisol (Pagan *et al.*, 1995; Mowat, 1996). Ward and coworkers (1995) interpreted their variability in response across trials to have resulted from differences in animal stress. As mentioned earlier, chromium has improved immunocompetence in stressed feeder cattle in several trials. Although we generally think of pets living with us in our houses, there are many areas of the world where heat stress is a potential problem. Several breeds which are brachycephalic such as Pugs and Pekinese in the Toy Group and Boxers in the Working Group are thought to be less tolerant to heat stress. Additionally, during the reproductive cycle, dams are often subjected to heat stress to provide an optimum environment for the litter. A final factor to consider is feeding management. Swine are fed in many ways, including limit feeding (gestating sows), scale feeding (programmed growth in Europe), feeding multiple times per day (lactation, stress diets) or *ad libitum*. Meal pattern obviously influences hormonal responses as well as the pattern of nutrient intake. Both meal feeding and *ad libitum* are common in pet food applications. The former is common with canned pet foods and in kennel situations while the latter may be more common where dry products are fed to household pets.

Pet food applications for chromium

Based on the previous discussion of the types of products present in the pet food industry and the functions of chromium, it would appear that there are numerous immediate applications for chromium to the pet food industry requiring consideration.

OBESITY/WEIGHT LOSS

Similar to the case with humans, obesity is extremely prevalent in the pet population. The incidence of this condition has varied in surveys from 6 to nearly 50% (Hand *et al.*, 1989; Markwell *et al.*, 1990; Crane, 1991; Sloth, 1992). This variation likely was affected by species, breed, age, diet/owner lifestyle, gonadal status, owners, survey type and definition. Incidence for cats has been surveyed to be less than dogs, although today the gap has narrowed considerably. Among dogs, labrador retrievers, Cairn terriers, cocker spaniels, long haired dachshunds, Shetland sheepdogs, cavalier King Charles spaniels and beagles have been identified as breeds prone to obesity (Markwell *et al.*, 1990). Relative to age, it would appear that maximum obesity is present in animals in middle age, similar to the situation with humans (Kronfeld *et al.*, 1991, 1994; Armstrong and Lund, 1996). An average propensity of 25% may therefore represent as much as 50% in some age groups (Figure 2). Relative to diet and owner lifestyle, the incidence of obesity has arisen in concert with the availability of high fat, high energy 'superpremium diets'. This is not to say that the diets themselves are a problem but rather that they require more careful feeding management which is not always practiced. The incidence of obesity appears to be higher where the owners are more sedentary as well. Gonadectomized pets appear to have lower energy requirements (as much as 30% lower) than intact animals, again making feeding management critical in obesity prevention (Root, 1995; Flynn *et al.*, 1996). Relative to survey type it must be considered that those involving clinical submissions may or may not be representative of the general pet population.

Relating to definition, functionally obesity is defined as the state at which level of body fat impairs body function or health (Kirk and Toll, 1996). Practically this has meant weight exceeding the 'ideal' by 15 to 20% in both humans and pets. With the acceptance and validation of body composition measurement and body condition scoring systems, perhaps a standardization of measurement procedure will be possible. At any rate, recent articles using modern methodology have settled on an obesity incidence of approximately 25% (Laflamme *et al.*, 1994; Armstrong and Lund, 1996). This is not an insignificant

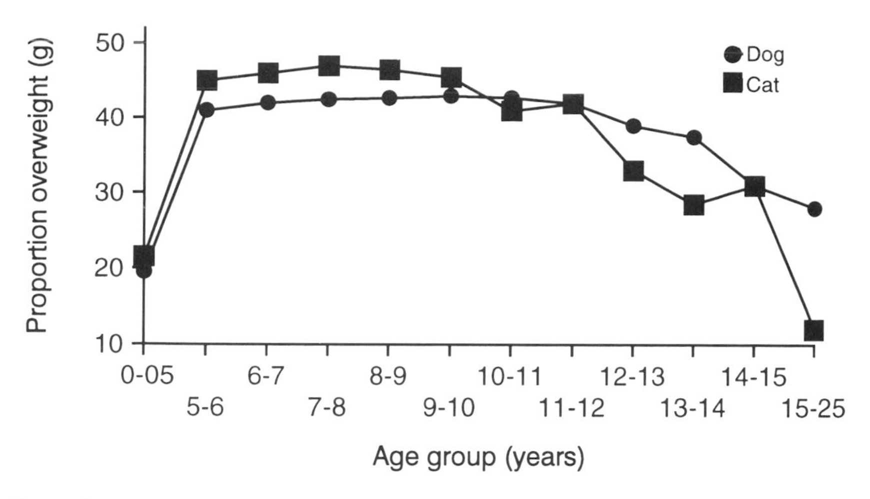

Figure 2. Incidence of obesity in cats and dogs of various age groups.

number. The presence of this condition can in turn exacerbate other conditions including osteoarthritis, heart disease, liver disease, diabetes, lung disease, cancer, pancreatitis, constipation, heat stress and exercise intolerance.

As mentioned previously, there is sound reasoning for considering chromium supplementation of diets designed for weight reduction. There is probably equal justification to consider adding chromium to puppy diets, particularly for those breeds prone to obesity as adults or because of osteoarthritis propensity need to remain in lean condition. Additionally, most organizations are recommending responsible pet ownership including spay and neuter of non-breeding animals. In the case of cats, this may represent nearly 87% of the adult population (Root, 1995). It is always easier to prevent a problem than to treat it.

GERIATRIC/MEDICAL DIETS

Similar to obesity, the consideration of the opportunity to supplement geriatric diets is confounded by the definition of old age. Many human nutritionists use a thumb rule of 50% of the maximum lifespan or 75% of the expected life of an individual based on genetic considerations. Based on the former (Kirkwood, 1992), humans, cats and dogs would be considered geriatric at 57.5, 14 and 10 years of age, respectively. As an alternative approach, results of a veterinary survey considering age at which problems associated with aging would manifest themselves, a range from 7.5 years of age in giant breeds (over 40 kg) to 11.5 years for small breeds (below 9 kg) was obtained (Goldston, 1989), reflecting differences in life expectancy of individual breeds (Anderson, 1987).

These surveys would suggest that 15 to 20% of the pet population is geriatric. The penetration of special diets for geriatric pets is much lower than this, perhaps in part because of the lack of consensus regarding whether geriatric pets have different nutrient requirements compared to healthy adults. It must also be considered that some of these pets may also be receiving diets designed for less active pets. Additionally, as stated earlier, the conditions which necessitate medical diets are more prevalent in geriatric pets so this group may be fragmented. Certainly diets designed for diabetic pets or those with renal disease or hyperlipidemias may benefit from consideration of chromium supplementation. As mentioned previously, geriatric pets may benefit from supplementation both based on the animals themselves and the typical composition of the diets which they are fed.

SEMI-MOIST PET FOODS

Although this pet food category continues to shrink and now represents under 5% of the pet food volume in the United States, this product form is widely used in manufacture of palatability treats. As stated earlier, these products generally contain simple sugars to control water activity. Additional incorporation of supplemental chromium may be an excellent safeguard in these formulations.

REPRODUCTION

There has been a disappointing amount of published research evaluating nutrient requirements during the reproductive cycle. Part of the difficulty with this research lies in the generation of puppies and kittens which then require adoption. Additionally, in the United States, testing of reproducing animals is required as part of nutrition substantiation from a regulatory standpoint, further reducing available animal resources. Few if any marketed products are designed exclusively for the reproductive cycle and none to my knowledge differentiate gestation and lactation, where needs may be quite different. Similar to the case with humans, there is a wide disparity in recommendations regarding feeding management of reproducing animals. While some facilities manage gestation gain as would be done in swine reproduction, others feed *ad libitum* as is done with rodents to maximize gestation gain. As would be expected, this practice generally leads to reduced food intake and increased weight loss during lactation, perhaps resulting in changes in insulin sensitivity during the reproductive cycle. The results with swine are extremely exciting to us in the pet food industry. Results of recent research conducted with dogs demonstrated improvements in reproductive efficiency from dietary incorporation of copper, zinc and manganese proteinate as part of the total mineral supplement (Kuhlman and Rompala, 1995). Perhaps similar benefits will result from organic chromium supplementation.

PERFORMANCE/STRESS FORMULAS

Several diets exist for use in hunting, coursing or racing. Greyhounds and especially sled dogs are asked to perform incredible tasks. In contrast to reproduction, several studies have been performed evaluating these highly trained canine athletes. However, it is equally important to consider the untrained athlete which must change exercise intensity from sedentary to active during the hunting season. There is potential for chromium supplementation in both applications. Most performance formulas contain high levels of protein and fat compared to the conventional diet counterparts. There is ample opportunity for consideration of functional ingredients to improve canine performance. Additionally, there are many service dogs and show dogs which may benefit from supplementation because of the stress levels involved with this activity. Even boarding kennels and animal shelters may benefit from inclusion of this nutrient in or supplementation to their standard diets. A final consideration would be for recuperative diets designed for pets following surgery.

HEAT STRESS

Although not marketed as such today, diets designed to help pets cope with heat stress may benefit from chromium supplementation. This would especially be applicable to pets housed outside. As mentioned earlier, it may also apply to lactating females nursing under heat lamps.

Conclusions

Yeast is a natural source of chromium which appears to have a multitude of applications within both the animal feed and pet food industry. It is truly a functional food component with much promise. The major factor limiting its long term use may be over-selling benefits and setting up unrealistic expectations in consumers and pet owners. Additionally, as with any conditionally essential nutrient (affected by animal or diet type), it is important that research be carefully designed to fairly evaluate the nutrient to prevent false negative results from proliferating research publications. Such information would undoubtedly delay approval of this nutrient in pet foods.

References

AAFCO. 1996. Official Publication, Association of American Feed Control Officials, Inc.

Amoikin, E.K., J.M. Fernandez, L.L. Southern, D.L. Thompson, Jr, T.L. Ward and B.M. Olcott. 1995. Effect of chromium tripicolinate on growth, glucose tolerance, insulin sensitivity, plasma metabolites, and growth hormone in pigs. J. Anim. Sci. 73:1123.

Anderson, R.A. 1986. Chromium metabolism and its role in disease processes in man. Clin. Physiol. Biochem. 4:31.

Anderson, R.A. 1987. Chromium. In: W. Mertz (Ed.). Trace Elements in Human and Animal Nutrition (5th Ed.). Academic Press, Inc. San Diego, CA. pp. 225–244.

Anderson, R.A. and M.M. Polansky. 1981. Dietary chromium deficiency: Effect on sperm count and fertility in rats. Biol. Trace Element Res. 3:1.

Anderson, R.A., M.M. Polansky, N.A. Bryden and J. J. Canary. 1991a. Supplemental chromium effects on glucose, insulin, glucagon, and urinary chromium losses in subjects consuming controlled low-chromium diets. Am. J. Clin. Nutr. 54:909.

Anderson, R.A., N.A. Bryden, M.M. Polansky and J.W. Thorp. 1991b. Effect of carbohydrate loading and underwater exercise on circulating cortisol, insulin and urinary losses of chromum and zinc. Eur. J. Appl. Physiol. 63:146.

Anderson, R.S. 1987. Age distribution in dogs and its relation to life expectancy.

Anonymous. 1991. Insulin resistance in the elderly. Nutr. Rev. 49:257.

Aoyagi, S., D.H. Baker and K.J. Wedekind. 1993. Estimates of copper bioavailability from liver of different animal species and from feed ingredients derived from plants and animals. Poul. Sci. 72:1746.

Aoyagi, S., K.M. Hiney and D.H. Baker. 1995. Copper bioavailability in pork liver and in various animal by-products as determined by chick bioassay. J. Anim. Sci. 73:799.

Armstrong, P.J. and E.M. Lund. 1996. Changes in body composition and energy balance with aging. Vet. Clin. Nutr. 3:83.

Biourge, V., J.M. Groff, C. Fisher, D. Bee, J.G. Morris and Q.R. Rogers. 1994. Nitrogen balance, plasma free amino acid concentrations and urinary orotic acid excretion during long-term fasting in cats. J. Nutr. 124:1094.

Boleman, S.L., S.J. Boleman, T.D. Bidner, L.L. Southern, T.L. Ward, J.E. Pontif and M.M. Pike. 1995. Effect of chromium picolinate on growth, body composition, and tissue accretion in pigs. J. Anim. Sci. 73:2033.

Burton, J.L., B.A. Mallard and D.N. Mowat. 1993. Effects of supplemental chromium on immune responses of periparturient and early lactating dairy cows. J. Anim. Sci. 71:1532.

Canfield, W. 1979. Chromium, glucose tolerance and serum cholesterol in adults. In: D. Shapcott and J. Hubert (Eds). Chromium in Nutrition and Metabolism. Elsevier/North-Holland Biomedical Press. Pp. 145–161.

Chang, X. and D.N. Mowat. 1992. Supplemental chromium for stressed and growing feeder calves. J. Anim. Sci. 70:559.

Combs, G.F., Jr. 1994. Nutritional essentiality: a changing paradigm. Proc. Cornell Nutr. Conf. Cornell Univ., Ithaca, NY.

Crane, S.W. 1991. Occurrence and management of obesity in companion animals. J. Sm. Anim. Prac. 32:275.

Evans, G.W. 1989. The effect of chromium picolinate on insulin controlled parameters in humans. Int. J. Biosocial Med. Res. 11:163.

Evans, G.W. and T.D. Bowman. 1992. Chromium picolinate increases membrane fluidity and rate of insulin internalization. J. Inorg. Biochem. 46:243.

Evans, G.W. and L. Meyer. 1992. Chromium picolinate increases longevity. Age 15:134.

Evans, G.W. and L.K. Meyer. 1994. Life span is increased in rats supplemented with chromium-pyridine 2 carboxylate complex. Adv. Sci. Res. 1(Vol.1):19.

Evock-Clover, C.M., M.M. Polansky, R.A. Anderson and N.C. Steele. 1993. Dietary chromium supplementation with or without somatotropin treatment alters serum hormones and metabolites in growing pigs without affecting growth performance. J. Nutr. 123:1504.

Flynn, M.F., E.M. Hardie and P.J. Armstrong. 1996. Effect of ovariohysterectomy on maintenance energy requirement in cats. JAVMA. 209:1572.

Franssila-Kallunki, A., A. Rissanen, A. Ekstrand, A. Ollus and L. Groop. 1992. Effects of weight loss on substrate oxidation, energy expenditure, and insulin sensitivity in obese individuals. Am. J. Clin. Nutr. 55:356.

Goldston, R.T. 1989. Preface to geriatrics and gerontology. Vet. Clin. N. Amer. 19:ix.

Hand, M.S., P.J. Armstrong and T.A. Allen. 1989. Obesity: Occurrence, treatment and prevention. Vet. Clinics of North America. 19(3):447.

Harris, M.I. 1990. Noninsulin-dependent diabetes mellitus in black and white Americans. Diabetes Metab. Rev. 6:71.

Jackson, M.W., I.D. Kurzman, D.L. Panciera and E.G. MacEwen. 1994. Characterization of diet-induced obesity and subsequent weight loss in cats. Proc. 12th ACVIM Forum. p. 987.

Jacques, K. and S. Stewart. 1993. Does chromium have a future in feed? Feed Management 44(2):23.

Kegley, E.B. and J.W. Spears. 1995. Immune response, glucose metabolism, and performance of stressed feeder calves fed inorganic or organic chromium. J. Anim. Sci. 73(Suppl. 1):1.

Kirk, C.A. and P.W. Toll. 1996. Obesity treatment: Reasons for failure, the blueprint for success. Proc. 14th ACVIM Forum. p. 74.

Kirkwood, T.B.L. 1992. Comparative life spans of species: why do species have the life spans they do? Am. J. Clin. Nutr. 55:1191S.

Kronfeld, D.S., S. Donoghue and L.T. Glickman. 1991. Body condition and energy intakes of dogs in a referral teaching hospital. J. Nutr. 121:S157.

Kronfeld, D.S., S. Donoghue and L.T. Glickman. 1994. Body condition of cats. J. Nutr. 124:2683S.

Kuhlman, G. and R.E. Rompala. 1995. The influence of dietary sources of zinc, copper and manganese on canine reproductive performance. J. Anim. Sci. 73(Suppl. 1):186.

Laflamme, D.P., G. Kuhlman, D.F. Lawler, R.D. Kealy and D.A. Schmidt. 1994. Obesity management in dogs. Vet. Clin. Nutr. 1:59.

Lindemann, M.D. 1996. Organic chromium – the missing link in farm animal nutrition. In: Biotechnology in the Feed Industry. Proceedings of the 12th Annual Symposium (T.P. Lyons and K.A. Jacques, eds), Nottingham University Press, Loughborough, Leics. UK. pp. 299–314.

Lindemann, M.D., C.M. Wood, A.F. Harper, E.T. Kornegay and R.A. Anderson. 1995a. Dietary chromium picolinate additions improve gain feed and carcass characteristics in growing/finishing pigs and increase litter size in reproducing sows. J. Anim. Sci. 73:457.

Lindemann, M.D., A.F. Harper and E.T. Kornegay. 1995b. Further assessments of the effects of supplementation of chromium from chromium picolinate on fecundity in swine. J. Anim. Sci. 73(Suppl. 1):185.
Lyons, T.P. 1990. Pieces of the puzzle: Yeast cultures. Feed Management. 41(10):16.
Markwell, P.J., W. van Erk, G.D. Parken, C.J. Sloth and T. Shantz-Christienson. 1990. Obesity in the dog. J. Sm. Anim. Prac. 31:533.
Mertz, W. 1969. Chromium occurrence and function in biological systems. Physiol. Rev. 49:163.
Mertz, W. 1993. Chromium in human nutrition: A review. J. Nutr. 123:626.
Mooney, K.W. and G.L. Cromwell. 1995. Effects of dietary chromium picolinate supplementaion on growth, carcass characteristics, and accretion rates of carcass tissues in growing-finishing swine. J. Anim. Sci. 73:3351.
Morris, B.W., A. Blumsohn, S. MacNeil and T.A. Gray. 1992. The trace element chromium – a role in glucose homeostasis. Am. J. Clin. Nutr. 55:989.
Mossop, R.T. 1983. Effects of chromium III on fasting blood glucose, cholesterol and cholesterol HDL in diabetics. Cent. Afr. J. Med. 29:80.
Mowat, D.N. 1996. Twenty-five perceptions on trivalent chromium supplementation including effective fiber, niacin function and bloat control. Biotechnology in the Feed Industry. Proceedings of Alltech's Twelfth Annual Symposium. pp. 83–90.
Ohba, H., Y. Suketa and S. Okada. 1986. Enhancement of in vitro ribonucleic acid synthesis on chromium (III) in mouse liver. J. Inorg. Biochem. 27:179.
Okada, S., M. Suzuki and H. Ohba. 1983. Enhancement of ribonucleic acid synthesis by chromium (II) in mouse liver. J. Inorg. Biochem. 19:95.
Okada, S., H. Tsukada and H. Ohba. 1984. Enhancement of nucleolar protein synthesis by chromium (III) in regenerating rat liver. J. Inorg. Biochem. 21:113.
Olefsky, J.M. and O.G. Kolterman. 1981. Mechanisms of insulin resistance in obesity and non-insulin-dependent (type II) diabetes. Am. J. Med. 70:151.
Pagan, J.D., S.G. Jackson and S.E. Duren. 1995. The effect of chromium supplementation on metabolic response to exercise in thoroughbred horses. In: Biotechnology in the Feed Industry, Proceedings of the 11th Symposium (T.P. Lyons and K.A. Jacques, eds), Nottingham University Press, Loughborough, Leics. UK. pp. 249–256.
Page, T.G., L.L. Southern, T.L. Ward and D.L. Thompson, Jr. 1993. Effect of chromium picolinate on growth and serum and carcass traits of growing-finishing pigs. J. Anim. Sci. 71:656.
Press, R.I., J. Geller and G.W. Evans. 1990. The effect of chromium on serum cholesterol and apolipoprotein fractions in human subjects. West. J. Med. 152:41.
Purser, K.W. 1995. Effect of chromium picolinate on sow productivity. Proceedings of the fifteenth Prince feed ingredient conference. p.19.
Riales, R. and M.J. Albrink. 1981. Effect of chromium chloride supplementation on glucose tolerance and serum lipids including high-density lipoprotein of adult men. Am. J. Clin. Nutr. 34:2670.

Roginski, E.E. and W. Mertz. 1969. Effect of chromium III supplementation on glucose and amino acid metabolism in rats fed a low protein diet. J. Nutr. 97:525.

Root, M.V. 1995. Early spay-neuter in the cat: Effect on development of obesity and metabolic rate. Vet. Clin. Nutr. 2:132.

Schroeder, H.A. 1968. The role of chromium in mammalian nutrition. Am. J. Clin. Nutr. 21:230.

Schroeder, H.A. and J.J. Balassa. 1965. Influence of chromium, cadmium, and lead on rat aortic lipids and circulating cholesterol. Am. J. Physiol. 209:433.

Schwarz, K. and W. Mertz. 1959. Chromium (III) and the glucose tolerance factor. Arch. Biochem. Biophys. 85:292.

Sloth, C. 1992. Practical management of obesity in dogs and cats. J. Sm. Anim. Prac. 33:178.

Toepfer, E.W., W. Mertz, M.M. Polansky, E.E. Roginski and W.R. Wolf. 1977. Preparation of chromium-containing material of glucose tolerance factor activity from brewer's yeast and by synthesis. J. Agr. Food Chem. 25:162.

Trout, W.E. 1995. Hypothesis provides possible explanation as to chromium's effect on reproductive efficiency in swine. Feedstuffs. 67(53):12.

van Heugten, E. and J.W. Spears. 1994. Immune response and growth of stressed weanling pigs supplemented with organic and inorganic forms of chromium. J. Anim. Sci. 72(Suppl. 1):274.

Wang, Z., E.T. Kornegay, C.M. Wood and M.D. Lindemann. 1995. Effect of supplemental chromium picolinate on dry matter digestibility, nitrogen retention, and leanness in growing-finishing pigs. J. Anim. Sci. 73(Suppl. 1):18.

Ward, T.L., L.L. Southern and R.A. Anderson. 1995. Effect of dietary chromium source on growth, carcass characteristics, and plasma metabolite and hormone concentrations in growing-finishing swine. J. Anim. Sci. 73(Suppl. 1):189.

Wenk, C. 1995. Organic chromium in growing pigs: observations following a year of use and research in Switzerland. In: Biotechnology in the Feed Industry, Proceedings of the 11th Symposium (T.P. Lyons and K.A. Jacques, eds), Nottingham University Press, Loughborough, Leics. UK. pp. 301–308.

Wolter, R. 1995. Dietetic feeds and nutritional supplements. In: Biotechnology in the Feed Industry, Proceedings of the 11th Symposium (T.P. Lyons and K.A. Jacques, eds), Nottingham University Press, Loughborough, Leics. UK. pp. 143–150.

Wright, A.J., D.N. Mowat, B.A. Mallard and X. Chang. 1994. Chromium supplementation plus vaccines for stressed feeder calves. J. Anim. Sci. 72(Suppl. 1):132.

ORGANIC CHROMIUM: FROM THE KENTUCKY BLUEGRASS TO THE OLYMPIC ARENA

STEPHEN G. JACKSON and JOE D. PAGAN

Kentucky Equine Research, Inc., Versailles, Kentucky, USA

The effects of supplemental chromium in the diets of pigs, poultry, feeder cattle and humans have received a significant amount of attention. Effects on production have ranged from increasing lean body mass in an effect apparently similar to a repartitioning agent, to potentiating the immune system in cattle subjected to shipping stress. Pagan *et al.* (1995) reported that when chromium was fed to horses at 5 mg/day there was a decrease in glucose, insulin and cortisol compared to controls. More importantly, when horses fed supplemental chromium were subjected to a standard exercise tolerance test less lactic acid was produced in response to exercise compared to the untreated controls. Lactic acid is produced as a result of anaerobic glycolysis and is thought to be one of the factors involved in horses or other athletes becoming work intolerant. The greater the work intensity the higher the concentration of lactate since there is greater participation of anaerobic glycolysis in the generation of energy needed to support muscle metabolism. There are a myriad of factors besides work intensity which may affect the amount of lactate produced including cardiovascular fitness, diet, muscle fiber type and so on. The horses used in the experiment conducted by Pagan were in very similar states of fitness, were fed the same diet and subjected to the exact duration and intensity of exercise, the only variable being whether or not the horse received chromium. The results of that experiment stimulated a great deal of interest in our laboratory since it basically said that when the horses were fed chromium they were able to work more efficiently, at least in terms of energy yield. This work has led to the development of products which contain organic chromium and which are becoming widely accepted as effective nutritional adjuncts to a comprehensive feeding program for the equine athlete.

The use of chromium in the athletic horse

We have been using chromium supplementation in the Thoroughbred race horse for about three years. The specific supplement that we currently use is called 'Metaboleeze'. This product was designed for a friend who specializes in preparing two-year-old Thoroughbred horses for sale in what are referred to as 'Two-Year-Old-in-Training Sales'. His initial call indicated that he was having a severe problem with his two-year-olds becoming very nervous 'speed crazy', work intolerant, and going off feed, and was experiencing greater than acceptable numbers of horses which developed varying degrees of myositis. Myositis is a muscle disorder which is in some ways like severe cramps in the human. Horses get stiff, are unwilling to or cannot move. The muscles especially over the loin become very rigid and in extreme cases the horse may actually go down and be unwilling to rise. After a bout of 'tying-up' the urine becomes coffee-colored due to the presence of the muscle pigment myoglobin (hence another name frequently used is myoglobinuria), there is an elevation of the muscle enzymes CPK, SGOT (also called AAT and LDH). The presence of these muscle enzyme indicates muscle damage and is used clinically to diagnose the problem. Myositis, also called exertional rhabdomyolysis, has been very elusive in terms of prevention or treatment with consistent efficacy. Horses have responded to electrolyte therapy, to supplemental vitamin E and selenium, to reduction in the amount of grain (starch intake has been reduced), to treatment with mild tranquilizers (acepromazine), to sodium bicarbonate in the feed, and to a myriad of other treatments. However, not all horses respond to the same treatment regimens, and tying-up ranks as one of the more puzzling and frustrating clinical problems the horseman, nutritionist and equine practitioner have to deal with.

In evaluating the problems in the field we wanted to address two things. First, the nervousness and second the apparent problems that were occurring at the cellular level in the muscle. The product we put together was a combination of B vitamins and chromium from chromium yeast (Alltech Inc.). Organic chromium was added to address the possibility of insulin insensitivity and aberrant carbohydrate metabolism and B vitamins were added due to their role in energy metabolism and in the case of thiamin, nervousness. The trainer started using the product in 1994. From the beginning it was apparent that there was some beneficial effect. Whether the benefit was derived from the chromium or from the B vitamins or from a synergistic effect of the two has not been established. In this instance there is only the trainer's feeling and the reduction of the apparent problems of the horses in training and no solid data from a controlled experiment to offer as proof of the effectiveness of this supplement. Interestingly, last year the trainer took several horses to a sale and forgot his 'chrome'. Several days after arriving for the sale he felt that the horses were not 'looking well' and several of the fillies had elevated muscle enzymes indicative of horses that may be trying to tie-up. He immediately quizzed the stable manager and after some sleuthing discovered the 'chrome' had been left at home and the horses had been in a new and stressful surroundings for four days without supplemental chromium. I shipped 25 lb of Metaboleeze to him by next-day air and the horses were put back on the supplement and pretty quickly came right again. From this beginning Metaboleeze has grown into one of our major products. Trainers, literally all over the world, have heard of Metaboleeze and have started making it a part of their daily nutrition program. Metaboleeze is being used in

Dubai, Japan, England and extensively in the US by some of the major Thoroughbred trainers.

Even though it is our firm belief that the formula of Metaboleeze is the result of the effect of chromium and B vitamins, there is a body of evidence which suggests that adding 5 mg/day of organic chromium alone to the rations of horses in intense work has an effect as well. One of our clients, Rhone Poulenc Animal Nutrition, in Australia includes Alltech's chromium yeast in some of the rations designed for horses in rigorous training and sells chromium yeast for use as a top dress on race horse rations throughout Australia. Again, the major benefit that trainers see is a decrease in the incidence of tying-up and greater 'endurance down the stretch'.

The success of chromium on the race track has been more trial and error and word of mouth than a response of trainers to good, solid, scientific evidence. We continue to look at the science but until all of the answers are in there are some devotees out there who say 'don't confuse me with the facts I've made up my mind based on the performance of my horses'. Further evidence of the effectiveness of chromium in the diets of horses in intense training was seen this past summer during the Olympic Games. The following section will detail our involvement with the Atlanta Olympics; but while on the subject of chromium it bears mentioning that a large percentage of the horses at the Olympics were on Metaboleeze. Several horses that had a history of tying-up and work intolerance were immediately put on Metaboleeze when they arrived in Georgia upon the advise of Dr Pagan who directed Kentucky Equine Research involvement at Atlanta. The team veterinarians commented that the horses that were on chromium were the best they had ever been and several of those veterinarians have continued to order Metaboleeze for horses after the Olympics.

Since the original experiment in 1994 when we got positive results in the exercise tolerance test, we have explored opportunities for the introduction of chromium supplementation in the diets of the athletic horse. Data from human experiments have indicated that exercise increases the excretion of chromium and therefore results in an increased dietary requirement for this nutrient. We suspect that this is even more important in the horse which sees a great deal more carbohydrate and therefore blood glucose when in training than does the sedentary horse. The grazing horse may derive a majority of energy needs from the fermentation of fiber in the hindgut and the production and absorption of the volatile fatty acids acetate, propionate and butyrate. Of these, acetate and butyrate are ketogenic (fat-forming) and propionate is glucogenic. As such, resting plasma glucose concentrations in the hay- or pasture-fed horse range from 50–70 mg/dl depending on breed with that in ponies being lower and that in horses higher. These glucose concentrations are much lower than the concentrations seen in the grain-fed horse after a meal. Resting postprandial glucose concentrations in the grain-fed horse may be 85 mg/dl and peak glucose concentrations following a grain meal may reach a peak of 160 mg/dl. It is probably due to these high levels of plasma glucose in the grain-fed horse that we see an effect in the lowering of insulin, glucose and cortisol when chromium is fed. The chromium potentiated the action of insulin, resulting in lowered peak values for both glucose and insulin.

There is growing evidence that some horses may be insulin insensitive. This is particularly true of the chronic laminitic horse, the pony and the 'diabetic' horse. Thc future direction of research at KER with respect to chromium is in the areas of clinical anomalies in glucose metabolism and tissue response or lack of response to insulin. If in fact there is a potentiating effect of chromium on insulin, it may be that horses at risk due to the use of high carbohydrate meals may benefit from the routine administration of insulin in the feed.

The apparent success of the use of chromium yeast in the diets of performance horses is well documented in the minds of the trainers who have used the product. I guess our challenge as scientists is to document the metabolic and physiological reasons for this success.

Kentucky Equine Research feeds the Olympics

Kentucky Equine Research, Inc., an equine nutrition research and consulting company, recently served as the official nutritionists and feed supplier for the 1996 Summer Olympic Games held July 21–August 4 in Atlanta, Georgia. Dr Joe D. Pagan, president of KER, supervised the formulation, manufacture and distribution of horse feed to the 230 equine athletes representing 27 different countries. Dr Steve Duren and others on the staff of KER served in support roles to insure that the nutritional needs of all of the Atlanta participants were met.

This undertaking provided several unique challenges to KER. First, strict quarantine regulations established by the USDA to prevent the spread of the protozoal disease piroplasmosis required that all hay supplied to the Olympic venue be free of ticks. Ticks are the vector for the spread of this contagious disease which has been eradicated in the US, but which is still quite common in Europe and South America. To satisfy this regulation, hay was imported over 3000 miles (4800 km) from the Pacific northwest states of Washington and Idaho where there are none of the Dermacentor ticks in the environment. Maintaining strict quarantine requirements for the horses coming to Atlanta was part of the responsibility of KER in that all of the feed to the quarantine facility had to be quarantined as well.

Secondly, KER was required to provide a wide variety of horse feeds that would be suitable for horses from all over the world. KER supplied a range of hays that included timothy (the overwhelming favorite), alfalfa and a mixture of orchard grass and alfalfa. Compound feeds included three types of sweet feed and two pelleted rations in addition to a number of whole grains and feed ingredients. KER does consultation all over the world, and prior to the Olympics had a good idea of the types of feeds that Olympic caliber horses were used to eating. Formulated feeds ranged from a basic fortified grain mix to a more exotic type of textured feed using beet pulp and corn oil. Most of these feeds were manufactured by Hallway Feeds in Lexington, Kentucky. Additional feeds were manufactured by Banks Mill in Aiken, South Carolina and Flint River Mills in Bainbridge, Georgia. The feeds were stored in refrigerated trailers to insure that they remained fresh.

The feeds manufactured for the Atlanta Olympics all contained the KER vitamin-mineral fortification premix, KER 5X, which features Bioplex copper, zinc and manganese. The fortification package used and the formulation of the feeds was a key ingredient in trying to insure that the nutrient requirements of all of the horses at the Olympics were met. In addition to a full complement of trace minerals and macrominerals, fat and water soluble vitamins and a variety of energy and protein sources, Yea-Sacc, a live yeast culture, manufactured by Alltech, Inc., was used in all of the feeds provided to the Olympic venue as well as to all of the training centers prior to the Olympic Games. The research previously conducted at KER has shown that digestibility of fiber, phosphorous and dry matter in rations fed to horses is increased with the addition of Yea-Sacc. Also, Yea-Sacc is very palatable to horses and was one of the factors we used to insure that horses did not go off feed after the stress of air travel and being moved frequently in the months leading up to the Olympics.

Also, feed formulas were modified to include higher levels of corn oil. High fat intakes reduce the level of grain intake required to meet energy requirements and also reduce the total heat load placed on the horse. Additionally, many of the horses competing in the three-day event received chromium supplementation. Research at KER has demonstrated that chromium supplementation can improve metabolic response to exercise; and chromium supplementation has been shown to reduce heat stress in other species of animals.

Finally, the hot, humid weather in Atlanta was of great concern to everyone since most of the horses competing normally trained in more temperate climates such as found in Europe. To prepare for the heat, most international teams arrived in Georgia several weeks before the Games to allow their horses an opportunity to acclimatize to the Atlanta summertime weather. KER also made several nutritional modifications to prepare the horses for the heat. Most importantly, they provided a special electrolyte mixture which was formulated to replace the key electrolytes which are lost by horses exercising in the heat. This formula (appropriately named *Summer Games Electrolyte*) resulted from the consolidation of data gathered by a number of research groups from around the world. Much of this research was conducted specifically to address the conditions that were expected at the Atlanta games.

Overall, the horses tolerated the heat very well and there were no feed-related problems. The performances of the Australian, New Zealand, and American horses in the three-day event was particularly satisfying. These three countries swept both the team and individual medals in this discipline and each of these teams utilized much of KER's nutritional technology during their stays in Atlanta. Three-day eventing is one of the most strenuous equine competitions and it was very gratifying to see KER feeds and nutritional technology and supplements perform under such difficult conditions.

Although the 1996 Games have just been completed, KER has its sights set on the 2000 Olympics to be held in Sydney, Australia. KER has already had conversations with potential suppliers for the Sydney Games and is very interested in participating in the 2000 Olympics. Atlanta went very well and KER has a great deal of expertise to offer the Sydney organizing committee in meeting the nutritional needs of the horses at the next Olympic Games.

References

Pagan, J.D., T. Rotmensen and S.G. Jackson. 1995. Effect of chromium supplementation on metabolic response to exercise in Thoroughbred horses. In: Proceedings of the 14th Equine Nutrition and Physiology Symposium. Ontario, CA, Jan 19–21, 1995.

RESPONDING TO THE GROWING CHALLENGES OF CONSUMERISM FROM SALMONELLA TO ANTIBIOTICS

LIFE WITHOUT ANTIBIOTIC DIGESTIVE ENHANCERS

JOHN GADD

International Pig Consultant, Shaftesbury, Great Britain

Antibiotic digestive enhancers

Antibiotic digestive enhancers (also called growth promoters) have been with us since the 1950s. They have provided major benefits to pig producers, the welfare of the pigs under their care, and the reduction of environmental pollution, as well as to the consumer of pork. There is much evidence in support of antibiotics in all four areas that can be summarized as follows.

BENEFITS TO THE PRODUCER

The return to extra outlay ratios (REO) for antibiotic digestive enhancers vary in the literature from 1:2 to 1:8 (between 2 and 8 monetary units of increased income for each one monetary unit invested) with an average of 1:5. In Europe at present, a £2 per tonne investment in antibiotic should recoup £10 in increased return per tonne of feed, or 40 pence per pig securing £2 per pig extra income. The benefits are from better food conversion (typically 6% lower) faster growth (typically 5% better) and lower veterinary costs (typically 4% lower).

BENEFITS TO THE PIG

In Sweden following the ban of antibiotic use in feed there were 1.6% higher death losses post-weaning and a 400% rise (from 2 to 8%) in scouring incidence (Figure 1; Best, 1996).

BENEFITS TO THE ENVIRONMENT

The environmental advantages offered by antibiotic use are considerable, and at present tend to be overlooked. In the UK alone, producing 15 million finishing pigs per year, the reduction in slurry as a result of improved digestion due to antibiotics is around 45.5 liters per pig per year. This is equivalent to

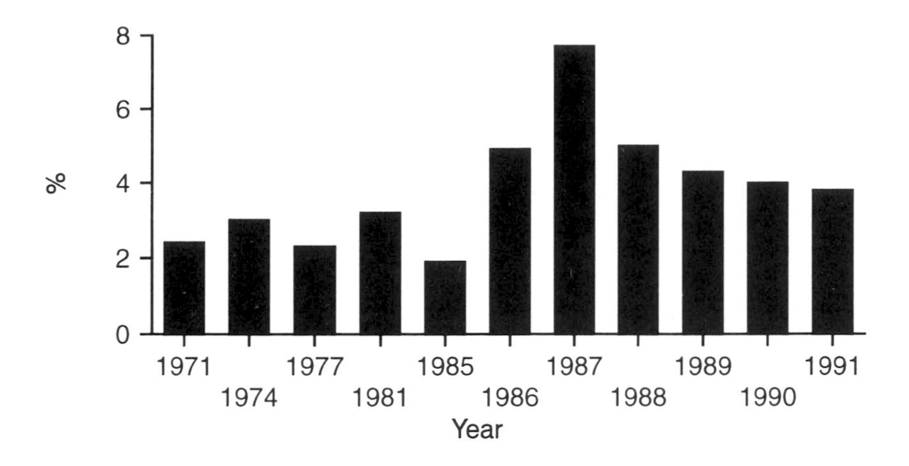

Figure 1. Incidence of post-weaning diarrhea in Sweden before and after the ban on in-feed antibiotic growth promoters (percentage of pig population affected) (Best, 1996).

a slurry lake one foot or more (305 mm) deep with an area of 15 square miles (Elanco, 1995). In terms of pollutants, assuming pig slurry diluted 1:1 with water contains 2 kg N_2 and 1 kg P_2O_5 per m^3 (MAFF, 1983), this is a reduction of over 127 tonnes of N_2 and 32 tonnes of P_2O_5 run-off per year in Britain alone. Lawrence (1992) quotes a reduction for the EU (12 countries at that time) of 48,750 tonnes N_2 and 16,250 tonnes phosphate from the use of digestive enhancers alone.

BENEFITS TO THE CONSUMER

If digestive enhancers were not available, then 42% of the added pork and bacon production costs in Europe (2.5 billion ECUs per year) would have to be met by the consumer. Production costs for pig meat would rise by up to 8.2%, or about £5 to £7 per head (Survey, by Viane and de Graene cited in Elanco, 1995). Thus any move to ban digestive growth enhancers, especially those based on antibiotics, would have a serious impact and on no account should be taken lightly. In Europe, approved antibiotic-based digestive enhancers in use now are licensed, carefully tested for safety, efficacy and freedom from residues in meat and thus can be considered safe in all respects. Why then, are we talking about the possibility of life without them? Is this a real threat or are we jumping at shadows?

The food safety scare

Listeriosis in cottage cheese, salmonella in eggs, *E. coli* 0157 and, most damaging of all, BSE have all provided those opposed to modern livestock farming methods with the opportunity to attack scientific advances in the productivity of animal production. Whether we agree with their arguments or not, the public at large is becoming very concerned indeed about food safety and the methods and materials we use to produce food. We are now seeing this concern being taken up by the retailers of food and by bureaucrats responsible for food safety. The retailers could be running scared of consumer concerns (e.g. recent requests to ban fishmeal from certain diets); and governments across Europe are lobbied increasingly by consumer groups to tighten regulations and control of modern farming methods (e.g. genetic engineering). There are signs that banning antibiotic growth enhancers could be next on the agenda.

AVOPARCIN: THE FIRST OF MANY TO BE BANNED?

The European Commission has now banned the use of the antibiotic avoparcin in animal feeds. It follows fears that its use could lead to antibiotic resistance being transferred from animals to humans. Denmark, Germany, Sweden and Austria had already introduced national bans on avoparcin and the European Commission has decided to press ahead with an EU-wide ban despite a report from its own Scientific Committee on Animal Nutrition that concluded there was no evidence of a link between avoparcin and increased antibiotic resistance in humans. But the Scientific Committee did recommend that more research was needed, and, at the time of writing, this seems to have been enough to encourage the Commission to press its case for an outright ban on this product.

Whether a ban of other antibiotic growth promoters will follow is arguable, and it is certainly not my intention in this paper to fuel any flames in this direction. However, we do need to look at what the pig-rearing world would be like without recourse to these valuable and safe, approved and well-proven products, now used on some billions of pigs over two whole generations of consumers eating pork and bacon from these pigs.

THE SUPERMARKETS' ATTITUDE

British supermarkets are the best in the world. I say this not because I am British myself, but because I visit every supermarket I can in the 14 main countries I work in. The British variety do an exceptionally good job at retailing food, and their meat departments are superb in dealing as they must with a perishable and relatively costly product. Prior to writing this paper, I have talked to representatives of four of the (nine) main UK supermarket chains. It is clear that all four are thinking along the same lines:

1. After the BSE scare, which is ongoing, the consumer is concerned about the safety of meat – not only of meat, but of all foods. It would be going too far to say that the public is apprehensive about food safety; but retailers claim that consumers look to them to ensure that what is on offer is, as far as humanly possible, safe to eat. The degree of trust is high.
2. It seems that the public's concern is far higher right now about food safety than for example animal welfare, a topic on which the British pig industry has been possibly overly-concerned for the past 10 years. Price, food safety and quality are now uppermost in the British meat buyer's mind while pig welfare seems to be a rather lame fourth.

3. The supermarkets I talked with were cautiously enthusiastic about 'farm safety assurance' schemes, ie. the co-ordination and supervision of clean and healthy meat from farm to plate. They are cautious because such protocols are costly to arrange and police. Retailers feel such programs are desirable altruistically, of course, but also must maintain sales volume and profits. For example, British supermarkets are now checking that British *feed mills* are adhering to agreed protocols, not just pig farms. One chief nutritionist reports that '20% of my time is spent talking to the retailers who audit our mills' (Hazzledine, 1997).
4. When directly asked about antibiotic-fed pork, the retailers all realized the value of digestive enhancers to all sectors of the marketplace. They had no intention of disparaging them, and were satisfied that safety checks are in place on the products licensed for that purpose. However, they were very sensitive to reaction on two counts: a. Any anxiety from the consumer that might lead to a drop in sales or profit. b. The stance a competitive supermarket chain might take. I gathered that even if such a stance were solely to gain a marketing advantage, if the competitor's strategy was successful other supermarkets could well follow suit. Of all things supermarkets may fear, what they fear most is another supermarket!

THE BUREAUCRAT'S ATTITUDE

It is exceptionally difficult to get a bureaucratic decision maker from one's own country to provide an informal comment on technical matters impinging on a farmer's profit or livelihood, let alone among the faceless hordes in Brussels! However, I was fortunate to sit among a gaggle of bureaucratic decision makers recently at an EU function and was able to ask questions, start a discussion and make some mental notes.

The senior bureaucrats who make decisions on new courses of action take their cues from their political masters who in turn are influenced by public opinion on the one hand (i.e. votes in the next election) and a variety of scientific committees on the other, with lobbyists bringing up rather a poor third, I was interested to note.

Once the strategy is proposed, bureaucrats have their own advisory committee(s) as a sort of cross-check – a wise precaution when dealing with any government or politician, apparently! They then decide on whether the proposal is sound enough to be mandated, or put up for general discussion within the Commission, which may or may not reach the public domain. With regard to antibiotic digestive enhancers, there seem to be two factors worrying the bureaucrats: antibiotic resistance and drug residues in the food.

Resistance and residues

It is generally accepted that digestive enhancers are unlikely to be involved to much degree (or any extent at all) in the residue problem due to the careful screening of this effect under the current UK/EU licensing arrangements. The area of concern lies in the curative use of antibiotics and withdrawal times. The recent alleged residue discoveries raised in Ireland could be an example of this, unless any unlicensed growth promoters were being used, of course.

The antibiotic resistance issue is less clear cut than the residue question, and the medical profession and veterinary scientists are not entirely in accord over it. Whenever bureaucrats, who exist due to support of their political faction, sense scientific disagreement or unease they will always err on the side of safety. Hence the withdrawal of avoparcin.

Unfortunately, the bureaucrats I talked with seem to consider that every additive molecule should be considered toxic unless proved otherwise! This attitude is as grotesque as the public's mistaken belief that farmers use the same in-feed drugs to further livestock performance as are used by doctors and hospitals in the treatment of human diseases and disorders. While the consumer is a decision *influencer*, the bureaucrat is a decision *maker*, which is much more immediate and serious in its long-term effect on perfectly viable and safe procedures and commercial products some of which, like antibiotic digestive enhancers, are of great value. For example, the decision makers I talked with seem unimpressed that 'to consume the same amount of amoxycillin as is in a single tablet prescribed by a doctor for a sick patient, for example, someone would have to eat 500 grams of meat containing the maximum acceptable level of residue every day for 55 years' (Ghesquière, 1996).

It is in this disquieting – I might almost say irrational – scenario that makes possible a ban on antibiotic digestive enhancers today.

Pig production without antibiotics

I now turn to the second part of my paper; should the worst come to pass what can pig producers do to mitigate the disaster?

THE SWEDISH EXPERIENCE

We are fortunate, at the expense of Swedish pig (and poultry) farmers who have had to suffer the problems of drastically-reduced availability of antibiotics for pig production, to be able to examine what happened and how they have managed to cope. I work regularly in Sweden at farm level so can provide background and inside information.

How did it happen?

In 1981, the Swedish media revealed that 30 tonnes annually of antibiotics were added to feed of pigs and chickens, solely to promote growth. Other reports followed, and the public were surprised to learn that medically-familiar drugs were given to *healthy* animals. The Swedish farming organizations responded to this heavy criticism at once by drawing up a voluntary policy of restricted and controlled use of antibiotics in all animal production, including banning of antibiotic growth promoters.

This was not enough. In 1985, a law was passed 'restricting antibiotics and other chemotherapeutics in feed solely for the prevention, easing or curing of disease'. Immediately afterwards a directive stated: 'Feed containing antibiotics or other chemotherapeutic substances may be sold *or used* only on a veterinary prescription *in each single case*'. This became law on January 1, 1986. With hindsight, the main fear leading to this draconian measure was that of antibiotic resistance.

What were the results?

Pork producers were worst affected, followed by calf rearers. Poultry units were least affected as alternative vaccinations were already available though at a higher cost. Farmers, of course, objected strongly, as did some veterinarians. Ståhle (1996) reported 'The first years were tough. Many veterinarians were not used to the situation. Some veterinarians were too strict and some too generous in prescribing medicine'.

Some reports have suggested that banning antibiotic growth promoters soon resulted in an overall increase in the use of antibiotics, legally administered on prescription, to cure the disease outbreaks previously held at bay by the growth enhancers. Figure 2 clearly shows that this was not the case. The reality is that antibiotic use is still less than half what it was in 1985. Written veterinary approval was needed last year for only 2% of grower/finisher pig feeds and 5% for piglet creep and post-weaning diets (Inborr, 1996, as cited by Best, 1996). Only a very few feed mills are licensed to make medicated feeds. A figure of US $45 per order (not per tonne) is charged by a typical manufacturer for additional mixing and separation charges.

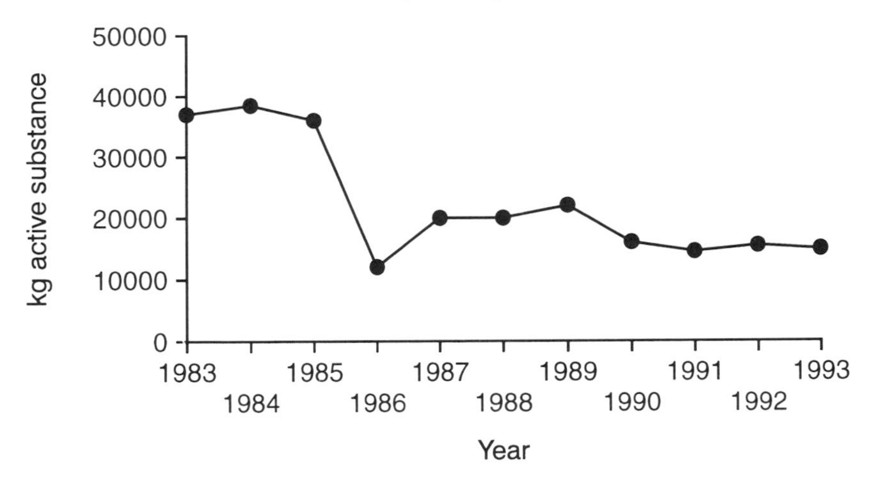

Figure 2. Total use of antibiotic growth promoter chemicals in Sweden, including those used as feed additives (1983–1985) and on a prescription only basis (1986–1994} (Best, 1996).

How did performance fare?

Productivity of pigs in Sweden has now recovered and surpassed the pre-1985 levels, although still somewhat below British, Irish, Danish and French (Brittany) average productivity (Tables 1 and 2).

Table 1. Swedish piglet production, 1993 to 1995.

	1993	1994	1995	1995 (best 20%)
Number of farms	417	419	417	84
Number of sows	30,274	32,179	35,237	7,476
Piglets per sow per year	18.5	19.0	19.4	22.8
Weight at delivery, kg	26.7	27.0	27.6	27.9
Weaning age, days	40.4	40.0	39.6	37.5
Litter size	10.8	10.9	11.0	11.5
Mortality, %				
Birth to weaning	15.2	15.1	14.7	12.4
Weaning to delivery	4.0	3.4	3.4	1.8

Table 2. Swedish slaughter pig production, 1993 to 1995.

	1993	1994	1995	1995 (best 20%)	1995 (best 20% based on FCR)
Number of pigs	228,000	189,000	241,000	33,400	
Slaughter weight, kg	76.9	78.5	79.6	81.4	80.2
Daily gain, g	800	818	820	902	852
Feed conversion	2.90	2.89	2.91	2.75	2.60
Mortality, %	1.6	1.5	1.6	0.9	1.3
Meat, %*	60.0	60.0	60.2	60.1	60.3

* Head and feet included.

The Swedish experience from the inside

It is important to take the Swedish figures in context. In the immediate post-ban period, much investigation was done on nutritional and alternative product uses along with weaning age and management changes (Table 3). Lactic acid bacteria, yeast cultures and organic acids were found to have a role when used at appropriate physiological stages. Mannanoligosaccharides (MOS), dietary fiber sources and other sugars had beneficial effects. Zinc oxide, used in Sweden at 2000 ppm for 14 days, was found very useful in controlling post-weaning scours (Table 4). There has been little evidence of the 'Zinc oxide fallaway' phenomenon experienced in some countries where the post-weaning scour seems to be held off while zinc oxide was in the diet, but returns once the product is withdrawn. This seems to me to be due to the fact that the Swedes now wean much later (4 to 5 weeks pre-1985; about 6 weeks now). Thus zinc oxide can be taken out in the eighth or ninth week rather than at the sixth or seventh week as in the UK. I have found that MOS have a scour-preventive action if used when zinc oxide is removed at 45 days of age; however, the Swedes report no difference in scour incidence with or without other oligosaccharides at the normal weaning time of 40 days or more (i.e. zinc oxide has to be withdrawn at 54 days). To my knowledge MOS have not been tested in this regard. Should zinc oxide use be banned due to soil pollution fears, this could markedly affect post-weaning scour control, thus raising the importance of other nutritional countermeasures, including that of MOS use.

Table 3. Alternatives investigated following the Swedish ban on antibiotic growth promoters.

Topic or product	Conclusions
Lactobacilli, yeast cultures, organic acids	Effective when used at the right physiological stage (i.e. young pigs)
Oligosaccharides, digestible fiber, special sugars	Beneficial results for oligosaccharides were positive but modest, the others (e.g. 5% dried sugar beet pulp) are useful in older finishers.
Zinc oxide (2000 ppm for 14 days)	Very useful in controlling post-weaning scours, little evidence of the 'Zinc oxde fallaway' phenomenon due to later weaning (4 to 5 weeks pre-1985; about 6 weeks now).
Nutrient density reduced	Post-weaning and creep diets were made more digestible and less antigenic, lower protein post-weaning, coarser grinding, use of feed enzymes both beneficial.
Later weaning	Stronger pigs with more developed digestion, better control of thermodynamics, and a better immune threshold reduce need for medication. Disadvantages lie in lower numbers of slaughter pigs produced per sow per year.
Management factors	Later weaning, all-in/all-out facilities, batch farrowed sows and careful control of stocking densities are important precursors to having to use far fewer antibiotics.

Table 4. The value of zinc oxide in controlling scour where routine in-feed medication is denied (10 pig farms)*.

	Without zinc oxide	With zinc oxide
Mortality, %	4.1	1.3
Scouring incidence, %	10–20	5
Daily liveweight gain, % change	–	+30

* Kavanagh (1993).

Nutrient density has been reduced by 10 to 20%. Post-weaning and creep diets were made more digestible and less antigenic well before the general move in these directions outside Sweden. Both are now generally accepted by pig nutritionists, if not every pig producer in the rest of the world. Coarser grinding helps, and of course the use of feed enzymes has made a positive contribution (Inborr, as cited by Best, 1996). Protein levels were also reduced in post-weaning diets with amino acid intake maintained (Table 5).

At later weaning dates the Swedes are certain of stronger pigs with more developed digestive systems, better control of thermodynamics within the

Table 5. Performance on lower protein post-weaning diets*.

	Crude protein (%) 18.6%	16.5%
Liveweight gain, g/day	785	802
FCR	2.77	2.68

*Goranssen (1997).

body and a better immune threshold to help counteract the need for peri-weaning medication. Still, the 14 day lag must cost them two slaughter pigs produced per sow per year.

Management factors (suggestions made at the end of this paper) have also contributed to reducing the need for antibiotics in Sweden. The Swedes tell me that later weaning, all-in/all-out facilities, batch farrowed sows and careful control of stocking densities with spare accommodation to take care of production bulges are important precursors to having to use far fewer antibiotics on the farm.

THE BRITISH EXPERIENCE

Over the past eight years I have had two clients who were engaged in 'green' or 'clean meat' (additive-free, plus many other restrictions), pork production, and three who subscribe to a supermarket's antibiotic-free pork contract. Thus, I have been privileged to assist all five farmers in the struggle to make a profit from these more specialized outlets.

The closely controlled 'clean meat' scenarios are described in a detailed paper (Gadd, 1992). However, both 'green' producers have reverted to being 'conventional' producers. Despite the supermarket offering their pork at a 30% (later 26%) premium and offering the producers 13% above a conventional contract sale price, the two producers found their increased costs were between 11 and 12% more and could easily be 20% more if standards were relaxed or they experienced a run of bad luck outside their control (Table 6). The supermarket, by the way, found that the public, while enthusiastically espousing meat produced to 'green' standards, would not pay 26% more for it, so the exercise failed.

The three other specialized-market producers have been more successful. These producers cannot use antibiotics except when their specialist pig veterinarian agrees it is wise; and even then antibiotics can only be administered by water medication or injection, never in the feed. They get a modest premium of from 2 to 5 pence per kg (about another 1.5 to 4% on the deadweight price); but that is not their objective. Their goal is to secure a firm, guaranteed market

Table 6. Estimated costs of compliance with a real meat contract.

	Cost (per pig)		Cost (per lb meat)	
	£	US	£	US
No growth promoters or copper	£3.00	$5.40	1.70p	3.06¢
Use of straw (extra handling)	£1.30	$2.34	0.80p	1.44¢
Cost of Veterinary Inspection Scheme	10p	18¢	0.05p	0.09¢
Cost of packers' inspection	Nil	Nil	–	–
Extra cost of post-weaning problems due to restriction in routine medication	£1.70	$3.06	0.77p	1.39¢
Extra cost of compliance to packers' inspection (extra hygiene, repairs, etc. mostly labor)	50p	–	0.25p	0.45¢
Total	£6.60p	$11.88	3.57p	6.43¢

Value of premiums offered = £9/pig, or +3.7 percent of sale price.

outlet. The premium is paid for better quality pork, not necessarily antibiotic-free meat, though the supermarket makes a claim for this.

British pioneer's advice
Similar 'green' contracts exist in Holland (e.g. Scharrel pigs) based on 'free' range but the market is small, about 0.5% of total production. The expectation is that it will grow to at least 2.5% (Den. Hartog, personal communication). A similar situation exists in France (Produits Label Rouge) but the uptake is low. Both countries regard the abandonment of licensed antibiotic-based growth enhancers as being without foundation. The three producers I visit have been raising pigs in these programs successfully for eight years, so we can learn the following from them:

1. Average weaning age is 28 to 32 days, but all three multi-suckle from 14 to 18 days. This is a difficult, highly-skilled procedure, and all three producers (averaging 180 sows and expanding) are excellent stockmen. Multi-suckling lessens weaning trauma and the weaker piglets are better able to withstand final separation.
2. As in Swedish operations, sows are batch farrowed in groups and re-arranged into service dates by the use of P.G. 600, a product which does not seem to be proscribed.
3. All pigs are all-in/all-out up to the final finishing move at 35 kg.
4. The services of a pig specialist veterinarian are considered essential and on average he routinely visits the farms every 5½ weeks.
5. Vaccinations and hygiene protocols are strictly followed, especially in the weaner accommodation, which is low-roofed with weaners in small (less than 25) batches. As a result I am sure the herds have a high immune status.
6. Food, except the creep and immediate post-weaning food, is bought from one source and comes from a baby pig food specialist company. While costing more, these feeds pay (Table 7).
7. Breeding stock (females) have mainly female-dominant traits, especially appetite, milkiness, litter size and motherliness, and also come from one source (i.e. one multiplier only). Boars (and artificial insemination) come from another source and are chosen on EBV, male-dominant traits. As a result, the individual price of the boar selected is of minor importance. His likely impact on profit is the cost-effective item of choice.
8. Strict quarantine (6 to 8 weeks) is practiced, off-farm in one case, and at a separate induction unit in the others).
9. Water is supplied after weaning via the dome-shaped turkey drinker instead of via the bite drinker (Tables 8 and 9).

What have we learned from all this?

1. Life will go on if no antibiotic digestive enhancers are permitted, but it will be another heavy burden, particularly for weanling pig producers.
2. All producers I have met over the past five years will suffer some loss of profit. I estimate a loss of £1 per pig for the best producers and up to £3 for average producers. Losses may be more for those below average, which may be unsustainable.

Table 7. The value of the link feed concept. UK costings of a specially-designed highly digestible post-weaning feed compared to a conventional formula. Fed for seven days post-weaning (the cost of the special feed is 50% higher).

	Conventional	Highly digestible link feed
Piglet growth rate/day		
0–7 days post-weaning	100 g	200 g
Days to slaughter	171	163
Food eaten/week, kg	1.05	2.10
Cost of food, % increase/week, pence	24.15	72.14 (+199%)
Food saved to slaughter at 2 kg/day, pence	–	16 g (256p)
Overhead saved (8 days) value in pence	–	8 days (50p)
Summary		
Savings in costs to slaughter, pence/pig		306
Less: extra cost of post-weaning food, pence/pig		48
Value of highly-digestible post-weaning feed, pence/pig		258

Table 8. Oasis turkey drinkers versus conventional drinkers for pigs.

	Oasis drinker	Bite drinkers
Weight in, kg	6.2	6.2
Weight out, kg	23.7	22.1
Days on trial	41	41
Average daily gain, g	427	388
Food eaten pig, kg	30.27	28.14
FCR	1.73	1.77
Pigs per tonne of feed, kg	578	565

Table 9. Effect of oasis drinkers on performance to slaughter.

	weaner stage oasis drinkers	weaner stage bite drinkers
Weight in, kg	23.7	22.1
Weight out, kg	88.6	88.2
Average daily gain, g	818	809
FCR	2.41	2.48
Killing out percentage	74.2	73.9
Saleable meat per tonne of feed, kg	420	398

3. Most average producers can regain 50% of this loss by attending to various management and production factors such as later weaning or changing to a full, (not watered-down) segregated early weaning regime.
4. The main problems encountered are post-weaning diarrhea and respiratory infection, both at present well-contained by several in-feed and perfectly safe antibiotic additives. Diarrhea can reduced by better attention to nutrition and the adoption of an all-in/all-out strategy (Gadd, 1994). Many of the respiratory infections can be addressed by recourse

to vaccination and concurrent environmental attention. To do one without the other jeopardizes the chances of success.

5. Herd immunity needs reinforcement. Building immunity is an amalgam of:
   a. Attending to hygiene, biosecurity and adoption of all-in/all-out management.
   b. Culling more intelligently to raise herd age profile and thus having more breeding females in the peak immune ages of third to sixth parities.
   c. Using vaccination protocols advised by the veterinarian.
   d. Adopting longer and more disciplined quarantine and induction regimes as advised by the veterinarian.
   e. Reduce stress, both acute (e.g. fright, fear) and low level (e.g. worry, anxiety, discomfort).
   f. Use any supplementary nutrients thought to raise immunity (e.g. organic chromium, selenium, proteinated trace elements, maybe some enzymes). New methods of watering weaners can help considerably.
   g. Attend to reduction in mycotoxin residues.

6. Concentrate on baby pig and weaning nutrition. This means paying more for the right feed (highly-digestible, non-antigenic, fresher), feeding management care (clean, low-waste receptacles), feeding special pre-weaning and for a varying length of time post-weaning (3 to 10 days) diets dependent on the digestive challenges present. The additional cost per tonne is more than compensated by performance to slaughter reduction (Table 7).

CAN NUTRITIONAL BIOTECHNOLOGY HELP?

Certainly, there are two attractive features of such products and procedures. First, they are nutrients derived from natural products, like yeast. Second, REOs for these products are high as they are generally cheap to add per tonne of feed. Amounts needed are relatively small (e.g. organic chromium added at 200 mg/tonne), but benefits can be high. Organic selenium and chromium have REOs estimated at 12:1 and as high as 26:1, respectively. Even products used at higher usage rates like Allzyme Vegpro for soya, MOS and acidulants have REOs at or better than antibiotic growth enhancers which can only usually manage 3:1 to 5:1.

The biotech products that should be considered by any pig farmer faced with an antibiotic digestive enhancer ban, especially at the post-weaning stage, from the author's experience, are listed in Table 10 along with some REOs from the literature and from my own clients' experience on-farm.

Performance reports of the biotech products demonstrate their benefits. There are excellent reports on MOS where they are used when other digestive enhancers are removed. It is recommended that MOS be used following withdrawal of 3 kg per tonne zinc oxide (after the UK mandatory permitted period of two weeks use for pigs up to 10 weeks old), especially if the pigs are weaned 'early' (i.e. 18 to 24 days).

An organic acidifier combined with enzymes, probiotics and an electrolyte

Table 10. Return to extra outlay ratio for Alltech products.

Product	Use	REO
De-Odorase	As a growth promoter	3 to 5:1
(Yucca extract)	In the mating house	4 to 7:1
Mannanoligosaccharide	On weaner pigs	
Bio-Mos	Modest conditions	Up to 15:1
	Good conditions	5 to 8:1
	In older finishers	Up to 6:1
Organic chromium	Typical breeding sows	Up to 26:1
Biochrome	Top breeding herd	12:1
	Growing/finishing pigs	14:1
Bioplex Mn	Average grading	11:1
(compared to $MnSO_4$)	Good grading	4:1
Acidifier/electrolytes/ probiotics/enzymes (i.e. Acid Pak 4-Way)	Young pigs scouring	5 to 9:1
	Not scouring	Up to 5:1
Organic selenium	Breeding sows and gilts	
Sel-Plex 50	In general	4:1
	In Se deficient situations	up to 8:1

mixture (Acid Pak 4-Way) is particularly valuable in cases of milder post-weaning scour where rank nutritional indigestibility or very poor environmental conditions are not involved.

An enzyme targeted to soya, Allzyme Vegpro, is useful in cases of post-weaning indigestibility leading to a hostile microbial increase.

Nutrients which enhance immune response are also available. Very small amounts of organic chromium, selenium and proteinated trace elements (Bioplexes) will go some way towards increasing natural immunity. While none of these (except possibly MOS) will replace a good antibiotic digestive enhancer *per se*, their concurrent use will help mitigate the effects of antibiotic unavailability, especially if used with other strategies suggested in this paper. Used wisely, the cost comes out about the same, if only a little more.

ECONOMETRICS

As we have seen, removal of antibiotic growth enhancers could save between £2 and £5 per tonne of feed (40 pence to £1 per slaughter pig) but the cost in gross margin terms, if a realistic choice of the best options is taken, is somewhere between £1 and £5 per pig, thus needing a 60 pence to £4 per pig recovery after the savings from not having to use a growth enhancer are deducted.

I have attempted to calculate the costs and benefits (based on European figures) of the defensive measures which the dedicated pig producer could put in place of antibiotic digestive enhancers currently available (Table 11). The calculations are based on current UK costs (December, 1996) and from

Table 11. The cost of a blanket-ban counter measures.

Savings	About 40–60p a pig from dispensing with a routine digestive enhancer.	
Cost of doing nothing once antibiotics are banned	At least £2.00 per pig, maybe up to £5.00 per finished pig sold	
Cost of advised or necessary countermeasures per pig sold:		
Countermeasure	Comment†	Cost
1. Wean later*	Lowered income less lowered costs on 22 pigs sold per sow per year – deficit	1p
2. Zinc oxide*	For 14 days	3p
3. Use of non-antibiotic replacements	Optional	8p
4. Organic acids*	In creep and weaner feeds	Up to 15p
5. Oligosaccharides, more digestible fiber, special sugars, etc.	Optional	Up to 20p
6. Multi-suckling	Probably useful if feasible. Cost of conversions varies from 20 to 80p per finished pig over 7 years progeny of 100 sows, but can with luck cost nothing.	0 to 80p
7. Destocking	15% destocking increases housing costs by 15%.	35p
8. All-in/all-out	Very variable depending on the refurbishment needed for farrowing and nursery. On the few cases known to me where this was done properly (i.e. approved by a veterinarian), cost was between £200 – £600 per sow, including interest, over 7 years.	£1.29 to £3.89
9. Improved post-weaning diet*	No cost, all recouped in better performance.	Nil
10. Increased use of vaccines*	Respiratory disorders	£1.60
11. 4% more scouring to be treated		15p
12. Use turkey drinkers in the nursery	Benefit payback quickly	Nil

* Strongly advised by the Swedes.
† Comment: Excluding the cost of conversion to all-in/all-out, the total costs of all these essential measures is around £3.28 per pig. After deducting 50p for non-use of a growth enhancer, the extra cost of maintaining respectable performance is well over £2.50 a finished pig. Note, however, that there are certain items you should be doing already – namely 2, 4, 7, (8), 9 and 12 that will improve performance and could reduce the necessary cost increase.

published evidence of benefits. The picture suggests that, over two years after an antibiotic digestive enhancer ban was imposed, the measures put in place would still provide a £2 per pig shortfall, at least in the short term. Incidentally, this is what the Swedes also found. Not taking defensive action (i.e. just take antibiotics out and carry on as before) could cost up to £6 per pig in the worst scenario. I know of one man who suffered an immediate run of problems which eventually was equivalent to £9 per pig. Both are unsustainable deficits in profit and cash flow terms.

The cost of delaying weaning from 24 days (average in northwest Europe in 1996) to 35 to 40 days in lost productivity has not been included in Table 10 because I feel very few British, Irish, Danish, Dutch and French (Brittany) producers would even countenance such a deliberate scaling down in potential performance. Mathematically, 40-day weaning must be costing the Swedes two slaughter pigs per sow per year (£20 gross margin), or around £1 on each finished pig to add to the £2 pig shortfall on current gross margin.

Conclusions

A ban on all antibiotics except on prescription for each and every case would have a severe impact on the performance and profit of pigs if no alternative strategy or products were adopted. Even so, gross margins could fall in the first two years until producers found the time to put these strategies in place. There are alternative products available, particularly in the biotechnology field, which are nutrients, not drugs, and offer a means of cushioning the blow. Further exploration of these possible solutions should be a priority from EU research establishments now that there is talk of antibiotic restrictions.

References

Best, P. Production without antibiotics; the Swedish experience. Feed International, April 1996, p. 8.

Elanco Animal Health. 1995. Digestive Enhancers – Answers to Some Commonly Asked Questions.

Gadd, J. 1994. One Out – All Out. Pig Farming, July 1994, pp. 16–17.

Gadd, J. 1992. The Green Animal, with Special Reference to Pig Production and Marketing. In: Proceedings Alltech Asia-Pacific Lecture Tour, pp. 39–58.

Goransson, L. 1997. Swedish experiences from banning in-feed antibiotics in: Recent Advances in Animal Nutrition, Proceedings Nottingham Feed Manufacturer's Conference (W. Haresign and P.C. Garnsworthy, eds). Nottingham University Press, Loughborough, Leics, UK. Repeated in Farmers Weekly Jan 10 1977 p. 34.

Ghesquière, 1996. Residue rules threaten drug supplies. Pig International, Aug 1996. pp. 35–36. Watt Publishing Co. Petersfield, England.

Hazzledine, M. 1997. Food Safety: Nutritionist Urges Common Sense Approach. Pig Farming, Farming Press, Ipswich, England. January. p. 20.

Kavanagh (1993) The case for zinc oxide. Pig International, Feb 1993, pp. 11–12. Watt Publishing Co. Petersfield, England.

Lawrence, K. 1992. Europe's View of Growth Promoters. Pig International, May, pp. 6–7.

MAFF. 1983. Profitable Uses of Livestock Manures. Advisory Booklet 2081.

Ståhle, G. 1996. Healthy Animals Without Antibiotics – Is It Possible? Proceedings: Antibiotics in Animal Health Sessions, Brussels.

BIOCHEMICAL AND PHYSIOLOGICAL BASIS FOR THE STIMULATORY EFFECTS OF YEAST PREPARATIONS ON RUMINAL BACTERIA

KARL A. DAWSON and IVAN D. GIRARD

Department of Animal Sciences, University of Kentucky, Lexington, Kentucky, USA

Introduction

Yeast products prepared from active cultures of *Saccharomyces cerevisiae* have become important strategic tools in many livestock production systems. While the magnitude of their beneficial effects can often be quite variable, it is clear that both the performance and health of livestock can be improved through the applications of specific yeast preparations and yeast cultures in feeds (Williams and Newbold, 1990; Wallace and Newbold, 1993; Harris, 1994; Yoon and Stern, 1995). In the last decade, the interest in using yeast as feed supplements for ruminant animals has resulted in considerable research on mechanisms which can be used to explain their beneficial effects on animal production. Many studies have suggested that some of the measurable production responses to yeast culture supplementation can be related to induced changes in the normal gastrointestinal bacterial population which result in improved digestion and protection from physiological dysfunctions and disease (Martin *et al.*, 1989; Dawson, 1992; Wallace, 1996). Several investigators have attributed these microbial changes in the rumen to the demonstrated stimulatory activities of viable yeast cells which enhance the growth of specific beneficial groups of bacteria in the digestive tract (Williams and Newbold, 1990; Dawson, 1992; Wallace 1996). Specifically, yeast preparations have been shown to increase the concentrations of cellulolytic (Wiedmeier *et al.*, 1987; Harrison *et al.*, 1988; Dawson *et al.*, 1990; Newbold and Wallace, 1992; Kim *et al.*, 1992), lactic acid-utilizing (Edwards, 1991; Girard *et al.*, 1993) and proteolytic (Yoon and Stern, 1996) bacteria in the rumen. In addition, yeast has been shown to enhance the activities of the bacteria that convert molecular hydrogen to acetate in the rumen (Chaucheyras *et al.*, 1995b). The ability of specific yeast preparations to increase the concentrations and activities of the beneficial bacteria in the rumen is believed to be key to their overall effects on ruminant production (Dawson, 1992; Newbold *et al.*, 1996; Wallace, 1996). Despite this basic understanding of some of the beneficial effects of yeast cultures on the bacterial populations in the rumen, the physiological basis for this enhanced microbial growth has not been completely described.

As a result, it is not possible to define the specific characteristics of yeast strains or cultures which make them useful as feed supplements.

Models explaining stimulatory effects of yeast cultures

The realization that yeast culture preparations can stimulate the growth and activities of ruminal microorganisms has led many investigators to examine and describe some basic biochemical and physiological mechanisms which may account for the stimulatory activities of yeast culture supplements and specific yeast preparations in the rumen (Table 1). Nisbet and Martin (1991) were the first to suggest that some of the stimulatory activities of yeast preparations can be related to the presence of the dicarboxylic acids in aqueous extracts from cultures of *S. cerevisiae*. This proposed mechanism is based on the observation that malic acid can stimulate the growth and activities of one of the predominant lactic acid-utilizing bacteria from the rumen, *Selenomonas ruminantium*. These investigators were able demonstrate that yeast preparations containing malic acid could enhance growth and lactate utilization in cultures of these ruminal anaerobes. Similarly, Rossi *et al.* (1995) observed that extracts from yeast culture preparations could enhance the growth of a second lactic acid-utilizing ruminal organism, *Megasphaera elsdenii*. These two organisms play an important role in processes which prevent lactic acid accumulation in the rumen. Selective stimulation of these organisms could moderate ruminal pH and prevent ruminal dysfunction. Data from these studies are consistent with other studies that have demonstrated an increased potential for lactic acid utilization when mixed ruminal populations are supplemented with yeast (Girard *et al.*, 1993). The metabolic basis for this stimulatory mechanism has not been clearly elucidated but may be related to a biochemical role for malate as an electron sink for lactate metabolism in these anaerobic organisms (Martin and Park, 1996). This particular mechanism is attractive from a practical point of view because it can explain the role of yeast cultures in controlling lactic acidosis in the rumen of animals fed high energy diets (Williams *et al.*, 1991). However, in studies that directly compared malic acid with active yeast preparations, malic acid was not able to provoke the same stimulatory response as a yeast culture preparation in the rumen of mature sheep (Newbold *et al.*, 1996). These *in vivo* studies suggest that malate may have an important physiological role in lactate-metabolizing ruminal bacteria, but the main stimulatory effects of yeast culture supplements on ruminal bacteria may not be mediated by malic acid.

More recent studies suggest that other naturally occurring metabolites from yeast cells may have a role in stimulating the growth of beneficial strains of ruminal microorganisms. Chaucheyras *et al.* (1996) provided evidence that suggests amino acids and vitamins derived from *S. cerevisiae* cultures can stimulate lactate uptake by *M. elsdenii*. These studies describe a nutritional role for specific components in yeast culture preparations which can stabilize the physiochemical conditions in the rumen and provide for improved fermentation efficiencies by moderating ruminal pH. Interestingly, these investigators also suggest that the ability of yeast to stimulate lactic acid utilization may be augmented by the ability of live yeast cells to compete with lactic acid-producing bacteria like *Streptococcus bovis* for substrates. This type of activity would also prevent the build up of detrimental concentrations of lactic acid in the rumen and prevent the associated ruminal dysfunctions. This same group of investigators (Chaucheyras *et al.*, 1995a) have demonstrated that vitamins such as thiamin derived from yeast cell preparations can enhance the growth and activities of anaerobic fungi in the rumen. These fungi are actively cellulolytic and may play a key role in the digestion of some of the more recalcitrant fibrous substrates in the rumen. Stimulation of these fungi could enhance fiber digestion and could support the growth of greater concentrations of ruminal bacteria through cross-feeding mechanisms or by increasing substrate availability. These nutritional roles for yeast culture preparations could explain some of the variability associated with the use of yeast supplements in practical situations, since the nutritional needs of ruminal microorganisms are often influenced by the animal's diet. However, these mechanisms are inconsistent with other studies which suggest that active yeast cells are needed for optimal stimulatory effects in the rumen and that these optimal stimulatory activities are not seen with nutrients found in simple yeast extracts.

The ability of yeast to stimulate the growth of ruminal bacteria by removing trace amounts of oxygen from the environment has also been proposed as a basic mechanism which can be used to explain its stimulatory activities (Newbold *et al.*, 1996). This hypothetical mechanism is based on the observation that many of the organisms that contribute to normal microbial activities in the gastrointestinal tract are strict anaerobes and are very sensitive to oxygen exposure. Small amounts of oxygen entering the rumen as swallowed air or in feed materials could depress the activities of many of these important anaerobes. Newbold *et al.* (1996) demonstrated that the ability of specific yeast strains to stimulate the growth of anaerobic bacteria in rumen-simulating cultures is partially dependent on the respiratory activities of the yeast and that this activity is absent in respiratory mutants which have lost their ability to utilize oxygen. These investigators have also demonstrated that yeast strains selected for their superior ability to metabolically remove oxygen from the environment can also increase the concentrations of total anaerobes and cellulolytic bacteria in rumen-simulating continuous cultures. These observations are all consistent with the demonstrated effects of successful yeast cultures on the ruminal bacterial population. This oxygen-scavenging mechanism remains controversial because facultative bacteria in the rumen have a tremendous potential for removing oxygen from the environment. This activity normally prevents oxygen from reaching toxic levels in the rumen. In addition, attempts to remove oxygen by substituting specific chemical reducing agent for yeast cultures have been unsuccessful (Chaucheyras *et al.*, 1996).

Several laboratory methods have been used to demonstrate the stimulatory effects of yeast preparations on ruminal bacteria in both mixed and pure culture systems. Rumen simulating continuous cultures have routinely been used to successfully demonstrate the stimulatory effects of yeast cultures on the total anaerobic microbial populations, cellulolytic bacteria and lactic acid-utilizing bacteria (Dawson *et al.*, 1990; Girard *et al.*, 1993; Newbold *et al.*, 1995, 1996). In addition, a limited number of studies have examined the effects of various yeast-derived preparations on the growth characteristics, cell yields and physiological activities of certain ruminal bacteria (Nisbet and Martin, 1991; Rossi *et al.*, 1995; Chaucheyras *et al.*, 1995a,b). Most of these studies with isolated ruminal bacteria have specifically focused on the effects of yeast on a small group of lactic acid-metabolizing bacteria from the rumen and have not yet examined the effects on other important physiological groups. One of the major concerns with many of these assays for stimulatory activities has been the large amount of material needed to elicit a measurable response. While only a limited number of dose-rate studies have been reported, most of the assay systems have only demonstrated stimulatory activities with massive doses of yeast culture (Table 1). When these *in vitro* dose rates are translated into effective dose rates for use in animal feeding programs, the yeast culture supplements examined in these systems would have to be added to ruminant diets at levels between 60 g and 3 kg per day. This requirement for large amounts of yeast casts some doubt about many of the proposed physiological mechanisms based on studies of stimulatory responses because the required doses of yeast are far above those provided in normal feed supplementation strategies. As a result, the requirement for large quantities of specific supplements make many of these mechanisms untenable in practical feeding situations.

Stimulation with low concentrations of yeast culture

The stimulatory effects of small concentrations of yeast culture and yeast cells can be readily examined in batch cultures of isolated ruminal bacteria grown on well-defined anaerobic media which do not support the growth of the yeast. In these assay systems, the stimulatory effects of the yeast supplement are reflected in altered bacterial growth patterns and can be evaluated over a rather short period (< 24 h). For example, the addition of low concentrations (2.4 10^4 CFU/ml) of active yeast cells resulted in a significant decrease in the amount of time required for the cellulolytic bacterium *Ruminococcus albus* to initiate active growth in the basal medium (Figure 1). This type of assay appears to provide a sensitive test for evaluating the stimulatory activities of specific yeast and yeast-derived preparations on specific ruminal bacteria (Table 2). It is possible to use these types of assays to demonstrate that yeast cells can beneficially alter the growth characteristics of many of the major cellulolytic bacteria (*R. albus*, *R. flavefaciens* and *Fibrobacter succinogenes*), and lactic acid-utilizing bacteria (*S. ruminantium* and *M. elsdenii*), while supplementation has little influence on some of the bacteria associated with lactic acid production (Lactobacilli and *Streptococcus bovis*). These stimulatory activities are consistent with the population changes observed in mixed cultures of ruminal bacteria in rumen-simulating continuous cultures given yeast culture supplements.

Studies with isolated ruminal bacteria clearly suggest a mode of action which can provide beneficial effects when very low concentrations of yeast are included in animal diets and does not rely on massive additions of yeast in

Table 1. Proposed mechanisms used to explain the stimulatory and inhibitory effects of yeast preparations on gastrointestinal bacteria.

Reference	Key component or activities	Method for measuring effects	Concentration required for optimal activities	Dose responses measured	Estimated daily dose required for a ruminant*(g)
Nisbet and Martin (1991)	Malic acid in yeast preparations enhances the activities of lactic acid-utilizing bacteria	Enhanced growth and lactate uptake by *Selenomonas ruminantium*	2.5 mg of yeast culture/ml of culture	Yes, between 1 and 10 mg/ml	125–250
Rossi *et al.* (1995)	Malic acid in yeast preparations enhances the activities of lactic acid-utilizing bacteria	Enhanced growth, lactate uptake and activities of *Megasphaera elsdenii*	5.0 mg of yeast culture/ml of culture	Yes, between 1 and 5 mg/ml	125–250
Chaucheyras *et al.* (1995a)	Vitamins from yeast enhance the activities of cellulolytic fungi in the rumen	Enhanced growth, cellulolytic activity and germination of *Neocallimastix frontalis* in vitamin-deficient media	2×10^7 cells/ml	Yes, between 2×10^5 and 2×10^7 cells/ml	100–1000[†]
Chaucheyras *et al.* (1996)	Amino acids produced by viable yeast cells enhances the activities of lactic acid-utilizing bacteria in the rumen	Enhanced growth and lactate uptake by *Megasphaera elsdenii*	6×10^7 cells/ml	Yes, between 10^7 and 10^8 cells/ml	300–3000[‡]
Chaucheyras *et al.* (1996)	Glucose metabolism by yeast cells competes with lactate producers in the rumen	Inhibition of glucose utilization by isolated strains of *Streptococcus bovis*	10^7 cells/ml	No	50–500[‡]

Table 1. Continued

Reference	Key component or activities	Method for measuring effects	Concentration required for optimal activities	Dose responses measured	Estimated daily dose required for a ruminant*(g)
Newbold *et al.* (1996)	Respiratory activities in yeast cells protect anaerobic gastrointestinal bacteria from oxygen damage	Increased oxygen uptake by *in* vitro culture of mixed ruminal bacteria	1.3 mg/ml of culture	No	66
Girard (1996)	Small peptides from metabolically active yeast trigger exponential growth of ruminal bacteria	Enhanced growth of *Ruminococcus albus* and other strains of ruminal bacteria	140 to 280 thousand CFU/ml	Yes, between 70 and 560 thousand cells /ml	7–14

*Based on the amount required to give an optimal response with a single daily dose in a 50 l rumen.
† May give measurable responses with as little as 1–10 g/day
‡ Assumes that the supplement contains between 1 and 10 billion cells/g.

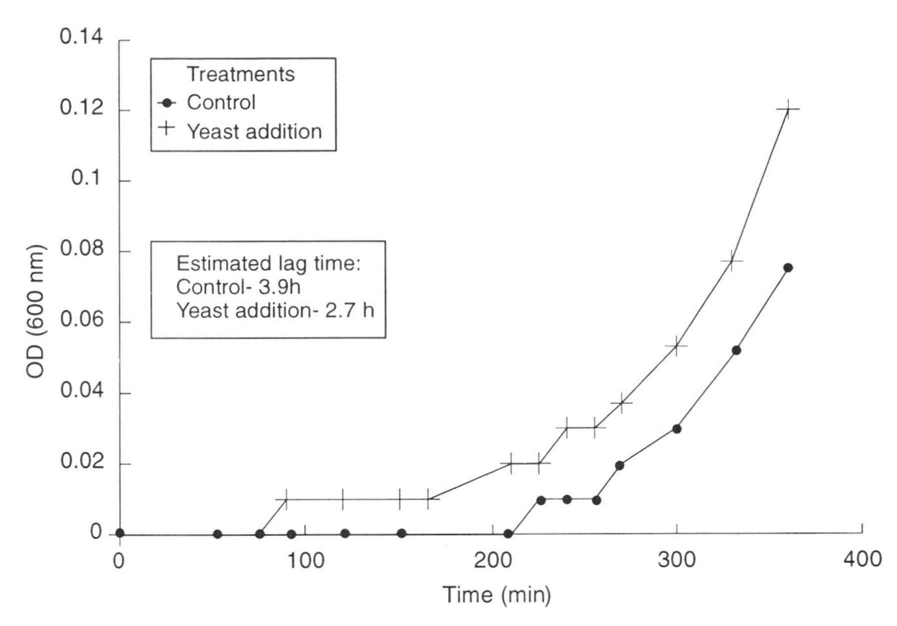

Figure 1. Effect of active yeast cells on the time required for *Ruminococcus albus* to initiate growth in baral medium.

Table 2. Effects of live yeast cell preparations containing *Saccharomyces cerevisiae* stain 1026 on the growth characteristics of representative strains of ruminal bacteria.*†

Organism	Effects on growth
Proteolytic/sacchrolytic bacteria	
Prevotella ruminicola	36% decrease in time required to initiate growth
Ruminobacter amylophilus	7% increase in maximum cell yield
Cellulolytic bacteria	
Butyrivibrio fibrisolvens	21% decrease in time required to initiate growth
Ruminococcus flavefaciens	58% decrease in time required to initiate growth
Ruminococcus albus	31% decrease in time required to initiate growth
Fibrobacter succinogenes	63% increase in maximum cell yield
Lactic acid-utilizing bacteria	
Selenomonas ruminantium	13% decrease in time required to initiate growth and 130% increase in maximum cell yield
Megasphaera elsdenii	35% decrease in time required to initiate growth
Lactic acid-producing bacteria	
Lactobacillus plantarum	<10% decrease in time required to initiate growth
Streptococcus bovis	<10% decrease in time required to initiate growth

* Yeast cells were added to fresh broth cultures of the ruminal bacterium from an overnight culture to provide final concentration of 2.8×10^4 yeast cells/ml.
† Adapted from Girard and Dawson (1994) and Nisbet and Martin (1991).

the rumen (Girard and Dawson, 1994). In addition, detailed studies of this type of stimulatory response have indicated that these stimulatory activities are independent of the respiratory activities in the yeast cells and have been observed in respiratory deficient mutants (Table 3). This suggests that the stimulatory activity is not influenced by the oxygen-scavenging activities described by other investigators.

Girard and Dawson (1995) have used these batch culture assay techniques to examine various fractions derived from whole yeast culture preparation and yeast cells. In these studies, a number of small stimulatory components (molecular weight <1000 Da) have been identified which stimulate the growth of *R. albus* and decrease the time required for a transition from a stationary growth phase to an active exponential growth by up to 50%. Some of these compounds have been identified as small short-chain peptides which appear to be metabolic products derived from the growth of specific yeast strains. These compounds can be isolated from culture fluids collected from yeast cultures grown on a semi-defined medium and are active in very small concentrations (<1 micromoles/l). The active concentrations are within the ranges which might be expected with low levels of yeast culture supplementation (<10 g/head/day).

It is not currently possible to assign a specific metabolic role to these stimulatory peptides in specific ruminal bacteria. However, the useful concentrations of these compounds are much lower than would be expected if these materials were acting as a specific nutrients or metabolic precursors for the bacterial cells. These active concentrations are also lower than would be expected if these materials are required as cofactors for specific metabolic reactions within the bacterial cell. This leaves the possibility that these small peptides serve as metabolic triggers or inducers which can serve to initiate the transition from a stationary growth phase to active exponential growth (Figure 2). Similar types of triggers have not been described in bacteria but may be useful for inducing protein synthesis and serve in a regulatory role at the level of nucleic acid transcription.

The ability of yeast culture to influence anaerobic bacteria concentrations in the rumen appears to be associated with the presence of live yeast cells or some heat labile component in the yeast culture supplement. Autoclaved yeast culture preparations were ineffective in increasing the concentrations of culturable bacteria in rumen-simulating cultures (Dawson *et al.*, 1990). However, actively growing or budding yeast cells may not be needed. Studies with irradiated yeast cells which can no longer reproduce but are

Table 3. Effects of *Saccharomyces cerevisiae* strain 1026 and selected respiratory mutants on the lag time in cultures of *Ruminococcus albus* strain 7.

Yeast addition*	Estimated lag time (hours)
None	3.16^{b}
Strain 1026	1.58^{c}
Respiratory mutant 1	1.23^{c}
Respiratory mutant 2	1.45^{c}

* Yeast cells provided to give a final concentration of 1.4×10^{5} CFU/ml.

[bc] Means within a column with different superscripts differ ($P<0.05$).

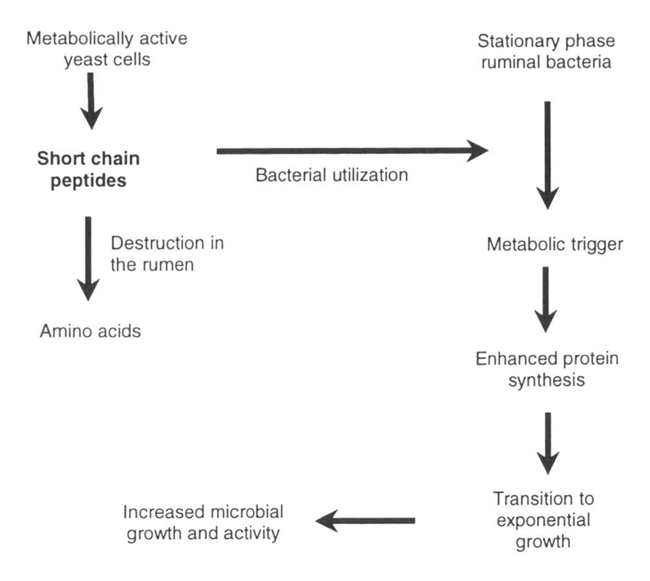

Figure 2. Possible metabolic role of yeast stimulatory peptides on ruminal bacteria.

metabolically active have shown that some of the stimulatory effects are associated with metabolic activities within the yeast cells and do not require actively growing yeast cells (El Hassan *et al.*, 1993). One feature of this short chain peptide model for the beneficial effects of yeast in the rumen is that it explains the need for metabolically active yeast in the rumen. Chromatographic analysis suggests that these small peptides are not readily found in the ruminal environment and may not be freely available even in the rumens of animals receiving a yeast supplement. Because of the peptide nature of these compounds, it is very likely that they have a very short residence time in the rumen and can be rapidly destroyed by the proteolytic activities of the bacterial population (Figure 2). As a result, the ability of viable yeast to replenish these peptides may be a key part of their activities and would explain the apparent need for actively metabolizing yeast in the feed supplements.

Conclusions

The interaction between metabolically active yeast cells and the microbial populations in the rumen is complex, but appears to play a key role in the beneficial effects of yeast cultures in ruminant production systems. A number of metabolic mechanisms have been proposed which could explain the stimulatory effects of yeast on ruminal bacteria. However, most of these models suggest that large doses of yeast are required for the beneficial effects on the ruminal microbial population. Recent studies suggest that low levels of yeast can stimulate the growth of specific strains of ruminal bacteria through the activities of a number of small molecular weight compounds which actively stimulate the growth of specific ruminal bacteria at extremely low concentrations. An understanding of the role of these compounds in metabolism and their production in yeast cells will provide a rational basis for selecting yeast strains and production systems on the basis of their abilities to stimulate metabolic activities in the beneficial microbial populations in the rumen.

References

Chaucheyras, F., G. Fonty, G. Bertin and P. Gouet. 1995a. Effects of live *Saccharomyces cerevisiae* cells on zoospore germination, growth and cellulolytic activity of the rumen anaerobic fungus, *Neocallimastix frontalis* MCH3. Current Microbiol. 31:201.

Chaucheyras, F., G. Fonty, G. Bertin and P. Gouet. 1995b. *In vitro* H_2 utilization by a ruminal acetogenic bacterium cultivated alone or in with an Archaea methanogen is stimulated by a probiotic strain of *Saccharomyces cerevisiae*. Appl. Environ. Microbiol. 61:3466.

Chaucheyras, F., G. Fonty, G. Bertin, J.M. Salmon and P. Gouet. 1996. Effects of a strain of *Saccharomyces cerevisiae* (Levucell SC), a microbial additive for ruminants, on lactate metabolism *in vitro*. Can. J. Microbiol. 42:927–933.

Dawson, K.A. 1992. Current and future role of yeast culture in animal production: A review of research over the last six years. In: Supplement to the Proceedings of Alltech's 8th Annual Symposium. Alltech Technical Publications, Nicholasville, KY.

Dawson, K.A., K.E. Newman and J.A. Boling. 1990. Effects of microbial supplements containing yeast and lactobacilli on roughage-fed ruminal microbial activities. J. Anim. Sci. 68:3392.

Edwards, I.E. 1991. Practical uses of yeast culture in beef production: Insight into its mode of action. In: Biotechnology in the Feed Industry, Vol.VI. Alltech Technical Publications, Nicholasville, KY.

El Hassan, S.M., C. J. Newbold and R.J. Wallace. 1993. The effects of yeast culture on rumen fermentations: Growth of the yeast in the rumen and requirement for viable yeast cells. Anim. Prod 56:463.

Girard, I.D. 1996. characterization of stimulatory activities from Saccharomyces cerevisiae on the growth and activities of ruminal bacteria. Ph.D. Dissertation. University of Kentucky, Lexington.

Girard, I. D. and K. A. Dawson. 1994. Effects of yeast culture on the growth of representative ruminal bacteria. J. Anim. Sci. 77(Suppl.1):300 (Abstr.).

Girard, I.D. and K.A. Dawson. 1995. Stimulatory activities from low-molecular weight fractions derived from *Saccharomyces cerevisiae* strain 1026. 23rd Biennial Conference on Rumen Function, Chicago, Illinois, p. 23.

Girard, I.D., C.R. Jones, and K.A. Dawson. 1993. Lactic acid utilization in rumen-simulating cultures receiving a yeast culture supplement. J. Anim. Sci. 71(Suppl. 1):288.

Harris, B. 1994. Direct-fed microbial update. Feed Management. 45:22–26.

Harrison, G.A., R.W. Hemken, K.A. Dawson, R.J. Harmon and K.B. Barker. 1988. Influence of addition of yeast culture supplement to diets of lactating dairy cows on ruminal function and microbial populations. J. Dairy Sci. 71:2967.

Kim, D.Y., M.R. Figueroa, D.P. Dawson, C.E. Batallas, M.J. Arambel and J.L. Walters. 1992. Efficacy of supplemental viable yeast culture with or without *Aspergillus oryzae* on nutrient digestibility and milk production in early to mid lactation dairy cows, J. Dairy Sci. 75(Suppl. 1):206 (Abstr.).

Martin, S.A., D.J. Nisbet and R.G. Dean. 1989. Influence of a commercial yeast supplement on the *in vitro* ruminal fermentation. Nutr. Rep. Int. 40:395.

Martin, S.A. and C.M. Park. 1996. Effect of extracellular hydrogen on organic acid utilization by the ruminal bacterium *Selenomonas ruminantium*. Current Microbiol. 32:327.

Newbold, C.J. and R.J. Wallace. 1992. The effect of yeast and distillery byproducts on the fermentation in the rumen simulation technique (Rusitec). Anim. Prod. 54:504.

Newbold, C.J., R.J. Wallace, X.B. Chen and F.M. McIntosh. 1995. Different strains of *Saccharomyces cerevisiae* differ in their effects on ruminal bacterial numbers in vitro and in sheep. J. Anim. Sci. 73:1811.

Newbold, C.J., R.J. Wallace and F.M. McIntosh. 1996. Mode of action of the yeast *Saccharomyces cerevisiae* as a feed additive for ruminants. Br. J. Nutr. 76:249.

Nisbet, D.J., and S.A. Martin. 1991. Effect of *Saccharomyces cerevisiae* culture on lactate utilization by the ruminal bacterium *Selenomonas ruminantium*. J. Anim. Sci. 69:4628.

Rossi, F., P.S. Cocconcelli and F. Masoero. 1995. Effect of a *Saccharomyces cerevisiae* culture on growth and lactate utilization by the ruminal bacterium *Megasphaera elsdenii*. Ann. Zootech. 44:403–409.

Wallace, R. J. 1996. The mode of action of yeast culture in modifying rumen fermentation. In: Proceedings of Alltech's 12th Annual Symposium on Biotechnology in the Feed Industry. Nottingham University Press, Loughborough, Leics. UK.

Wallace, R.J. and C.J. Newbold. 1993. Rumen fermentation and its manipulation: The development of yeast cultures as feed additives. In Biotechnology in the Feed Industry. Ed by T. P. Lyons, Alltech Technical Publications p 173.

Wiedmeier, R.D., M.J. Arambel and J.L. Walters. 1987. Effects of yeast culture and *Aspergillus oryzae* fermentation extract on ruminal characteristics and nutrient digestion. J. Dairy Sci. 70:2063.

Williams, P.E.V., and J.C. Newbold. 1990. Rumen probiosis: The effects of novel microorganisms on rumen fermentation and rumen productivity. In Recent Advances in Animal Nutrition W. Haresign and D.J.A. Cole, ed. Butterworths, London, England. p. 211.

Williams, P.E.V., C.A.G. Tait, G.M. Innes and C.J. Newbold. 1991. Effects of the inclusion of yeast culture (*Saccharomyces cerevisiae* plus growth medium) in the diet of dairy cows on milk yield and forage degradation and fermentation patterns in the rumen of steers. J. Anim. Sci. 69:3016.

Yoon, I.K. and M.D. Stern. 1995. Influence of direct-fed microbials on ruminal microbial fermentation and performance of ruminants: A review. Asian Ausralas J. Anim. Sci. 8:533.

Yoon, I.K. and M.D. Stern. 1996. Effects of *Saccharomyces cerevisiae* and *Aspergillus oryzea* cultures on ruminal fermentation in dairy cows. J. Dairy Sci. 79:411

PATHOGEN CONTROL IN THE HUMAN FOOD CHAIN AND THE NECESSITY FOR HAACP PROGRAMS BEGINNING AT THE FARM

MELISSA C. NEWMAN

Q Laboratories, Cincinatti, Ohio, USA

The 1990s have been a decade of food awareness. This awareness includes the search for fast, fat free, nutrient fortified, safe food. Unfortunately, most American consumers view food safety as someone else's problem. Education of the general public about the safe preparation and storage of food is just now beginning. The Centers for Disease Control estimates that 97% of food-borne illness is a result of improper food handling in the home or restaurant. The predominant factors associated with food-borne illness in meat and poultry are described in Table 1.

The United States Department of Agriculture Food Safety and Inspection Service (USDA-FSIS) estimates that the contamination of meat and poultry products with pathogenic bacteria results annually in as many as 4000 deaths and 5,000,000 illncsscs (USDA/FSIS, 1996). The primary bacterial pathogens associated with these numbers are salmonella, campylobacter, *E. coli* 0157:H7 and *Listeria monocytogenes*. The accuracy of the statistics for meat and poultry-associated illness has been highly disputed, although the USDA contends that the estimates may actually be conservative. The tragic deaths of four children and 600 others who were made ill by eating improperly cooked hamburgers on the west coast drew public attention to the potential danger of food pathogens. This incident is a primary reason for the meat inspection regulations released on July 25, 1996.

The purpose of the FSIS is to ensure that meat, poultry, and egg products are safe, wholesome, and properly marketed. This is accomplished by an inspection program of meat and poultry slaughter and processing operations. FSIS has developed the Pathogen Reduction/Hazard Analysis Critical Control Point (HACCP) regulations to reduce the risk of food-borne illness associated with the consumption of meat and poultry products. These regulations replace the somewhat antiquated visual inspection program utilized by USDA in the past. The modernization of the old inspection system incorporates microbiological testing to monitor the presence of potentially pathogenic bacteria. The regulations confirm that meat and poultry facilities are responsible for producing products that are not adulterated or misidentified, and that achieve all performance criteria and standards. These regulations will only be achieved

Table 1. Factors that contribute to outbreaks of meat and poultry-borne disease*.

Contributory factor	Percentage
Improper cooling	48
Foods prepared a day or more before serving	34
Inadequate cooking or thermal processing	27
An infected/colonized person touching cooked foods	23
Inadequate reheating of cooked foods	20
Improper hot storage of cooked foods	19
Cross contamination of cooked foods from raw foods	15
Inadequate cleaning of equipment	11
Ingesting raw products	8

* Bryan (1980).

by continuous improvement in hazard identification and prevention. The five major elements of the Pathogen Reduction/HACCP plan are: (1) establishment of critical control systems; (2) target efforts to control pathogenic bacteria; (3) adoption of food safety performance standards; (4) removal of regulatory obstacles to innovation; (5) address food safety hazards from the farm to the table. The primary tools required to follow these regulations are the development of Standard Operating Procedures for sanitation and the implementation of a HACCP plan customized for each production facility.

History and purpose of HACCP

Hazard Analysis Critical Control Point systems were first utilized by NASA to ensure that only the highest quality materials were used as components of space vehicles. The first application of HACCP in the food industry was a result of providing food for the astronauts in 1971. HACCP systems identify the potential hazards associated with food production from the very beginning through consumption. Hazards can be defined as biological (including microbiological), chemical, or physical. A successful HACCP program should identify only the critical control points (CCP) associated with producing a safe product, independent of quality control or regulatory requirements. Selection of CCPs depends on the nature of the product being produced. Ready-to-eat cooked products should be free of pathogen contamination and the CCPs should focus on preventing re-contamination of the product. Raw products that have not undergone irradiation will contain potentially pathogenic bacteria and the goal would be to minimize the contamination level and the opportunity for growth. The initial responsibility to minimize the contamination levels of pathogenic bacteria and all other potential hazards on raw meat and poultry products starts with the farm.

Biological hazards include parasites and bacteria that may or may not cause disease in the host animal, but can cause illness in humans. *Trichinella spiralis, Toxoplasma gondii,* and *Balantidium coli* are parasites that enter the human food supply primarily in undercooked or raw pork. The prevalence of these parasites may increase in the animal population through direct contact, poor management practices or contamination of the water supply. Microbiological hazards associated with meat and poultry products may include organisms that have no apparent pathological effects on the host animal. Microorganisms have been determined to cause either food-borne infections or toxicity. An infection results when pathogenic bacteria are consumed in sufficient numbers to cause an illness, e.g. salmonellosis. A food-borne intoxication is caused by the consumption of already formed toxins, which are produced by some bacteria during growth, e.g. staphylococcal enterotoxin. The nine pathogens of common concern for the meat and poultry industry are found in Table 2.

Chemical hazards include compounds that could result in toxic residues in the meat. These chemicals include natural toxins, metals, drug residues, sanitizers, cleaners, pesticides, fertilizers and environmental contaminants (Ahmed et al., 1993; Winter et al., 1994). Physical hazards include a wide variety of foreign materials that could cause illness or injury upon consumption. These hazards include: metal, glass, insect or pest parts or wood.

Table 2. Prominent pathogens of meat and poultry.

Organism	Illness	Characteristics/source
Bacillus cereus	Intoxication: diarrheal and emetic	Sporeformer found in raw milk, cereal products
Campylobacter jejuni	Infection: enteritis or gastroenteritis, possible association with Guillain-Barre syndrome	Wild and domestic animal reservoirs, swine herds/ poultry flocks
Clostridium botulinum	Intoxication: neurotoxin, extremely potent	Sporeformer, found in sausage, improperly canned foods
Clostridium perfringens	Intoxication: enterotoxin produced during sporulation in the human intestine causing gastroenteritis	Sporeformer found in raw milk (not commonly associated with elevated mastitis test scores)
Escherichia coli O157:H7	Intoxication: verotoxin causing hemorrhagic colitis, hemolytic uremic syndrome	Apparent inhabitant of the GI tract of some domestic animals, found in undercooked meat
Listeria monocytogenes	Infection: listeriosis, a hospital-associated pathogen causing septicemia, central nervous system disease, increased risk for immune compromised individuals	Psychrophile, multiplication at 3°C, in beef, lamb, milk, pork, sausage
Salmonella species	Infection: gastroenteritis, septicemia, increased risk for immune compromised individuals	Isolated from the intestinal tract of most animals, the pathogen most frequently associated with food-borne illness
Staphylococcus aureus	Intoxication: enterotoxin causing enterotoxicosis, enterotoxemia	Heat-stable toxin found in raw milk

HACCP implementation on the farm

All HACCP programs are identical in that they are a common sense approach to identification of hazards and to monitoring control over these hazards. Table 3 provides a generalized approach to implementation of a HACCP program (Deeth and Fitzgerald, 1983). The initial step is to develop a flow chart that diagrams movement of the animal through all stages of production: air, water, feed, housing, livestock background and health, storage, equipment, pest control, chemical storage and personnel are all areas that must be defined and evaluated to identify potential hazards. Once identified, the risk to consumer health must be determined. If a hazard does present a risk to consumer health then the point at which it appears becomes a CCP. Obviously in animal production there are CCPs associated with animal health and safety. In most cases these would also be CCPs for the production of food for human consumption. To reduce the risk of food-borne illness associated with the consumption of meat and poultry products, the health of the animal is of primary concern.

Salmonella has been a pathogen of concern to animal producers for many years. The effects on animal health, which generally result in decreased production and death, have made a significant economic impact on the industry. The FSIS has established pathogen reduction standards for salmonella in meat and poultry products (USDA/FSIS, 1996). These standards will be used to monitor the effectiveness of HACCP programs in meat and poultry production facilities. Salmonella was selected because it is the most common cause of food-borne illness in meat and poultry products. Control of salmonella in animals and birds has been extensively evaluated, and many of the methods of control are also effective for other biological hazards. However, it would be naive to expect all food-borne illness to disappear if salmonella was completely eradicated. The intention of the FSIS is to focus on salmonella at this point and add other pathogen reduction standards in the future.

A primary objective of a farm-based HACCP program should be to reduce or eliminate biological hazards from animal or bird production. Salmonella can be used as an indicator of the program's effectiveness. It is important to realize that our knowledge of the growth and control of many pathogenic organisms is still evolving, and therefore so must HAACP programs. The control of biological hazards in animal production must include practices that

Table 3. The seven principles of HACCP.

1.	Hazard analysis inspection, including biological, chemical, and physical conditions that could jeopardize the safety of foods.
2.	Identify CCPs at which control can be applied and a food safety hazard can be prevented, eliminated, or reduced to an acceptable level.
3.	Establish critical limits for each CCP.
4.	Establish CCP monitoring requirements.
5.	Define the corrective action when monitoring indicates a problem.
6.	Establish effective record-keeping procedures to document the HACCP system.
7.	Establish verification procedures for the HACCP system.

will eliminate or reduce exposure of healthy, non-infected animals to sources of contamination, and decontaminate any animals previously exposed (Table 4). The USDA initiated a program in February 1990 which was designed to trace reported human illness due to salmonella back to the farm producing the fresh grade A shelled eggs which were implicated (Louis, 1988). Salmonella-infected hens can pass the organism through the ovaries and oviducts to the yolk resulting in illness or contamination of chicks prior to hatching (ICMSF, 1980). Salmonella may also be on the exterior of the shell from fecal contamination and may penetrate the egg after laying (Williams *et al.*, 1968). The flocks were then tested for salmonella, and if positive, eggs were then diverted to pasteurization plants until the hens were depopulated. Flocks would then be reestablished with chicks form hatcheries following the National Poultry Improvement Plan for salmonella eradication. This plan relies on testing, sanitation and flock management techniques to detect and exclude salmonella from breeding flocks (Mason, 1992).

The FDA has recommended a zero tolerance of salmonella in animal feeds, indicating that no level of salmonella contamination is acceptable. This recommendation has contributed to an awareness in the feed industry that management of potentially pathogenic microorganisms requires diligence on all levels. Feeds produced for food-producing animals have traditionally contained the by-products of human foods. These by-products include animal proteins, wheat, corn, barley, sunflower, oats, and milk. Once defined as animal feeds these ingredients often undergo storage and production conditions that increase the potential for microbial contamination. Elimination of these conditions with the implementation of HACCP principles and procedures will increase the possibility of controlling salmonella contamination. The non-medicated feed industry and many renderers and blenders do not have Good Manufacturing Practices standards and are not familiar with HACCP programs, although meat, bone, feather, and fish meals are rendered at temperatures that are lethal to most pathogenic organisms. Salmonella contamination in rendered product and finished feed is generally associated with re-contamination after processing (Clark *et al.*, 1973). The contamination rate for rendered product ranges from 0–100% depending on the specific product and producer (Graber, 1991).

Table 4. Measures to reduce biological hazards on the farm.

1. The origin of new animals/birds should only be from facilities that also follow HACCP programs designed to reduce the risk of biological contamination. New animals/bird should be quarantined for an adequate amount of time.
2. Animal/bird holding facilities should be sanitized and monitored for contamination prior to housing new animals.
3. Transportation of animals/birds should be in thoroughly sanitized vehicles.
4. Efficient rodent and insect control programs should be established.
5. Use of only salmonella-free feed and maintenance of its integrity by proper storage.
6. Maintain all equipment in clean working order.
7. Provide only potable water for consumption.
8. Reduce exposure to animal/bird waste.
9. Train employees in sanitation and personnel hygiene
10. Evaluate vaccines, therapies, and biological control procedures for reducing biological contamination in your facility.

A true estimate of the extent of salmonella contamination in finished feeds and ingredients is not available. However, there are methods available that will maintain an unadulterated product or eliminate salmonella from animal feeds. The addition of chemicals or preservatives demonstrated to be safe and effective can be included in the animal feed. Proper pelleting of feed (e.g. 15% moisture, 180°F) reduces the incidence of salmonella contamination; and the use of pelleted feed has been encouraged in poultry operations. A new option for the eradication of salmonella and many other potentially harmful organisms is irradiation. Doses from 10 to 40 kGy have been used successfully with no apparent adverse effect on animal performance (Leeson and Maecotte, 1993).

The objective of zero salmonella in feeds is an immense one, particularly when you realize that 75–85% of all animal feeds are grown and fed on the farm and receive no heat or chemical treatment. These feeds include hay, silage, and many grains. Treatment of these feeds for the elimination of salmonella and other pathogens is not always practicable or possible. Implementation of control measures to reduce the severity of possible salmonella contamination via silage treatments, incorporation of competitive exclusion cultures, and water treatment may facilitate the process (Figure 1). Routine chlorination of drinking water to a minimum of 1–1.5 ppm free chlorine has been reported to prevent the spread of salmonella in poultry.

Rodent and insect control are vital to the control of biological hazards on the farm. Rodents and insects have been implicated in the transmission of salmonella and many other diseases. Rodents spread disease into uncontaminated areas via their droppings, feet, fur, urine, saliva, or blood. Additionally, rodents are a food source for foxes, coyotes, raccoons, skunks, dogs or cats which may also carry disease. Rodent control will require the elimination of hiding spots in and around the facility, e.g. high grass, garbage, broken equipment, cardboard boxes. Feed should be stored in secure containers to reduce the attraction of rodents and avoid contamination. Installation of rodent barriers and guard strips on doors and the use of rodenticides may be required. Insect control requires the use of good sanitary habits and the removal of animal waste. Excessive amounts of moisture and garbage must be eliminated. The insects of primary concern include roaches and flies. The use of biological control methods such as fly parasites or insecticides would be recommended.

Salmonella and other pathogens can survive in litter and bedding for periods from a few days to several weeks (Mitscherlich and Marth, 1984). The purpose of cleaning is to remove the soil and debris from a surface in preparation for sanitization. Initial cleaning is performed by sweeping, dusting, or scraping in the absence of water. The objective is to reduce the amount of material that adheres to floor, walls, and cages. A chemical cleaner or surfactant can then be used to lift the remaining residues to be rinsed away. Detergents and surfactants are generally not antimicrobial. Efficient reduction of pathogenic bacteria requires the use of a sanitizing agent or disinfectant after thorough cleaning. The most common chemical sanitizers are chlorine-containing sanitizers such as sodium hypochlorite or chlorine dioxide, iodine-based

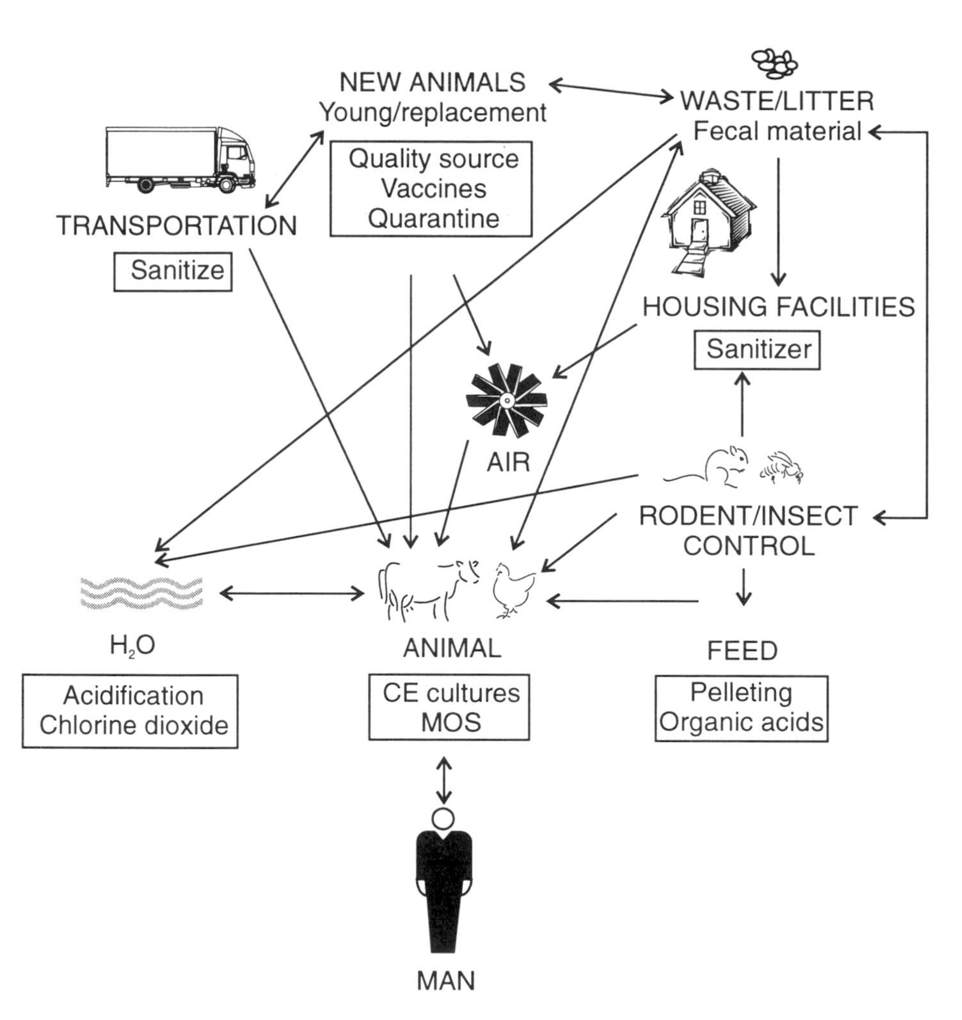

Figure 1. Control measures to reduce salmonella contamination at various points in feed and animal production.

sanitizers, cresol, phenol, and quaternary ammonia compounds. In production facilities with dirt floors application of 1 gallon diluted chlorine-based product or formaldehyde per square foot has been effective for sanitization.

Summary

It is impossible to eliminate all food-borne illness or even all pathogens from our food supply. We have traditionally tested food for specific pathogens, often ignorant of the presence of other potentially pathogenic organisms. To address the presence of these organisms in food we must direct our focus at the source – which unfortunately is the farm. We know very little about the ecology of most food pathogens, particularly when the livestock and poultry show no clinical signs of infection. An example of this is *E. coli* 0157:H7. This organism has been isolated from some herds while others are not affected. More information about the source of this organism and its effects on cattle is required before we can attempt to reduce the incidence of this pathogen in the food supply. In addition, it may be necessary to change some of our management practices such as overcrowding of livestock and poultry. Intensive rearing of animals where thousands of animals are kept together is conducive to the spread of pathogens (McCoy, 1975). Under these conditions one infected animal can contaminate the entire group. Improved management practices in animal production and food processing will reduce the risk to the consumer. However, the consumer must also accept ultimate responsibility to prevent food-borne illness by following safe food preparation practices.

References

USDA/FSIS. 1996. Federal Register of July 1996, 9CFR Part 304.

Ahmed, F.E., D. Hattis, R.E. Wolke and D. Steinman. 1993. Risk assessment and management of chemical contaminants in fishery products consumed in the USA. J. Appl. Toxicol. 13:395.

Bryan, F.L. 1980. Food-borne diseases in the United States associated with meat and poultry. J. Food Prot. 43:140.

Deeth, H.C. and C.H. Fitzgerald. 1983. Lipolytic enzymes and hydrolytic rancidity in milk and milk products. In: Developments in Dairy Chemistry – lipids (P.F. Fox, ed.). Applied Science Publishers, London. p. 195.

Graber, G. 1991. Feeding Animal Protein, Concerns of FDA. International Poultry Trade Show, Southeastern Poultry and Egg Association, Atlantic, GA.

ICMSF (International Commission on Microbiological Specifications for Foods). 1980. Microbial Ecology of Foods. Vol. 2. Food Commodities. Academic Press, New York, NY.

Leeson, S. and M. Maecotte. 1993. Irradiation of poultry feed I. Microbial status and bird response. World Poultry Sci. J. 49.

Louis, S.T., M.E. Morse and D.L. Potter. 1988. The emergence of Grade A eggs as a major source of *Salmonella enteritidis* infections. JAMA 259(14):2103.

Mason, J. and E. Kbel. 1992. APHIS *Salmonella enteritidis* control program. Salmonella Task Force, USDA, APHIS, Hyattaville, MD. 78.

Mitscherlich, E. and E. H. Marth. 1984. Microbial survival in the Environment: Bacteria and Rickettsiae Important in Human Health. Springer-Verlag, Berlin.

Williams, J.E., L.H. Dillard and G.O. Hall. 1968. The penetration patterns of *Salmonella typhimurium* through the outer structures of chicken eggs. Avian Dis. 12:445.

Winter, H., A.A. Seawright and H.J. Noltie. 1994. Pyrrolizidine alkaloid poisoning of yaks: identification of the plants involved. Vet. Rec. 134:135.

UNDERSTANDING THE DEVELOPMENT OF THE AVIAN GASTROINTESTINAL MICROFLORA: AN ESSENTIAL KEY FOR DEVELOPING COMPETITIVE EXCLUSION PRODUCTS

PETER SPRING

Institute of Animal Sciences, Nutrition Biology, ETH Zurich, Switzerland

Introduction

Perhaps the most important function of the indigenous intestinal microflora to the host is its ability to inhibit the colonization of invading pathogens in the intestinal tract (Rolfe, 1991). Many complex bacterial control mechanisms are involved in regulating the composition of the gut microflora and in excluding intestinal pathogens. Imbalances in the gastrointestinal ecosystem can weaken the protective effect of the indigenous microflora which gives enteric pathogens a better chance to colonize in the gut. Imbalances in the ecosystem mainly occur in the young animal, during periods of stress, digestive disorders, changes in the diet or periods of nutritive or therapeutic application of antibiotics. The young bird combines all these situations. It is therefore no surprise that the young bird is prone to colonization with enteric pathogens such as salmonellae. Profound knowledge of the development and the composition of the avian microflora and its regulatory forces is essential to understand failures of the innate microflora to exclude pathogens and to take appropriate measures to minimize the occurrence of such failures.

Competitive exclusion

The term competitive exclusion (CE) is used to describe the process by which beneficial bacteria exclude pathogens. Competitive exclusion implies the prevention of entry or establishment of one bacterial population into the gastrointestinal tract because that niche is already occupied by a competing bacterial population. To be able to succeed, the latter population must be better suited to establish or maintain itself in that environment or must produce compounds inhibitory to its competition (Bailey, 1987). The mechanisms which are involved in CE are very complex, since bacterial populations have a variety of different approaches to competition with invaders (Table 1). These approaches can be divided into direct and indirect mechanisms. Indirect mechanisms are the result of the normal microbial flora altering the physiologic response of the host, which in turn affects the interaction between the host and the microorganism (Rolfe, 1991). Direct mechanisms are exerted by different bacterial populations on each other. The complexity of the interactions makes it extremely difficult to study the effects of one isolated factor. We still do not understand most mechanisms involved in CE in detail, nor do we know the exact function of each bacterial population in the gut. However, comparison of the newly hatched and the adult bird reveals that some of the elements in the gut microflora are missing in a young bird and can therefore give clues as to what is important when developing CE products.

Table 1. Indirect and direct bacterial regulatory mechanisms which affect the composition of the microbial flora in the gastrointestinal tract [a].

Regulatory mechanism	Control factors
Indirect mechanisms:	
Induction of immunological process	Intestinal Ig
Modification of bile salts	Unconjugated bile acids
Stimulation of peristalsis	Flow rate
Direct mechanisms:	
Nutrient utilization	Competition for nutrients or growth factors
	Synergistic nutrient utilization
Attachment	Competition for receptor sites
	Stimulation of epithelial cell turnover
Creation of a restrictive environment	pH
	Lactic acid
	VFA
	Hydrogen sulfide
	Eh
	Modification of bile salts
	Induction of immunologic process
Production of antimicrobial substances	Ammonia
	Hydrogen peroxide
	Hemolysin
	Bacterial enzymes
	Bacteriophage
	Bacteriocins
	Antibiotics

[a] Adapted from Miles (1993), Rolfe (1991) and Savage (1987)

Development of the gastrointestinal microflora in the bird

The alimentary tract of the newly hatched, healthy chicken is usually sterile (Mead and Adams, 1975; Savage, 1987). Soon after hatching, young birds naturally develop a mature intestinal microflora through contamination with fecal material from mature birds. With coprophagic birds, such as the domestic fowl, the transfer of bacteria from parent to chick occurs very efficiently and allows the young animal to establish a protective intestinal flora within the first couple of days after hatching. With young animals such as rats or puppies the colonization of the alimentary tract with microorganisms takes longer as they are not active at birth and are kept in the nest under comparatively clean conditions (Smith, 1965). The same delay in establishing a protective microflora is observed in chickens reared in a modern production system under hygienic production standards. These birds have to establish their microflora by ingesting bacteria from hatching debris, the hatchery environment, on transport, in the production facility and through feed or water.

The contact with bacteria from the environment allows the chick to develop bacterial concentrations in the gut very similar to the concentrations found in adult birds within the first three days of life. However, the early flora is very simple and the change in complexity, especially the establishment of obligate anaerobic bacteria, takes several weeks since those microorganisms are scarce in the immediate environment of the young bird. From the microbiological standpoint, the avian gastrointestinal tract can be divided into five sections: crop; proventriculus and gizzard; small intestine; ceca; colon and cloaca (Figure 1). The microbial populations mainly differ due to differences in pH, Eh and flow rate in these sections. *Escherichia coli* and enterococci are the dominant organisms found in the gut of the newly hatched chick. Lactobacilli colonize much slower than those two non-fastidious organisms but eventually outnumber them in the upper part of the tract. In the crop, lactobacilli become the dominant organism within the first five days of life while in the small intestine it can take up to two weeks for lactobacilli to become dominant (Smith, 1965). Bacterial colonization is slower in the small intestine than in other parts of the gut, with concentrations found often below 10^5 CFU/g on day 1 (Hutanen and Pensack, 1965). As in the upper gastrointestinal tract, the cecal population of the 1-day-old chick is predominated by enterococci and coliform bacteria. Those early invaders are outnumbered by obligate anaerobes after about two weeks and the proportion of obligate anaerobes increases until the age of four weeks. At this time bifidobacteria, bacteroides, eubacteria, peptostreptococci and clostridia predominate in the ceca (Hutanen and Pensack, 1965; Barnes, 1977; Salanitro *et al.*, 1974). Cellulolytic organisms have never been isolated from the chicken ceca at levels above 10^3 CFU/g (Barnes *et al.*, 1972). Differences in the development of the gut microflora can occur due to differences in bacterial loads in the immediate environment of the young bird. Hygienic rearing conditions which minimize the risk of pathogen infection will also reduce the environmental concentrations of beneficial bacteria. Hygienic rearing conditions will therefore increase the time it takes the bird to develop a complex climax gut microflora.

The most striking difference between an early and a climax microflora in commercial birds is the lack of complexity of the microflora, in particular the lack of obligate anaerobes in the lower intestinal tract during the first weeks of life. In addition, lactobacilli become the dominant organism in the upper part of the gut only a couple of days after hatching. Therefore, natural CE mechanisms to exclude pathogens are prone to fail in the entire gastrointestinal tract during the first days of life and in the ceca during the first weeks of life. Strategies to improve CE in the gut have to focus on improving those two weaknesses of the immature microflora.

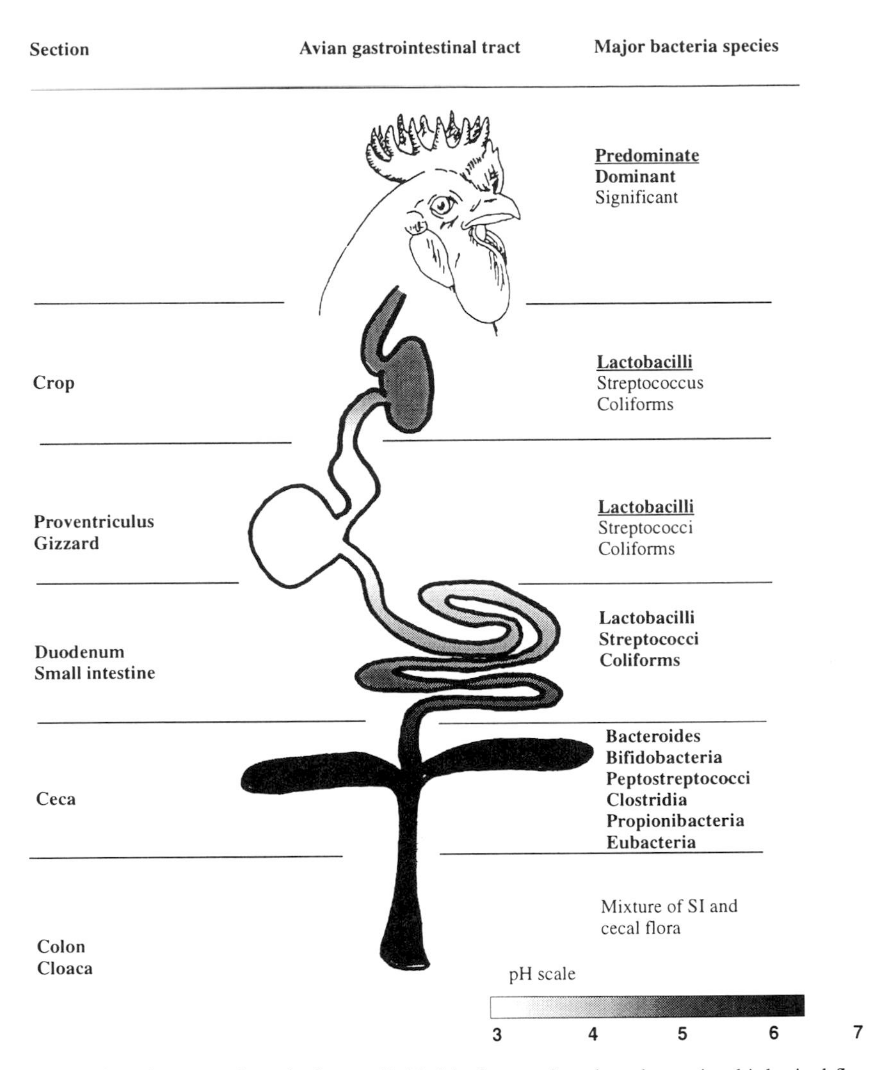

Figure 1. Avian gastrointestinal tract divided in five sections based on microbiological flora: crop, proventriculus and gizzard, duodenum and small intestine, ceca, colon and cloaca.

Bacterial cultures

Competitive exclusion cultures (Nurmi cultures) must focus mainly on diversifying the composition of the intestinal microflora, particularly in the cecum. Stavric *et al.* (1985) and Impey *et al.* (1982, 1984) developed relatively complex defined mixtures of approximately 50 bacterial strains containing species of *Escherichia, Enterococcus, Bacteroides, Fusobacterium, Lactobacillus, Eubacterium, Propionibacterium, Clostridium, Bifidobacterium* and unidentified Gram-positive rods. These mixtures gave protection against

Salmonella typhimurium comparable to that obtained with an undefined cecal culture. When the number of strains in the mixture was gradually reduced by deletion of individual strains, the protective activity of the mixture decreased. This shows the importance of the complexity of a mixture. Schneitz *et al.* (1993) convincingly demonstrated the importance of lactobacilli with the ability to adhere as CE microorganisms in poultry. A wild strain of *Lactobacillus acidophilus* that adhered strongly to chicken epithelial cells was tested for its ability to displace *Salmonella infantis.* When the same strain was subcultured several times on a selective agar it lost its adhering ability. This change was accompanied by a considerable loss of protective activity against colonization of newly hatched chicks with *S. infantis.*

Three important conclusions for the production of a CE culture can be drawn from those findings. First, the number of bacterial generations between the harvest of intestinal content from the adult bird and the final CE product should be as few as possible in order to minimize losses of important bacterial characteristics such as ability to adhere. Second, special care must be taken to optimize anaerobic conditions during the entire production process and during storage to maintain viability of obligate anaerobic bacteria. Third, since experiments failed to determine a few single strains that are the main players involved in CE, the production process must be conducted in a manner that maintains ratios among the major organisms in the culture similar to those found in the gut.

The effect of CE cultures on salmonella colonization has been well documented over the last two decades. Limited work has been done on CE of other enteric pathogens such as *Campylobacter jejuni* or *E. coli*. However, some of the work with CE cultures looking beyond salmonella is promising. Stavric *et al.* (1992) reported improved resistance of chicks against *E. coli* O157:H7 with a CE culture (Table 2). Competitive exclusion of campylobacter must receive increased attention due to the problems in the poultry industry caused by this microorganism. Due to its unique niche in the crypt mucus of the intestine, campylobacter seems to be harder to exclude than salmonella. The selection for anti-campylobacter cultures will have to focus on finding bacteria that can compete with campylobacter in its niche. Aho *et al.* (1992) reported some very promising results of campylobacter exclusion when spiking a commercial CE culture with beneficial bacterial strains containing similar growth properties as campylobacter. Positive results against campylobacter were also reported from Schoeni *et al.* (1992; 1994) with a CE culture composed of bacterial strains with anti-campylobacter factors. These data indicate that specifically designed cultures might be able to help the poultry industry overcome the problems with this organism.

Mannanoligosaccharide and other specific inhibitors of bacterial attachment

Understanding the gastrointestinal microflora is essential not only for the development of probiotics but also for developing new pre-biotics. Specific inhibitors of adhesion which have been discovered throughout the last decades

Table 2. Effect of a CE culture on colonization of *E. coli* O157:H7 in chicks[a]

Challenge dose (CFU/chick)	Control (% colonized)	CE culture (% colonized)
10^3	50	0
10^5	89	10
10^7	60	25
10^9	63	15

[a] adapted from Stavric *et al.*, 1992

have not only provided the best evidence that adhesion is important, but also enabled the design of new approaches to the prevention of infections. Only the knowledge that mannose sensitive type-1 fimbriae are involved in attachment of pathogens and that mannose can inhibit their attachment has allowed us to specifically look for carbohydrates such as mannanoligosaccharide (MOS) with properties similar to mannose. In laboratory scale trials MOS has been shown using cecal material from 10-day-old chicks to reduce prevalence and concentration of salmonella expressing type-1 fimbriae *in vitro* (Figure 2; Spring, 1996). Ability of MOS to bind type-1 fimbriae and therefore block bacterial attachment is thought to be the major mode of action for CE with MOS. Immune stimulation with MOS might also play a role in the effects seen against salmonella. However, a possible involvement of immune stimulation in excluding salmonella has not yet been studied. It has been shown with other sugars such as lactose that results of short-term trials do not always reflect the effect of CE products in trials conducted throughout the growth period. Data are now available from a salmonella challenge trial conducted through grow out (Figure 3; Caskey personal communication). Mannanoligosaccharide (Bio-Mos, Alltech Inc.) has been shown to be effective at lowering *Salmonella enteritidis* colonization steadily over the growing period. Chicks treated with Bio-Mos were therefore free of salmonella much earlier than control chicks. Freeing infected chicks of salmonella as fast as possible is essential because this will help to reduce environmental contamination with the pathogen and will therefore reduce the risk of recontamination. These data suggest that MOS, when combined with proper management practices, can be an effective anti-salmonella tool in the field.

While MOS is able to prevent attachment of bacteria that express type-1 fimbriae, many other non-mannose specific fimbriae are present in the bacterial kingdom. Some strains of campylobacter for instance use fimbriae specific for fucose to adhere to the intestinal wall. Cinco *et al.* (1984) showed that the addition of L-fucose could inhibit *C. jejuni* attachment to intestinal cells *in vitro* (Figure 4). Inhibition was a function of fucose concentration and was not complete. The authors therefore hypothesized that other receptors or non-specific interactions may cooperate in cell attachment by campylobacter. Mouricout *et al.* (1990) isolated glycoconjugates from the non-immunoglobulin fraction of bovine plasma and administered the glycoconjugates in drinking water to colostrum-deprived calves. It was observed that the animals drinking glycoconjugates became more resistant to infections caused by $K99^+$ *E. coli.* Furthermore, adhesion of the bacteria to the intestines of the calves was reduced by two orders of magnitude.

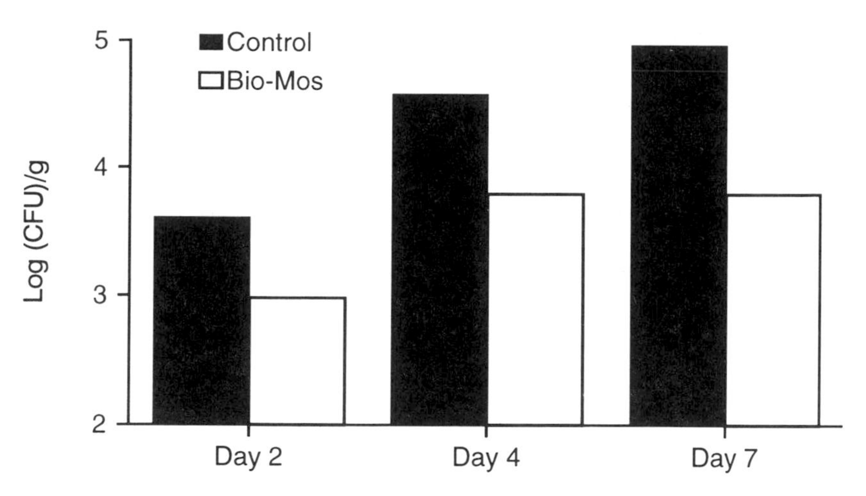

Figure 2. Effect of dietary Bio-Mos on cecal *Salmonella typhimurium* 29E concentrations of chicks maintained in microbial isolators at 2, 4 and 7 days after salmonella challenge (Spring, 1996).

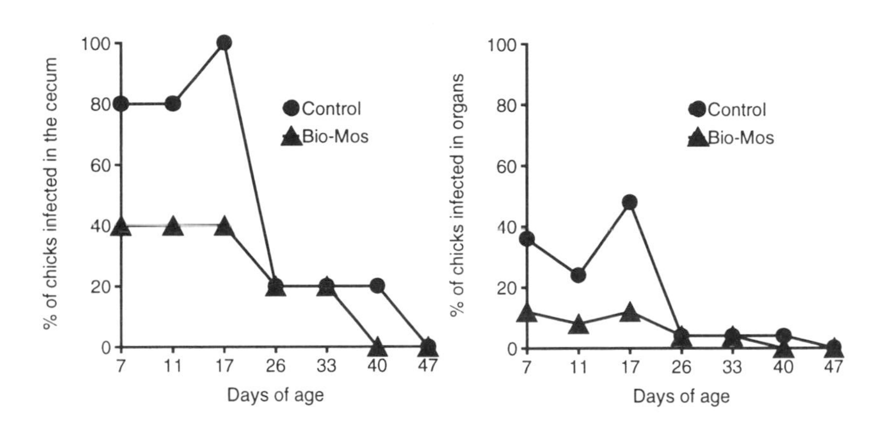

Figure 3. Effect of Bio-Mos on *Salmonella enteritidis* colonization in the gut and liver of broiler chicks.

Those findings indicate a potential to block bacterial attachment mediated by a variety of fimbriae. This ecological approach to fight bacterial colonization offers a new, promising tool for the future.

Interactions of mannanoligosaccharide with the gut mucosa

Research with different pro- and pre-biotics is mainly focusing on direct bacterial mechanisms as modes of action (Table 1). However, there is reasonable evidence that both the indigenous gut microflora and dietary additives enhance the immunologic mechanisms at the mucosa and systemic

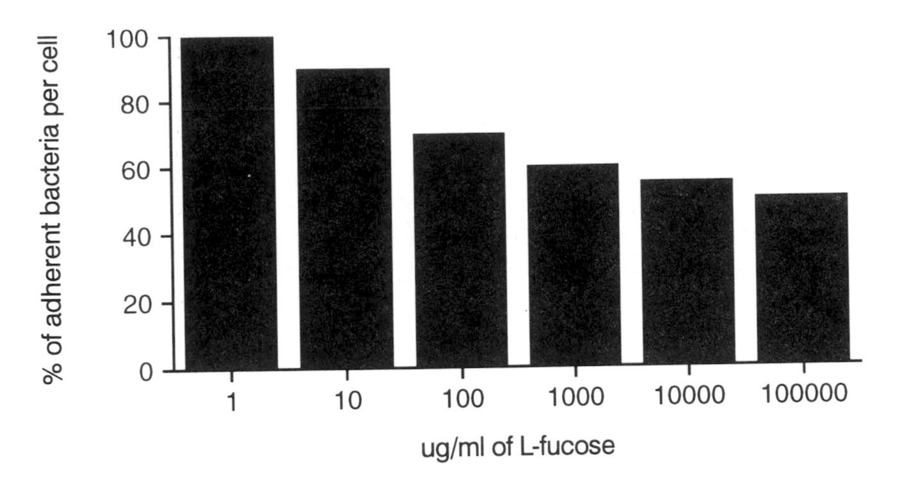

Figure 4. Inhibition of adhesion of *Campylocabacter jejuni* to epitheial cells *in vitro* by different concentrations of L-fucose (Adapted from Cinco *et al.*, 1984).

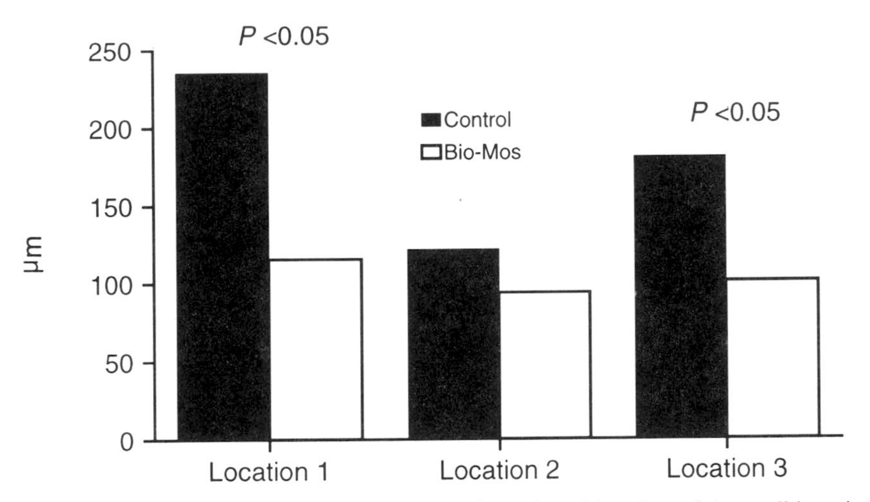

Figure 5. Effect of dietary Bio-Mos on crypt depth at selected locations of the small intestine in 8-week old turkeys.

level. The importance of indirect bacterial mechanisms as the mode of action of pro- and pre-biotics should therefore not be underestimated. Components of the microbial cell wall seem to play a key role in this stimulation. Immune stimulation has been show in different trials with MOS derived from yeast cell wall material.

Interactions of microorganisms (indigenous or added via diet) with the gut not only play a role in the development of the immune system but also in the overall development of the intestine. Great differences in the intestinal development of conventional and germ-free animals have been reported. Among other differences, conventional animals have been shown to have a greater mucosal surface area and a deeper crypt. However, while intestinal bacteria serve to stimulate intestinal development, certain pathogenic bacteria can also destroy the intestinal mucosa. Nabuurs *et al.* (1993) reported shorter intestinal villi and deeper crypts (decrease in villi length to crypt depth ratio fast cell turnover) in pigs from herds with post-weaning diarrhea when compared to pigs from healthy herds. A fast cell turnover will reduce the age and maturity of the absorptive cells and increase the energy requirement for gut maintenance. Since metabolic activities of the gut are known to be high relative to that of other tissues, factors that influence the energy efficiency of the digestive system may have a profound influence on the efficiency of energy utilization of the whole body (Ferrell and Jenkins, 1985; Jin *et al.*, 1994).

Different feed additives might exert some of their positive effects on animal performance through improving the energy efficiency of the gut. For instance subtherapeutic levels of antibiotics in the diet have been shown to reduce the weight of the intestinal tissue and to decrease the turnover of mucosal cells (Visek, 1978). Slow cell turnover requires lower protein synthesis. This can be associated with low rates of energy expenditure in the gut and could explain some of the performance improvements mostly seen with subtherapeutic dietary levels of antibiotics. The effect of yeast products on intestinal morphology has just recently started to be investigated. Yeast culture has been shown to affect intestinal mucosa development in turkeys. Bradley *et al.* (1994) reported reduced crypt depth and decreased numbers of goblet cells with dietary addition of *Saccharomyces cerevisiae* var. *boulardii.* These investigators suggest the energy conserved by reduced production of epithelial cells (decreased crypt depth) may be utilized for lean tissue mass synthesis and might explain some of the improvements seen in body weight gain with dietary *S. cerevisiae* var. *boulardii.* Two trials have been conducted which took a closer look at intestinal morphology with dietary Bio-Mos. In broilers, an increase in both villus length and crypt depth has been noted (Spring, 1996). An increase in villus length would indicate a more absorptive surface but an increase in crypt depth can be taken as an indicator of greater energy expenditure to produce the absorptive surface. It is therefore questionable if these changes might make the gut more energy efficient. Savage *et al.* (1997) on the other hand found 0.1 % dietary Bio-Mos decreased crypt depth and increased villi width with no change in villi length in turkeys (Figure 5). The mechanisms leading to those changes have not yet been revealed. Nabuurs *et al.* (1993) reported decreased villi length to crypt depth ratio in pigs from herds infected with *E. coli.* MOS might alleviate the negative effects of enteric pathogens by adsorbing them and their toxins and might therefore have increased villi length to crypt depth ratio in turkeys. A reduction in crypt depth might indicate energy savings for gut maintenance, and might explain some of the very significant performance improvements which have been reported with Bio-Mos in turkeys. The reasons for different effects of MOS on intestinal morphology in turkeys and chickens cannot be explained at this point. Differences in animal species and experimental conditions, in particular the composition of the diet, might account for some of the differences.

Summary

Improving our understanding of the development and the regulatory mechanisms of the avian microflora is essential for the development of effective CE products. The most striking difference between an early and a climax microflora in commercial birds is the lack of complexity of the microflora, particularly the lack of lactobacilli in the upper part and the lack of obligate anaerobes in the lower part of the tract. This lack will weaken the CE effect of the early flora and CE products have to focus on overcoming those two weaknesses. Bacterial cultures and MOS have both been shown to be effective at improving CE and decreasing the prevalence of enteric pathogens in poultry. Bio-Mos is particularly interesting as a CE product because it has also been shown to improve animal performance under various conditions. In certain situations changes in conditions in intestinal morphology with dietary MOS might be involved in improvements seen in animal performance.

References

Aho, M., L. Muotio, E. Nurmi and T. Kiiskinen. 1992. Competitive exclusion of campylobacters from poultry with K-bacteria and Broilac. Int. J. Food Technol. 41:88–92.

Bailey, J.S.. 1987. Factors affecting microbial competitive exclusion in poultry. Food Technol. 41:88–92.

Barnes, E.M. 1977. Ecological concepts of the anaerobic flora in the avian intestine. Am. J. Clin. Nutr. 30:1793–1798.

Barnes, E.M., G.C. Mead, D.A. Barnum and E.G. Harry. 1972. The intestinal flora of the chicken in the period 2 to 6 weeks of age, with a particular reference to the anaerobic bacteria. Br. Poult. Sci. 13:311–326.

Bradley, G.L., T.S. Savage and K.I. Timm. 1994. The effect of supplementing diets with *Saccharomyces cerevisiae* var. *boulardii* on male poult performance and ileal morphology. Poult. Sci. 73:1766–1770.

Cinco, M., E. Banfi, E. Ruaro, D. Crevatin and D. Crotti. 1984. Evidence for L-fucose- (6-deoxy-L-galactopyranose) mediated adherence of *Campylobacter* spp. to epithelial cells. FEMS Microbiol. Lett. 21:347–351.

Ferrell, C.L. and T.G. Jenkins. 1985. Cow type and the nutritional environment. Nutritional aspects. J. Anim. Sci. 61:725–741.

Hutanen, C.N. and J.M. Pensack. 1965. The development of the intestinal flora of the young chick. Poult. Sci. 44:825–830.

Impey, C.S., G.C. Mead and S.M. George. 1982. Competitive exclusion of salmonellas from chick caecum using a defined mixture of bacterial isolates from the caecal microflora of an adult bird. J. Hyg. Camb. 89:479–490.

Impey, C.S., G.C. Mead and S.M. George. 1984. Evaluation of treatment with defined and undefined mixtures of gut microorganisms for preventing *Salmonella* colonization in chicks and turkey poults. Food Microbiol. 1:143–147

Jin, L., L.P. Reynolds, D.A. Redmer, J.S. Caton and J.D. Crenshaw. 1994. Effects of dietary fiber on intestinal growth, cell proliferation and morphology in growing pigs. J. Anim. Sci. 72:2270–2278.

Mead, G.C. and B.W. Adams. 1975. Some observations on the cecal microflora of the chick during the first two weeks of life. Br. Poult. Sci. 16:169–176.

Miles, R.D.. 1993. Manipulation of the microflora of the gastrointestinal tract. Natural ways to prevent colonization by pathogens. In: T.P. Lyons (Ed.). Biotechnology in the Feed Industry. Nottingham University Press, Nottingham, UK. pp. 133–150.

Mouricout, M., J.M. Petit, J.R. Carias and R. Julien. 1990. Glycoprotein glycans that inhibit adhesion of *Escherichia coli* mediated by K99 fimbriae: treatment of experimental colibacillosis. Infect. Immun. 58:98–106.

Nabuurs, M.J.A., A. Hoogendoorn, E.J. Van der Molen and A.L.M. Van Osta. 1993. Villus height and crypt depth in weaned pigs reared under various circumstances in the Netherlands. Res. Vet. Sci. 55:78–84.

Rolfe, R.D. 1991. Population dynamics of the intestinal tract. In: L.C. Blankenship (Ed.). Colonization control of human enteropathogens in poultry. Academic Press, Inc., San Diego, CA. pp. 59–75.

Salanitro, J.P., I.G. Fairchild and Y.D. Zgornicki. 1974. Isolation, culture characteristics and identification of anaerobic bacteria from the chicken cecum. Appl. Microbiol. 27:678–687.

Savage, D.C. 1987. Factors influencing biocontrol of bacterial pathogens in the intestine. Food Technol. 41:82–87.

Savage, T.F., E.I. Zakrewska and J.R. Andreasen. 1997. The effects of feeding mannanoligosaccharide supplemented diets to poults on performance and the morphology of the small intestine. Presented at the Annual Meeting of the Southern Poultry Science Association. Atlanta, January 20.

Schneitz, C., L. Nuotio and K. Lounatma. 1993. Adhesion of *Lactobacillus acidophilus* to avian intestinal epithelial cells mediated by the crystalline bacterial cell surface layer (S-layer). J. Appl. Bact. 74:290–294.

Schoeni, J.L. and M.P. Doyle. 1992. Reduction of *Campylobacter jejuni* colonization of chicks by cecum-colonizing bacteria producing anti-*C. jejuni* metabolites. Appl. Environ. Microbiol. 58:664–670.

Schoeni, J.C.L. and A.C.L. Wong. 1994. Inhibition of *Campylobacter jejuni* colonization in chicks by defined competitive exclusion cultures. Appl. Environ. Microbiol. 60:1191–1197.

Smith, H.W. 1965. The development of the flora of the alimentary tract in young animals. J. Path. Bact. 90:495–513.

Spring, P. 1996. Effects of mannanoligosaccharide on different cecal parameters and on cecal concentrations of enteric pathogens in poultry. Diss. ETH no. 11897. ETH Zurich, Switzerland.

Stavric, S., T.M. Gleeson, B. Blanchfield and H. Pivnick. 1985. Competitive exclusion of salmonella from newly hatched chicks by mixtures of pure bacterial cultures isolated from fecal and cecal contents of adult birds. J. Food. Prot. 48:778–782.

Stavric, S., B. Buchman and T.M. Gleenson. 1992. Competitive exclusion of *Escherichia coli* 0157:H7 from chicks with anaerobic cultures of faecal material. Appl. Microbiol. 14:191–193.

Visek, W.J. 1978. The mode of growth promotion by antibiotics. J. Anim. Sci. 46:1447–1469.

THE NEW GENERATION OF PROTEIN SUPPLEMENTS: ULTIMATE PROTEINS

COMBINING THE BENEFITS OF ULTIMATE PROTEIN 1672 WITH THE CONCEPT OF AN IDEAL PROTEIN

D.J.A. COLE

Nottingham Nutrition International, East Leake, Loughborough, Leics, UK

The role of protein in animal nutrition has largely been considered on the basis of the requirements for maintenance and production. Such requirements are influenced by many factors, for example the nature and level of the productive function, genotype, etc. However, the role of proteins may be considered on the basis of functions quite separate from this. For example, the animal faces many and varied challenges from antigens (e.g. bacteria, viruses, parasites and a number of nutrients) and is able to respond by means of antibodies (immunoglobulins). These immunoglobulins (Ig) are a group of proteins secreted by lymphocytes. They are divided into several classes (e.g. of particular importance are IgM, IgA and IgG). IgA is produced by cells in the intestinal tissues and represents the first line of defence.

In some species, for example the human, antibodies are able to pass across the placenta so that the young are born with some degree of protection. This is not the case in all species (e.g. the pig) and the young are born with no circulating antibodies. Consequently the milk of the first few days (colostrum) which is very high in antibodies (IgA) is particularly important to the young pig. The immunoglobulins are large molecules and the newborn piglet has the ability to absorb them for several days after birth. IgA is well suited to the young animal as it is fairly resistant to digestive enzymes and can attach to the gut wall to give a defence against pathogenic challenge. By about the third week of life the pig is able to produce its own secretory IgA.

Thus, there has to be a consideration of the similarities and differences in the protein needs for:

- productive characteristics
- gut health and the immune system

In the immediate post-weaning period, there does appear to be some conflict in that to express its high genetic potential, the modern fast-growing pig has a high requirement for dietary protein while piglet well-being is more likely to be achieved by limiting dietary protein intake (particularly plant protein). The consequence of this is that great attention needs to be paid to providing high quality dietary protein in order to limit inclusion levels.

In a recent review Putzai *et al.* (1997) indicated the various disturbances that can occur in the young animal:

- poor nutritional performance
- reduced growth
- reduced digestion/absorption
- changes in gut motility
- structural damage in the small intestine
- diarrhoea

Protein requirement

A simple way of considering protein requirements and ensuring high quality dietary protein is to take a two-stage approach to the application of the concept of an ideal protein. This is done by defining the requirements for the most common first-limiting amino acid (generally lysine) and then relating to this the balance of the other essential amino acids. In normal cereal-based diets non-essential nitrogen should not be a problem.

Protein requirements have been comprehensively reviewed (e.g. Cole, 1996). They depend considerably on genotype, sex and liveweight. Genotype varies greatly and, in some countries rapid genetic gains are being made. Table 1 is an update of the lysine requirements of Cole (1992) by Van Lunen and Cole (1996). It incorporates values for modern fast-growing hybrids depositing protein at a rate of 175 g/day. Care should be taken to correctly identify the genotype by its protein deposition rate.

Having established the lysine requirements, the other amino acids should be supplied in the ratio of the ideal protein (Table 2).

Table 1. Optimum lysine/digestible energy (g/MJ) ratios in the diets of growing pigs*.

		Genotype			
Maximum PDR (g/d) Liveweight (kg)	Sex	Unimproved 100	Average 125	High 150	Hybrid 175
	Castrate	0.78	0.85	0.88	1.20
Up to 25	Gilt	0.80	0.85	0.90	1.20
	Boar	0.83	0.88	0.93	1.20
25 to 55	Castrate	0.73	0.78	0.83	1.10
	Gilt	0.75	0.80	0.85	1.10
	Boar	0.78	0.83	0.88	1.10
55 to 90 restricted fed	Castrate	0.55	0.55	–	–
	Gilt	0.65	0.65	–	–
	Boar	0.70	0.70	–	–
55 to 90 fed *ad libitum*	Castrate	–	0.58	0.63	0.95
	Gilt	–	0.60	0.65	0.95
	Boar	–	0.63	0.68	0.95

* Van Lunen and Cole (1996).

Table 2. Optimum balance of essential amino acids in the ideal protein for pigs (relative to lysine = 100).

Lysine	100
Methionine + cystine	50–55
Tryptophan	18
Threonine	66
Leucine	100
Valine	70
Isoleucine	50
Phenylalanine + tyrosine	100
Histidine	33

Extra nutritional effects

In looking at effects over and above those that are directly for maintenance and production, the weaned pig is a good example. The act of weaning is of great significance with the piglet moving from a situation where its gut was well protected from pathogens by the influence of sow's milk to the challenges of milk withdrawal and the introduction of solid food. Numerous changes including enzyme system development, new feed ingredients, reduction in lactic acid content of the gut and environmental challenge contribute to poor villus condition together with the reduction in productive performance (scouring, reduced growth and mortality). The poor villus height and *Escherchia coli* burden are at their worst in the two weeks after weaning (Figure 1). It is suggested that villus height (an indication of gut health) increases as feed

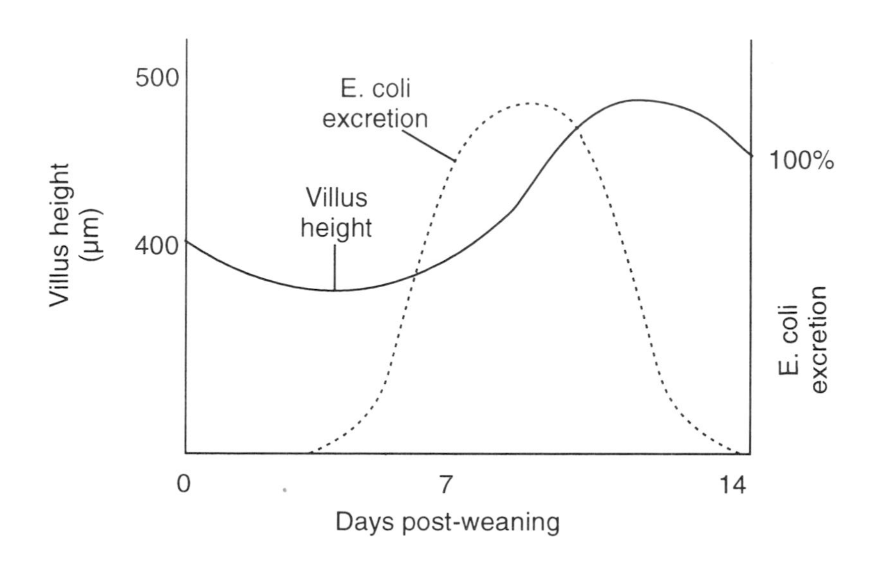

Figure 1. Changes in villus height and *E. coli* excretion (based on Welink, 1996 and Newby *et al.*, 1984).

intake increases (Figure 2). There is usually a drop in feed intake immediately after weaning and any aspect of nutrition which increases intake at this stage is of considerable importance.

The consequence of protein sources having extra nutritive effects is that it is not possible to assess their value in a conventional dose-response manner. Furthermore, the whole post-weaning period does not lend itself to simple evaluation. The value of increased weight gain in the immediate post-weaning period is not just the extra weight itself, but in the effect increasing weight gain has on reducing the days to final slaughter weight.

These benefits of increased growth are quickly manifest. For example, Campbell (1991) showed that piglets weighing 6.14 and 7.95 kg at 29 days of age were 30.4 and 35.6 kg, respectively, at 78 days of age, having grown at 454 and 529 g/day. Small differences in early life are magnified to greater benefits at the end of the production period (Figure 3). A heavier pig at any stage is further along the growth curve, has a bigger appetite and will reach maturity earlier. However, the situation is further complicated by behavioural considerations which may favour the larger pig in the hierarchy of the group: this would have further effects in increasing feed intake.

Furthermore, the period for which special precautions need to be taken is of short duration, ranging from 7 to 14 days after weaning. This is a period of low feed intake (Figure 4). It is important that the piglet quickly achieves an intake level which is adequate to meet its maintenance requirements (about 150 g/day). The healthy piglet does this by about the third day after weaning. A practical target to be achieved in the first week would be at least 200 g/day to ensure fat deposition rather than fat mobilization.

Starter diets giving protection to the piglet after weaning are expensive. However, their consumption is small, particularly in relation to total feed

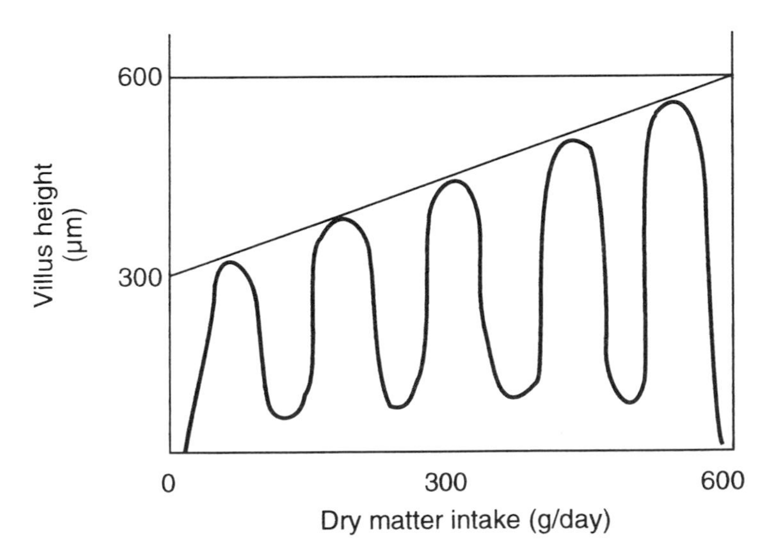

Figure 2. Relationship between feed intake and villus height (based on Weilink, 1996).

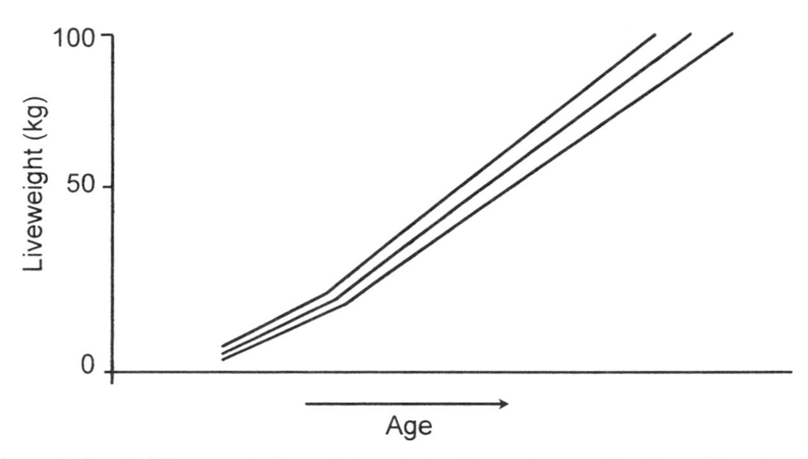

Figure 3. Small differences in liveweight early in life can be magnified later. (Based on data of Campbell, 1991, and P.R. English, personal communication).

Figure 4. Typical post-weaning feed intakes (Townson, unpublished).

consumption through the growing/finishing period. In the case of a pig growing from 7.5 to 100 kg liveweight:

- The intake of a special starter fed for the first seven days after weaning may be of the order of 1 kg compared with about 240 kg over the total period. This represents about 0.4% of the total food used.
- If the diet is fed for 14 days then intake will be of the order of 2.8 kg or about 1.1% of the total required for production.

As these special diets are such a small part of the total food requirement, the high cost necessary to give adequate protection to the young piglet adds little to total production costs.

There have been many investigations into the role of specific ingredients in the diet of the weaned pig to establish their effects over and above the direct supply of nutrients (Bolduan *et al.*, 1988). For example, weaned pigs are known to respond well to milk-based products. While these are good sources of highly digestible protein it is likely that their lactose content is of considerable importance. It is usual to encourage competitive exclusion in the intestine of the weaned pig. Specific strains of lactic acid bacteria are able to attach to the squamous epithelium and are thought to be able to give protection by inhibiting pathogenic microorganisms. As they require lactose as a nutrient, it is possible that the feeding of lactose-rich materials together with a *Lactobacillus* culture will afford a high degree of protection in terms of gut health. This has been well demonstrated in a large pig unit by Campbell (1991). In some countries plasma protein (at about 5% of the diet) is regarded as an essential constituent of post-weaning pig diets. There are numerous reports of plasma proteins helping to overcome the problems characteristic of this period.

A novel protein

Ultimate Protein 1672 (Alltech Inc.) is a very recent introduction to our range of ingredients available for the newly weaned pig. It is an interesting non-mammalian product, based largely on brewers yeast strains, wheat gluten and egg. The material is subjected to a unique production which yields, amongst other things, hydrolysed peptides.

A recent trial at Harper Adams College (Powles, unpublished) examined the use of this material in the diets of pigs in the first week after weaning at 21 days of age. All diets were balanced for energy (16.5 MJ/kg) digestible lysine (1.5%), methionine + cystine (0.9%) and threonine (0.95%). There were three treatments: high level milk protein, 5% inclusion of Ultimate Protein 1672, and 7.5% inclusion of Ultimate Protein 1672. Inclusion of Ultimate Protein enhanced both feed intake and growth rate (Figure 5).

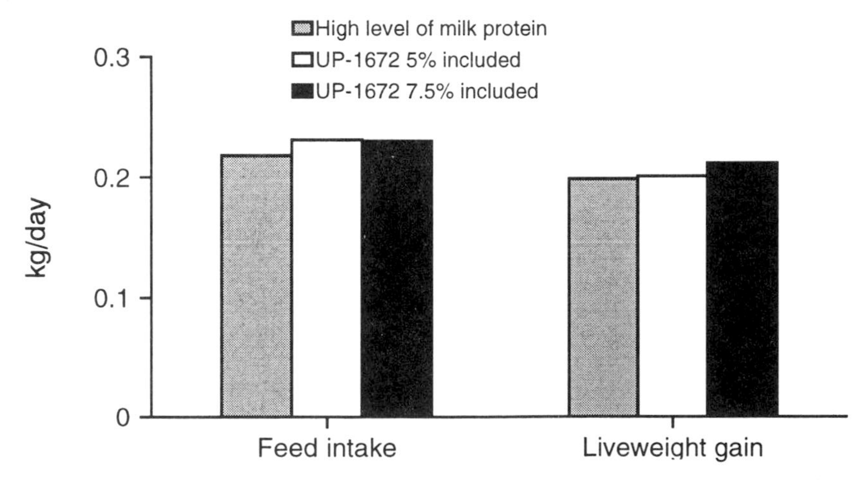

Figure 5. The response of piglets to Ultimate Protein 1672 in the first week after weaning.

Further work has been undertaken using complex commercial diets which already contained dried whey. The benefits of Ultimate Protein 1672 were again demonstrated (Table 3). Feed intake, growth and feed conversion were all improved.

Other work has shown the value of a brewers yeast based protein in replacing up to 75% of plasma protein (Trottier *et al.*, 1996) (Table 4).

It is too early to propose a distinct mode of action of any extra nutritional value of Ultimate Protein 1672. However, some of the important issues can be considered. Firstly, an excellent balance and level of digestible amino acids are provided by fortification of the basal yeast materials. This is enhanced by the unique production process which hydrolyses proteins.

It is interesting that Göransson (1997) has recently reviewed his concept of peptides being involved in gut protection. He stated that when the gut is challenged with a toxin, a substance is secreted from the pituitary gland which protects the gut. This substance was described as anti-secretory protein which is a peptide. He has developed the application of the anti-secretory protein concept by direct application in the diet.

Hydrolysis itself may be playing a further important role. For example, pigs receiving hydrolysed casein instead of casein showed increased villus crypt depth, height and brush border enzyme activity (secretion of lactase and sucrase) at some sites but not at others. Hampson (1986) suggested that the amino acids produced by hydrolysis might be able to stimulate crypt cell production directly, the so-called luminal nutrition hypothesis (Dowling and Booth, 1967; Diamond and Karasou, 1983). Being very labile, the intestinal mucosa adapts rapidly to a variety of changes in luminal conditions.

It can further be questioned to what extent the balance and sequence of dietary amino acids is of value. As far as the essential amino acids are concerned, many diets are formulated only to the first three or four. The better supply of some amino acids which are further down the order of limitation for growth characteristics may be more important when other aspects of response are considered. A change in requirement with improved genotypes or in stress situations may change the need for amino acids which have previously been regarded as semi-essential or non-essential. For example, Welink (1996) has suggested that some semi-essential amino acids

Table 3. A comparison of complex weanling diets, with or without 5% Ultimate Protein 1672.

	Without Ultimate Protein 1672	With Ultimate Protein 1672
Age at weaning, days	28	28
Weight at weaning, kg	7.39	7.25
Days on trial	20	20
Final weight, kg	12.32	13.00
Feed intake, g/day	349	379
Growth rate, g/day	247	288
Food conversion ratio	1.41	1.32

Table 4. The use of a dried brewers yeast based product to partially replace plasma protein in weanling pig diets*.

	Diet			
	1	2	3	4
Plasma protein	–	5%	50% replaced	75% replaced
Yeast protein	–	0	by yeast protein	by yeast protein
Feed intake, g/day	401	397	392	420
Growth rate, g/day	274	280	282	292
Food conversion ratio	1.46	1.42	1.39	1.44

* Figures are for 4 weeks; treatments were applied in week 1 and for the next 3 weeks 2.5 % plasma protein only was used (Trottier *et al.*, 1996).

that might be produced by the pig are not produced in sufficient quantities under stress situations. He says it is these semi-essential amino acids that help the villi recover.

It may also be postulated that some amino acids regarded as non-essential for production characteristics may have a role in gut defence.

References

Bolduan, G., M. Jung, E. Schnabel and R. Schneider. 1988. Recent advances in nutrition of weaner pigs. Pig News and Information 9(4):381:385.

Campbell, R. 1991. Digestive constraints in the young pig: implications on post weaning performance and health. In: Biotechnology in the Feed Industry. (Ed T.P. Lyons). Alltech Technical Publications, Kentucky, pp. 189–197.

Cole, D.J.A. 1992. Interaction between energy and amino acid balance. *International Feed Production Conference*, Piacenza, Feb 25–26, 1992.

Cole, D.J.A. 1996. Third generation proteins: an ideal protein for each species: Maximising performance while minimising excretion. In: Biotechnology in the Feed Industry, 1996. (Eds. T.P. Lyons and K.A. Jacques). Nottingham University Press, Nottingham, pp. 351–363.

Diamond, J.M. and W.H. Karasou. (1983). Nature 304:18.

Dowling, R.H. and C.C. Booth. 1967. Clinical Science 32:139–149.

Hampson, D.J. 1986. Attempts to modify changes in the piglet small intestine after weaning. Res. Vet. Sci. 40:313–317.

Göransson, L. 1997. Alternatives to antibiotics – the influence of new feeding strategies for pigs on biology and performance. In: Recent Advances in Animal Nutrition – 1997. (Eds P.C. Garnsworthy, J. Wiseman and W. Haresign). Nottingham University Press, Nottingham (in press).

Newby, T.J., B. Miller, C.R. Stokes, D. Hampson and F.J. Bourne. 1984. Local hypersensitivity response to dietary antigens in weaned pigs. In: Recent Advances in Animal Nutrition – 1984. (Eds W. Haresign and D.J.A. Cole). Butterworths, London. pp. 49–59.

Putzai, A., E. Gelencsér, G. Grant and S. Bardocz. 1997. Nutritional manipulation of immune competence in young animals. In: Recent Advances in Animal Nutrition – 1997 (Eds P.C. Garnsworthy, J. Wiseman and W. Haresign). Nottingham University Press, Nottingham (in press).

Trottier, H.L., J.L. Cline, R.A. Easter, G.R. Hollis, R. Bradfield and J.E. Sullivan. 1996. Dried brewers yeast as a partial replacement for porcine plasma in weaning pigs diets: Effect on growth performance. J. Animal Sci. 74 (Suppl. 1):172.

Van Lunen, T.A. and D.J.A. Cole. 1996. Energy-amino acid interactions in modern pig genotypes. In: Recent Advances in Animal Nutrition – 1996. (Eds P.C. Garnsworthy, J. Wiseman and W. Haresign). Nottingham University Press, Nottingham, pp. 233–261.

Welink, M. (1996). Aminozuur Tegen Groeidip, Boordenij/Vartcenshoeden 82 no 22. 22 October 1996.

SATISFYING THE DIGESTIBLE PROTEIN REQUIREMENTS OF HIGH PERFORMANCE RUMINANTS: THE DEVELOPMENT OF AN IDEAL BYPASS PROTEIN FOR DAIRY COWS

S.L. WOODGATE and J.E. EVERINGTON

Beacon Research International Ltd., Clipston, Leics, UK

Introduction

A description of feed proteins in terms of relative ability to produce growth, meat or milk of a certain quality would be of great economic benefit to the producer. In this world of both environmental and economic pressures, the efficient feeding of livestock to maximize production but minimize wastage through both physical and metabolic losses must be a major objective. To achieve this, diets would need to be formulated to meet the requirements of each species of animal for each specific physiological state, and hence the response of the animal to changes in substrate supply must be predictable. Thus, each nutrient must be able to be fully described in terms of both its chemical profile and its availability to the animal.

It is now possible to manufacture protein products with any particular blend of amino acids and hence there is the potential to produce amino acid mixtures that are ideal for a particular species in a particular metabolic state, e.g. a lactating cow or a fattening pig. In this respect an understanding of the individual species' requirements is a prerequisite to any attempt to supply an ideal protein. In ruminants the concept of an ideal protein is complicated by the fact that the amino acids presented for absorption at the small intestine are not the same as those in the diet.

This paper outlines the process of protein digestion in the ruminant, how estimates may be made of both the quantity and quality of protein that bypasses ruminal degradation along with the quantity and quality of microbial protein which enters the small intestine for absorption. Having described the supply of protein, progress towards an understanding of the cow's requirements will be considered and the development of a potential ideal protein for ruminants will be discussed.

Evaluating protein requirement and supply in ruminants

PRINCIPLES

There are several systems in use for evaluating protein requirement and supply for ruminant diets. They include the UK Metabolisable Protein System (MPS), the French Protein Digested in the Intestine (PDI) system and the Cornell Net Carbohydrate and Protein System (CNCPS) of North America.

For clarity, the MPS system (AFRC, 1992) will be used as a reference throughout. Metabolizable protein (MP) is the true protein available to the animal for metabolism which is used for maintenance and protein synthesis (i.e. lean tissue and milk protein) after digestion and absorption of nutrients. However, as protein digestion in the ruminant is rather a complex process, a simplified flow diagram is shown to assist in the explanation of the principles (Figure 1).

In essence, food crude protein (CP) entering the rumen is partially degraded by bacteria and protozoa to ammonia. This rumen degradable protein (RDP) together with available energy, is used for microbial protein synthesis. Microbial crude protein (MCP) and undegraded protein (UDP) then pass together into the small intestine for digestion and utilization.

The yield of metabolizable protein from any particular diet is the sum of absorbed amino acids from both microbial protein and undegraded feed protein and thus describes the supply of amino acids that is available for normal metabolic processes in the body. To determine the total MP yield it is necessary to be able to describe the ruminal degradation of feeds as fully as possible. The *in situ* technique described by Orskov and Mehrez (1977) is used to define the rumen degradable fraction within the MP and other systems.

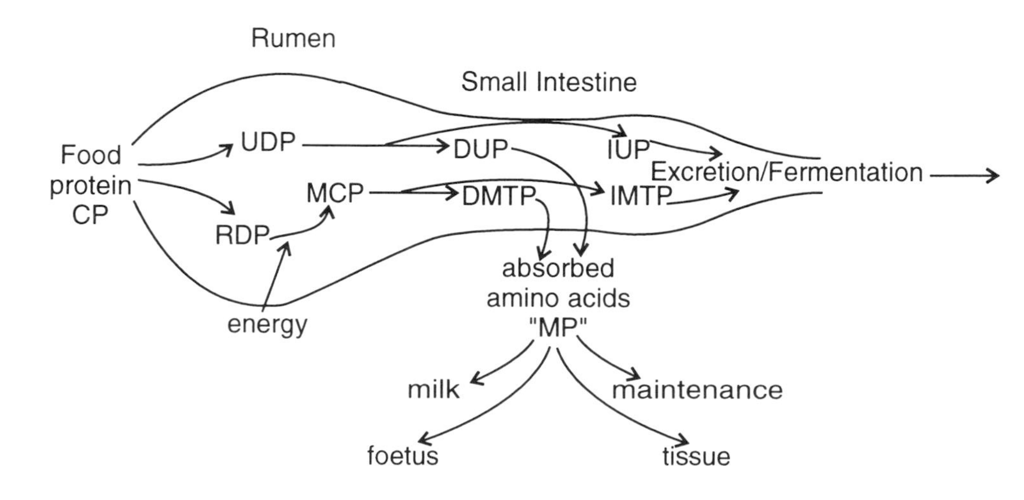

Figure 1. Protein digestion in the rumen.

Predicting protein supply: in situ technique or protein degradability

Protein degradability is determined *in situ* by incubating small nylon bags containing measured amounts of feed in the rumen for a range of time periods up to 48 h (concentrates) or 72 h (forages). After incubation the bags are washed and the feed residue is dried, weighed and analyzed. Losses of dry matter and nitrogen (N) from the bags can be plotted against time to obtain disappearance curves. Different feedstuffs produced differently shaped degradation curves (Figure 2).

The *in situ* data show that barley, soyabean meal (SBM) and fishmeal have varying rates of degradation. Soyabean meal, with 80% of the protein having disappeared in 24 h, compares with much slower and steadier disappearance of fishmeal with only half the protein degrading in 24 h. Interestingly, the immediate loss of water soluble material and fine particles at zero time is greater for fishmeal than SBM. An exponential equation may be fitted to the degradability data (Equation 1).

$$\text{Degradability} = a + b\,\{1\text{-}e^{-ct}\} \qquad \text{Equation 1}$$

where
a = immediate or water soluble fraction
b = potentially degradable fraction
c = rate of degradation of the 'b' fraction respectively.

Obviously the rate of outflow of rumen digesta will have a significant impact on the degree of degradation that takes place in the rumen. Orskov and McDonald (1979) derived an equation to relate effective degradability and rumen outflow rate for any given fractional rumen outflow rate per hour (Equation 2).

Effective degradability (p) at any given outflow rate (per hour, r) = $a+(b \times c)/(c + r)$ Equation 2

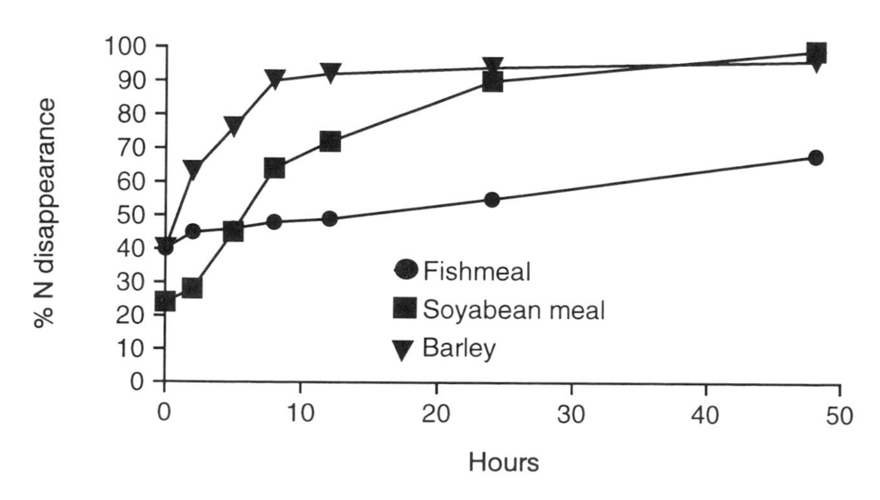

Figure 2. *In situ* disappearance of nitrogen over time from three feeds.

Outflow rates are highly correlated with the level of feeding. Higher feed intakes result in shorter rumen retention times and faster outflow rates. Different outflow rates have been described relative to three typical feeding situations:

1. 0.02/h for animals fed at maintenance, a low level of production,
2. 0.05/h for animals fed at higher levels, but less than twice maintenance, e.g. calves, beef cattle, sheep and low yielding dairy cows.
3. 0.08/h for high yielding cows fed at more than twice maintenance.

Using data from the degradability equation, quickly (QDP) and slowly (SDP) degradable protein fractions may be defined (AFRC, 1993). The immediately soluble or cold water soluble fraction of the total feed CP is called QDP and can be calculated by multiplying the constant value 'a' by the crude protein content (Equation 3).

Quickly degraded protein (QDP) = a (CP) Equation 3

The QDP fraction is rapidly released when feed enters the rumen. SDP is related to the 'b' fraction of feed CP and is determined by the time the feed remains suspended in the rumen exposed to bacterial digestion (Equation 4).

Slowly degraded protein (SDP) = $\{(b \times c)/(c + r)\} \times CP$ Equation 4

Effective rumen degradable protein (ERDP), which is the measure of total CP that is utilized by the microbes for growth and protein synthesis, may then be calculated (Equation 5). Some examples of proteins showing different degradability characteristics are given in Table 1. These data indicate that as rumen outflow rate increases, SDP content and hence ERDP decrease. This information can be used to calculate the UDP, for any outflow rate (Equation 6).

Effective rumen degradable protein (ERDP) = $0.8QDP + SDP$ Equation 5

Undegradable protein (UDP) = $CP - (QDP + SDP)$ Equation 6

Digestibility of undegraded protein

Going one step further, the digestibility of the UDP may be predicted from the acid detergent insoluble nitrogen (ADIN) fraction of the feed (Goering and Van Soest, 1970). ADIN is a measure of the amount of N associated with the acid detergent fiber (ADF) fraction of the feed; and as such is taken to represent that part of the undegradable N which is unlikely to be digested post-ruminally. The digestible undegraded protein (DUP) content of a feed can be derived if the UDP and ADIN are known (Equation 7). The ADIN content of some common feedstuffs given in Table 2 indicates a considerable variability across those feeds assayed.

Digestible undegraded protein (DUP) = $0.9UDP - 6.25ADIN$ Equation 7

Table 1. Degradable protein fractions of three feedstuffs at different outflow rates*.

Fraction		Fishmeal	Soyabean meal (g/kg dry matter)	Barley
Crude protein		714.0	462.0	119.0
QDP		221.0	37.8	37.3
SDP	r = 0.02/h	240.2	342.4	69.4
	r = 0.05/h	135.7	263.9	63.7
	r = 0.08/h	94.6	214.8	58.8
ERDP	r = 0.02/h	417.0	372.6	99.2
	r = 0.05/h	312.5	294.1	93.5
	r = 0.08/h	271.4	245.0	88.6

* Calculated from ADAS Feed Evaluation Unit Tables of Ruminant Degradability Values (ADAS, 1989).

Table 2. Crude protein and ADIN content of some common feeds.

Feedstuff	Grams/kg dry matter	
	Crude protein	ADIN*
Barley (ground)	129	0.4
Maize gluten feed	207	1.4
Fishmeal	694	—
Rapeseed meal	400	3.6
Distillers grains	317	13.0
Soyabean meal	497	2.2

* From AFRC (1993).

While ADIN appears to be a useful predictor of indigestible nitrogen, there is now evidence to show that in some cases part of the ADIN content may in fact be digestible. Therefore the calculated DUP contents of feedstuffs, particularly those with a high ADIN value, should be treated with caution.

Estimating digestible undegraded protein (DUP)

Using the equations described above, the ERDP, UDP and DUP contents of the three example feed proteins can be calculated (Table 3). All values are expressed for an outflow rate of 0.08/h relative to high yielding dairy cows at above twice maintenance.

Using the equations described previously, the amount of DUP entering the intestine may be estimated if the character of the feed is known. However, it

Table 3. The characterisation of three feed proteins at 0.08/h outflow rate.

	Fishmeal	Soyabean meal	Barley
Crude protein, %	71.4	46.2	11.9
ERDP, % protein	38.0	53.0	74.0
UDP, % of protein	55.7	45.2	19.3
DUP, % of protein	50.1	37.8	15.2

is normal that the amino acid profile of the DUP is not known. In fact, in most feeding situations it is unlikely to have exactly the same profile as that in the feed protein consumed. The reason for this is that when protein is degraded in the rumen, not all amino acids are deaminated at the same rate. Amino acids such as phenylalanine, which is more resistant to ruminal degradation, will be enriched in the DUP (Chen and Orskov, 1994). Of course not all of UDP is digestible (as DUP), so the undigested portion (indigestible undegraded protein or IUP) is either excreted or fermented in the hindgut.

Microbial protein

Between 35 and 66% of the amino acids entering the small intestine will be of microbial origin (Rulquin and Verite, 1993). As a result, the amino acid profile of the protein available for absorption in the small intestine is likely to be quite different from that in the original feed material. Studies on low yielding ruminants suggested that the amino acid profile in the intestine was almost constant, i.e. in the form of microbial protein (Cole, 1996).

The MP system calculates MCP supply for any particular diet and feeding level. Three quarters of the MCP is estimated to be microbial true protein (MTP) of which 85% is digestible, hence digestible MTP (DMTP) = 0.75× 0.85 × MCP. The indigestible portion (indigestible microbial true protein IMTP) is not absorbed for utilization and is excreted or fermented in the hindgut. Therefore, if the composition and yield of the DMTP and DUP entering the duodenum can be determined for a certain feeding situation, this should describe the protein supply to the site of absorption.

Predicting protein requirements

Given the complexity of ruminant protein digestion, how can the ideal balance of amino acids required by the ruminant be estimated? In the dairy cow, four functions need to be considered when looking at protein requirements. These are maintenance, tissue gain or mobilization, fetal development and milk production. In general, rumen microbial protein alone should meet the needs of a cow yielding less than 20 kg milk per day. Cows producing in excess of 20 kg/day will require feed protein (DUP) in addition to microbial protein.

Once microbial protein alone becomes insufficient to meet the cow's demands, opportunities arise to identify the first limiting amino acid in microbial protein. Hereafter, there is scope to use supplements to match or compensate for the imbalances of the amino acid composition of rumen microbial protein and hence minimize the need for supplemental protein.

The amounts of each essential amino acid needed to maintain a particular level of performance can, in theory, be calculated factorially by adding together estimated requirements for maintenance, for milk protein secretion and for tissue protein changes (i.e. weight gain or loss). Oldham (1994) has calculated the amino acid requirements for cows at four production levels (Table 4).

Given a knowledge of the profile and quantity of amino acids required at

Table 4. Amounts of amino acids required for absorption to meet the needs of a 600 kg cow for maintenance or when she is secreting 500, 1000 or 1500g milk protein per day.

	Amount of amino acid (g/day) to secrete milk at			
Amino Acid	Maintenance	500 g/day	1000 g/day	1500 g/day
Threonine	14	66	118	170
Valine	5	81	157	233
Methionine + cystine	5	46	87	128
Methionine	2	32	62	92
Isoleucine	4	68	132	196
Leucine	6	116	226	336
Phenylalanine	5	63	121	179
Lysine	9	53	97	132

the small intestine, those amino acids which appear to be in short supply can be supplemented. It is important to remember, however, that total protein requirements must be satisfied before optimizing individual amino acids. If this criterion is satisfied, limiting amino acids may be identified. In fact, under a range of conditions the first two limiting amino acids for milk protein secretion have been identified as lysine and methionine with histidine and leucine as potentially the third and fourth limiting (Sloan, 1997).

Thus it is theoretically possible to design an ideal protein for ruminants utilizing the amino acid requirements of the dairy cow. A comparison of supply and requirement for a cow secreting 1 kg milk protein per day is shown in Figure 3. The basal diet is in the form of silage, cereal, beet pulp and SBM which satisfies crude protein and energy requirements. However, some shortfalls of specific amino acids can be clearly seen. Under these feeding conditions it would be expected that the cow would not maximize milk protein yield.

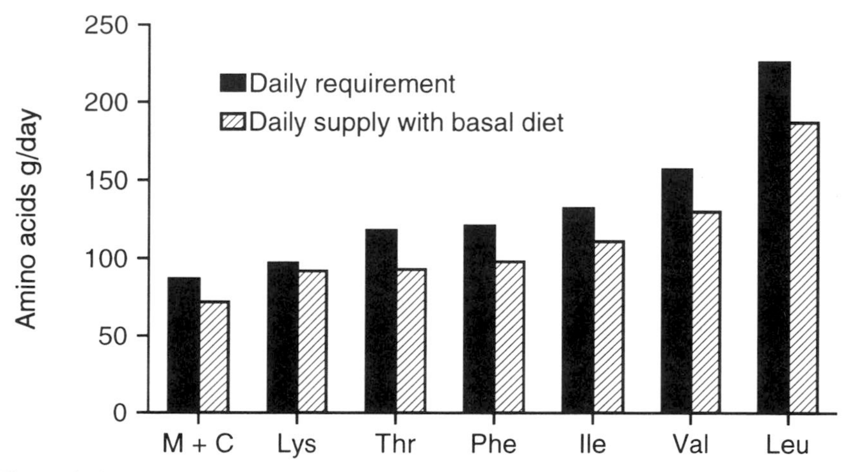

Figure 3. Comparison of amino acid supply and requirement in a cow secreting 1 kg milk protein/day.

Product development: the ultimate protein for dairy cows

To evaluate the theories described in the previous chapter, a new product in the Alltech Ultimate Protein range has been developed to provide an ideal supply of digestible undegraded amino acids to the high yielding dairy cow. Termed Ultimate Protein 1562, the product is designed to help meet the cow's total requirement for digestible undegraded amino acids. The Ultimate Protein 1562 is a wholly vegetable-based product, manufactured by a new processing technique. The protein incorporates vegetable fats during the process and is designed to have a high DUP content. Analysis of Ultimate Protein 1562 is shown in Table 5.

Table 5. The Ultimate Protein 1562: product analysis.

Nutrient	%
Protein	55
Fat	20
Ash	3
NDF	15
Metabolizable energy (ruminant), MJ/kg	17.5

DETERMINING UDP CONTENT

Evaluation of Ultimate Protein 1562 has proceeded ın stages. Initial rumen incubation tests at 0 and 8 h comparing it with fishmeal and soyabean meal suggested a high degree of rumen protection from degradation. These tests were followed by a full rumen degradability evaluation comparing a fishmeal control with Ultimate Protein 1562 as shown in Figure 4. The degradation characteristics obtained by fitting these values to the model of Orskov and McDonald (1979) are shown in Table 6.

If the data from Table 6 are converted to UDP (as a proportion of the protein feed itself), a comparison can be made between the relative bypass protein merits of a range of feed proteins (Table 7). Similar UDP values for fishmeal and Ultimate Protein 1562 can be seen, both of which are considerably higher than soya or rapeseed proteins.

DETERMINATION OF BYPASS AMINO ACIDS

The effective rumen degradability of Ultimate Protein 1562 at an outflow rate of 0.08/h (corresponding to a high yielding cow) was 41%. This figure is equivalent to the nitrogen loss after 12 h. To determine amino acid loss at this degradability, individual amino acid losses from samples incubated *in situ* were measured for the 12 h incubation samples. The values for amino acid disappearance were used to derive the bypass amino acid content of Ultimate Protein 1562 (Figure 5).

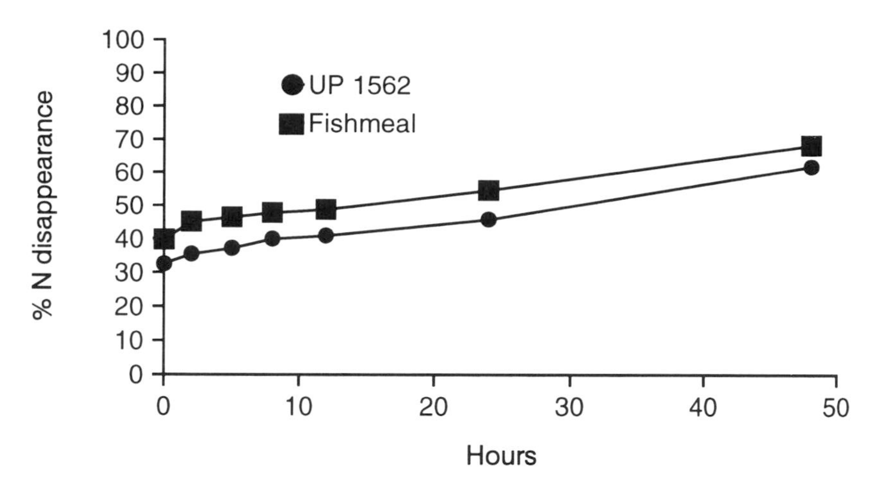

Figure 4. *In situ* disappearance of nitrogen for Ultimate Protein 1562 and fishmeal.

Table 6. Degradability characteristics*, ERDP and UDP of Ultimate Protein 1562 and fishmeal (0.08/h outflow rate).

Degradability characteristics	UP 1562	Fishmeal
a, %	34.0	40.6
b, %	56.8	48.1
a+b, %	90.8	88.7
c, per hour	0.012	0.017
ERDP, 0.08/h	41	49
UDP, 0.08/h	58	50

* See Equation 1.

Table 7. Comparison of UDP content* of a range of protein sources.

	UDP (%)
Fishmeal (66)	33
Soyabean (44)	18
Rapeseed (36)	12
Ultimate Protein 1562 (55)	32

* Measured *in sacco* and calculated at 8% outflow rate.

BALANCING SUPPLY WITH REQUIREMENT

In conclusion, these data show the Ultimate Protein 1562 to be protected through the rumen, to have a similar ERDP to fishmeal and therefore to be a good source of UDP. The profile of amino acids for absorption give a greater potential for matching requirement to supply. Figure 6 shows how a basal diet (described in Figure 3) supplemented with 500 g of Ultimate Protein 1562 may diminish the gap between supply and requirement.

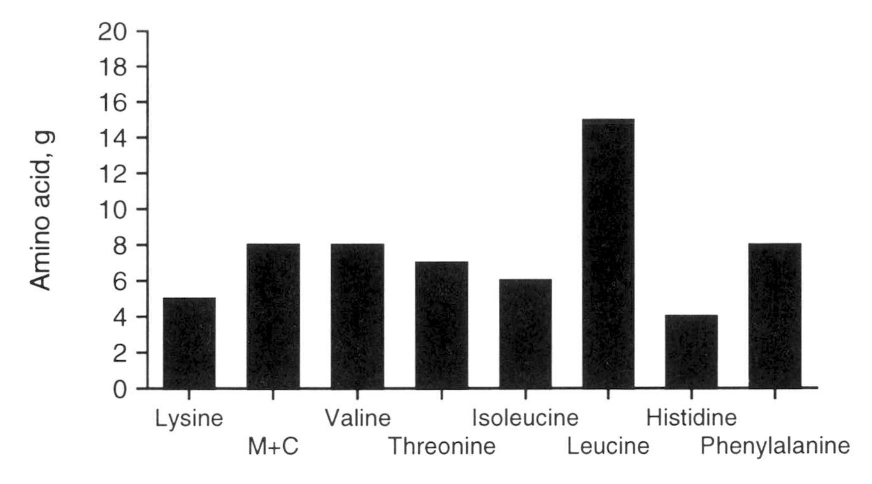

Figure 5. Supply of bypass amino acid, g per 500g Ultimate Protein 1562.

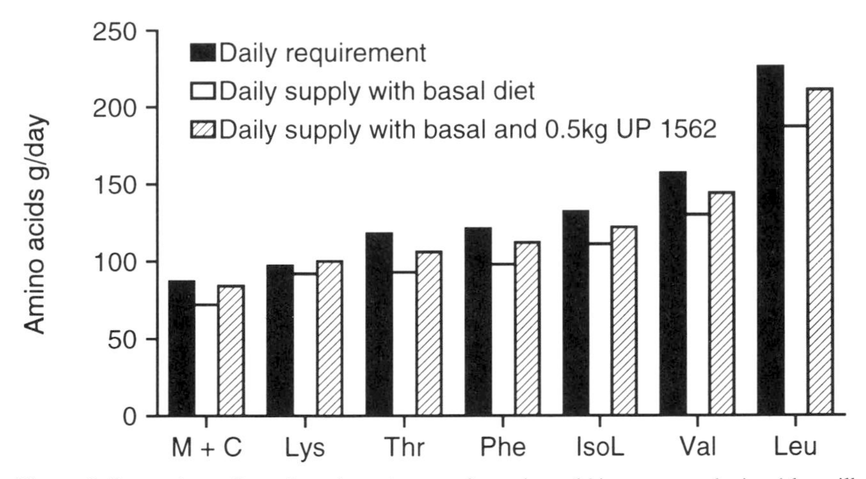

Figure 6. Comparison of supply and requirement for amino acid in a cow producing 1 kg milk.

DAIRY COW TRIALS

Two dairy cow production trials were designed to evaluate the principles of ideal protein supply and requirement under practical farming conditions. Both trials were designed to feed three groups of dairy cows a constant basal ration, sufficient to meet nutritional needs at the appropriate stage of lactation. The three dietary treatments evaluated were comprised of basal plus:

A. Degradable protein source 0.5 kg (SBM/rapeseed 50/50), or
B. UDP protein feed source 0.5 kg fishmeal, or
C. UDP protein feed source 0.5 kg Ultimate Protein 1562.

Milk records were obtained monthly and the preliminary data from the trials are shown in Figures 7 and 8.

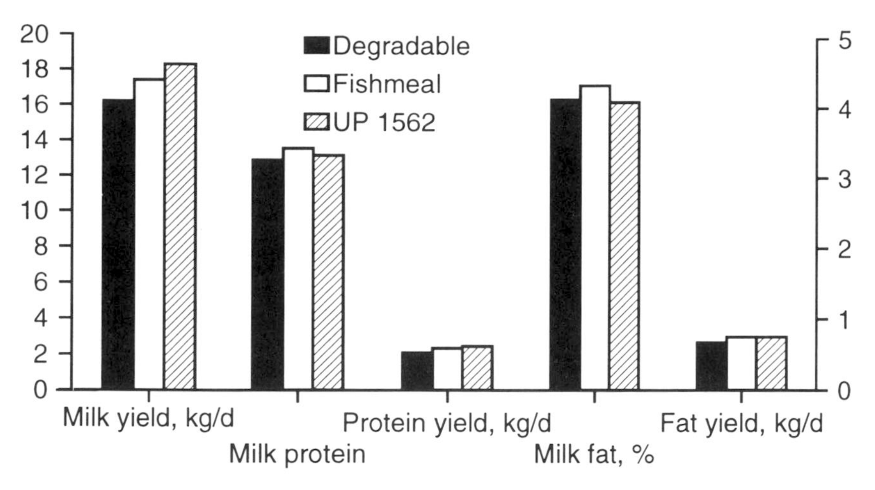

Figure 7. UK dairy cow production trial (A).

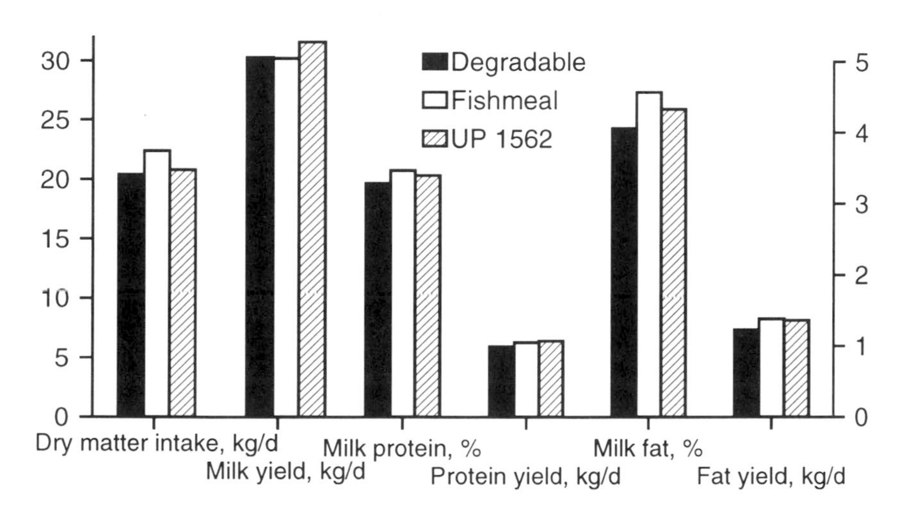

Figure 8. UK dairy cow production trial (B).

Production trial results

Preliminary data indicated that both medium and high yielding cows are responding to the addition of a UDP source (fishmeal and Ultimate Protein 1562) by increasing milk yield. Both UDP products appear to give a milk fat yield response compared with the control group. The cows consuming the Ultimate Protein 1562 supplement appear to achieve a higher milk protein yield when compared to both the control and fishmeal supplemented cows.

Conclusions

A protein feed material (Ultimate Protein 1562) which can satisfy the amino acid requirement in lactating ruminants has been designed by taking into account the current knowledge of the metabolizable protein system. Evaluation of Ultimate Protein 1562 using *in vivo* techniques and production trials have indicated that it is possible to balance the supply and requirement of amino acids in the dairy cow. The production responses in terms of increased milk yield and milk protein yield would tend to support the view that balancing supply and requirement is beneficial.

References

ADAS Feed Evaluation Unit. 1989. Tables of Rumen Degradability Values for Ruminant Feedstuffs.

AFRC. 1992. Technical Committee on Responses to Nutrients, Report No 9., Nutritive Requirements of Ruminant Animals: Protein. Nutr. Abs and Rev Series B 62 (12)787–835, CAB International, Wallingford, Oxon.

AFRC. 1993. Energy and Protein Requirements of Ruminants. An advisory manual prepared by the AFRC Technical Committee on Responses to Nutritents. CAB International, Wallingford, Oxon.

Chen, X.B. and E.R. Orskov. 1994. Amino acid nutrition in sheep. In : Amino Acids in Farm Animal Nutrition. (J. P. F. D'Mello, Ed) CAB International, Wallingford, Oxon. p.307.

Cole, D.J.A., 1996. Third generation protein: an ideal protein for each species. Maximising performance while minimising nitrogen excretion. In: Biotechnology in the Feed Industry. Proceedings of the 12th Annual Symposium. (T.P.Lyons, K.A. Jacques, eds). Nottingham University Press. Loughborough, Leics. UK p.351.

Goering, H.K. and P. J. Van Soest. 1970. Forage Fiber Analysis Agr. Handbook No 379. Agr. Res. Ser. USDA, Washington, DC.

Oldham, J.D. 1994 . Amino acid nutrition of the dairy cow. In: Amino Acids in Farm Animal Nutrition. (J.P.F. D'Mello, Ed) CAB International, Wallingford, Oxon.

Orskov, E.R. and A.Z. Mehrez. 1977. Estimation of extent of protein degradation from basal feeds in the rumen of sheep. Proc. Nutr. Soc. 36:78A.

Orskov, E.R. and I. MacDonald. 1979. The estimation of protein degradability in the rumen from incubation measurements weighted according to rate of passage. J. Agric. Sci. 92: 499.

Rulquin, H. and R. Verite. 1993. Amino acid nutrition of dairy cows: Productive effects and animal requirements. In : Recent Advances in Animal Nutrition. (P.C. Garnsworthy and D.C.A. Cole, eds). Nottingham University Press. Nottingham, UK p. 55.

Sloan, B.K. 1997. Developments in Amino Acid Nutrition of Dairy Cows. 31st University of Nottingham Feed Manufacturer's Conference. In Press.

THE GUT OF THE NEONATE, NURSING, AND WEANED PIG: DO DIETARY ELECTROLYTES AND CERTAIN HEAVY METALS MAKE IT FUNCTION MORE EFFECTIVELY?

DON MAHAN

Ohio State University, Columbus, Ohio, USA

Introduction

An ability to absorb and utilize large quantities of nutrients is essential for any young animal upon birth. The overall response to the maternal mammary secretions in the pig are particularly evident by the large increase in body weight that occurs within the first few weeks of life. After weaning, the young pig is expected to continue its rapid growth rate, but feeds that are effectively digested must be provided to meet the pig's nutritional needs. To achieve a high rate of growth not only is the nutrient source important, but the hydrolysis and absorption of these nutrients from intact macromolecules becomes extremely critical. The development of the gastrointestinal tract of the pig begins during fetal life, but continues to change post-partum. Changes in the intestinal tract upon weaning are perhaps the most dramatic in the pig's life, and how the pig responds to the dietary components during this transition may affect subsequent performance. Reviews by Cranwell (1995), Kidder and Manners (1978), and Varley (1995) have documented the rapidity of the digestive and microbial changes that occur in the intestinal tract as pigs mature, but more information is needed for better feedstuff utilization particularly as weaning age declines.

The young nursing pig can effectively digest the maternal milk supply, but the advent of early weaning and the changing dietary, environmental and microbial conditions that are imposed upon the young pig may alter the normal digestive process. The use of sodium chloride, copper sulfate and the recent incorporation of high dietary levels of zinc oxide during the early phases post-weaning have recently demonstrated post-weaning growth rate responses. The effects of these minerals appear to be largely in the digestive tract but the mode of action of each occurs in different ways. The use of specialty feeds during the early weaning period is important, but there is also the desire to make a quick transition toward using feeds of a lower cost in order to keep the cost of production at a minimum. Enzymes and specialty feed ingredients continue to be of prime importance in improving protein quality for early weaned pigs, but the role of minerals, particularly the electrolytes, copper and zinc during this period will greatly affect pig performance.

Pre-weaning digestive conditions

The stomach has a recognized storage role for ingested milk and feed, but it also has an important but unrecognized function in initiating the digestive process. The consumption of milk by the neonate results in the proliferation of *Lactobacilli* spp. These bacteria hydrolyze at least part of the milk lactose to lactic acid and a small amount of volatile fatty acids (Figure 1). The level of lactic acid in the stomach is particularly high while the pig nurses the sow, but declines greatly upon weaning when milk is excluded from the diet. The acidity which results from the action of the *Lactobacilli* organisms on milk results in a lowered stomach pH such that the growth of other microflora is inhibited (Table 1).

A low pH allows the pepsinogen enzymes which are secreted in the stomach to be closer to optimum pH for the clotting of milk and the subsequent hydrolysis of milk protein. Several enzymes are secreted in the stomach of the pig each having a different but optimum pH for protein hydrolysis activity (Table 2). Large quantities of mucous (Ito, 1987) and at least some sodium bicarbonate (Flemström, 1987) are secreted in the pyloric region of the stomach. These products serve to buffer the various gastric secretions and microbial products in the stomach. Milk protein and other dietary proteins can also serve as buffers for the acidic conditions generated.

Gastric secretion of hydrochloric acid (HCl) begins during late fetal development and is extremely low at birth. This is attributed largely to the immaturity of the parietal cells which are responsible for the secretion of this

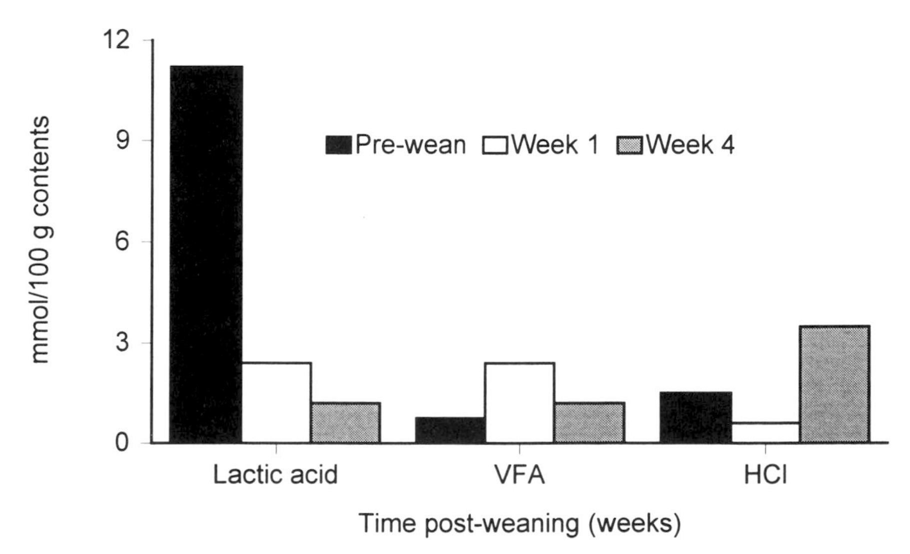

Figure 1. Concentrations of inorganic and organic acids in the stomach digesta of piglets weaned at 35 days (Schnabel, 1983).

Table 1. Approximate pH ranges for bacterial growth*.

Organism	Minimum	Optimum	Maximum
Clostridium perfringens		6.0–7.6	8.5
Escherichia coli	4.3–4.4	6.0–8.0	9.0–10.0
Pseudomonas aeruginosa	4.4–5.6	6.6–7.0	8.0–9.0
Salmonella sp.	4.0–5.0	6.0–7.5	9.0
Staphylococcus sp.	4.2	6.8–7.5	9.3

*Johnston (1991).

Table 2. Characteristics of porcine gastric proteases*.

IUBMB name and number	Optimum pH for GPA†	Relative MCA‡	
		Bovine milk	Porcine milk
Pepsin A	2.0	100	High
Pepsin B	3.0	10	–
Gastricsin (Pepsin C)	2.0–3.5	100	–
Chymosin (Rennin)	3.0–4.0	100	Very high

* Source: Cranwell (1991).
† GPA = general proteolytic activity.
‡ MCA = milk clotting activity.

gastric acid. Although HCl production increases during the first week of life, its production remains relatively low while the pig nurses the sow, increasing five to seven fold by five weeks of age. Because milk is easily hydrolyzed by the gastric and enzymatic conditions which exist in the stomach and small intestine, the need for HCl is lessened. The providing of creep feeds to nursing pigs will stimulate the production of HCl but weaning pigs by 10 days of age did not result in a lowered pH within the first 10 days post-weaning (Kidder and Manners, 1978). Consequently there appears to be very little HCl produced during the nursing and early post-weaning period (Figure 1). Because lactic acid production and HCl secretion is low in the weaned pig, protein digestive processes may therefore be hindered at least during the initial weeks post-weaning.

Post-weaning digestive conditions

Upon weaning not only is there a dramatic change in the composition and type of diet fed to young pigs, but there is also a change in the microbial population throughout the intestinal tract. Most pig starter diets have milk products incorporated at low dietary levels and a corresponding high percentage of vegetable proteins and starches. The intestinal hydrolysis of the protein and starch from these vegetable-based feeds is lower than milk proteins, and unless feeds of a high digestibility are used, feed intake and growth rate will be poor. Vegetable-based diets may also be retained in the stomach for a longer time than when the pig consumed milk from the sow. Because the population of *Lactobacilli* spp. declines post-weaning there is a subsequent increase in stomach acidity until HCl production increases. This condition could allow the growth of other microflora in the intestinal tract. Consequently, if a population of pathogenic bacteria proliferates during this period, severe diarrhea or disease may result within a few days of weaning. Efforts are therefore being made to prevent this condition from occurring by the proper formulation of starter pig feeds and the use of antibacterial agents, oligosaccharides and zinc and (or) copper supplements.

It is presumed by many that the responses noted with copper and zinc may be due to their antimicrobial properties. The exact mode of action of these minerals is due to oligodynamic action, i.e. metal ions combining with the -SH groups on cellular protein causing denaturation. In studies done in pure culture, Newman *et al.* (1995) found that the inhibitory concentrations of copper and zinc varied depending on the strain of bacteria examined. For example, concentrations as low as 16 ppm copper can be inhibitory to *Clostridium perfringens*, but higher levels of copper were necessary to inhibit beneficial bacteria such as *Lactobacilli*. Trials examining the effects of copper sulfate on ureolytic bacterial populations in the pig large intestine demonstrate reductions in the predominant intestinal ureolytic species and an overall reduction in intestinal urease activity (Varel *et al.*, 1987). The impact that this reduction in urease activity has on animal production is somewhat controversial. However, excessive ammonia in the intestinal tract can be toxic and lead to an increase in maintenance energy requirements due to epithelial cell turnover. With zinc, a similar antimicrobial activity exists, although higher concentrations of zinc are necessary to inhibit most bacterial strains tested (Newman *et al.*, 1995).

Not only is feed intake important for the weaned pig, but protein and carbohydrate sources are equally important. These feeds must be effectively digested or growth performance of the pig will be hindered. Because protein digestion is initiated in the stomach, and the acidic conditions are poorer in the stomach of the weaned pig, the role of dietary chloride was investigated.

Many of the specialty feeds have, however, high sodium and chloride concentrations largely because of the type of product and the processing methods used in production of these feeds (Table 3). Consequently, because of the dietary incorporation of one or several of these products into most weanling pig diets, it was generally considered that these minerals were provided at adequate levels to pig starter diets. These diets were therefore considered to provide adequate levels of these elements to meet requirements without further supplementation of these electrolytes. It was further speculated that because of the high levels of sodium in most of these specialty feeds, part of the diarrhea encountered post-weaning may be attributed to the upset of electrolyte balance in the intestinal tract.

The initial trial involved supplementing salt (NaCl) to the diets of three-week-old starter pigs. Diets were formulated with dried whey incorporated at 25%. The basal diet contained 0.20% Na and 0.35% Cl, substantially above the current NRC (1988) recommendations for these nutrients. A significant growth response occurred during the first two weeks post-weaning when added salt was provided, but no growth response thereafter (Table 4). Subsequent

Table 3. Electrolyte composition of commonly used specialty feeds in the Phase I diets of pigs.

Feed ingredient	Composition (%)		
	Na	Cl	K
Dried whey	1.30	1.50	1.11
Plasma protein	2.23	0.40	0.27
Ultimate Protein 1672	1.63	0.88	0.80

Table 4. Effectiveness of added salt (NaCl) to various starter feed ingredients used in Phase I nursery pig diets.

Feed ingredient	Daily gain (g) with added NaCl at				
	0%	0.2%	0.4%	0.6%	0.8%
7% Ultimate Protein 1672*					
0–7 days	37	44	53	68	61
8–14 days	201	215	210	212	216
15–28 days	359	361	365	399	393
7% Plasma protein (920)†					
0–7 days	168	188	214	210	—
7–14 days	329	374	411	414	—
25% Dried whey					
0–7 days	74	91	107	120	—
8–14 days	256	275	289	279	—
15–35 days	530	531	545	524	—

* Basal diet contained 0.10% Na, 0.22% Cl, 0.61% K.
† Basal diet contained 0.20% Na, 0.10% Cl, 0.68% K.
‡ Basal diet contained 0.20% Na, 0.35% Cl, 10.35% K.

trials using plasma protein and Ultimate Protein 1672 (Alltech Inc.) showed the same type of response to added salt (Table 4) demonstrating that a growth response to the added salt occurred in the initial weeks post-weaning, but no response was noted two weeks post-weaning.

Because the addition of salt provides two elements (sodium, chloride), two additional experiments were conducted. The first experiment evaluated the efficacy of added chloride on post-weaning pig performance responses, and a second evaluated the role of chloride on protein digestibility during the initial weeks post-weaning. The results presented in Table 5 demonstrated that pig growth rate increased each week post-weaning as the dietary chloride level increased, that nitrogen retention increased and that apparent protein digestibility was improved each week post-weaning. The dietary level of chloride that resulted in the maximum responses varied by week, but in general declined as the pig became older.

These results suggest that the young pig's ability to secrete adequate chloride for the protein digestion process is limited in the initial weeks post-weaning, but the addition of sodium chloride did not cause diarrhea and in fact resulted in improved performance responses. No evidence of upsetting

Table 5. The effectiveness of added chloride in the diets of weaned pigs on performance, nitrogen retention and digestibility*.

Item	Added chloride (%)					
	0	0.07	0.14	0.21	0.28	SEM
Week 1 post-weaning						
Daily gain, g	120	123	156	146	138	11
Daily feed, g	241	200	235	208	201	12
N retention, g/day	6.0	6.6	6.7	6.8	6.4	0.1
N digestibility, %	89.2	90.7	91.4	93.2	94.9	0.6
Week 2 post-weaning						
Daily gain, g	295	316	342	351	310	10
Daily feed, g	431	436	459	491	436	24
N retention, g/day	9.0	10.9	11.2	11.6	11.3	0.4
N digestibility, %	77.9	80.0	83.9	86.1	86.5	1.1
Week 3 post-weaning						
Daily gain, g	372	422	447	418	427	20
Daily feed, g	627	650	716	657	649	23
N retention, g/day	9.9	10.9	11.4	10.8	10.4	0.4
N digestibility, %	75.3	78.5	82.5	80.0	80.8	0.8

* Two separate trials were conducted; the performance data was a growth trial where *ad libitum* feed intake was allowed, whereas the digestibility trial restricted feed intake.

the electrolyte balance was evident during the course of these trials with any of the products used.

The use of copper sulfate and the more recent use of zinc oxide, both at high dietary levels, has resulted in improved pig performance responses during the nursery period. A recent starter pig study involving 1156 pigs weaned at 22 days of age was conducted at 10 universities in the United States. The results demonstrated an approximate 12 to 15% growth response to either copper sulfate (250 ppm Cu) or zinc oxide (3000 ppm Zn), but the effect was not cumulative (Figure 2).

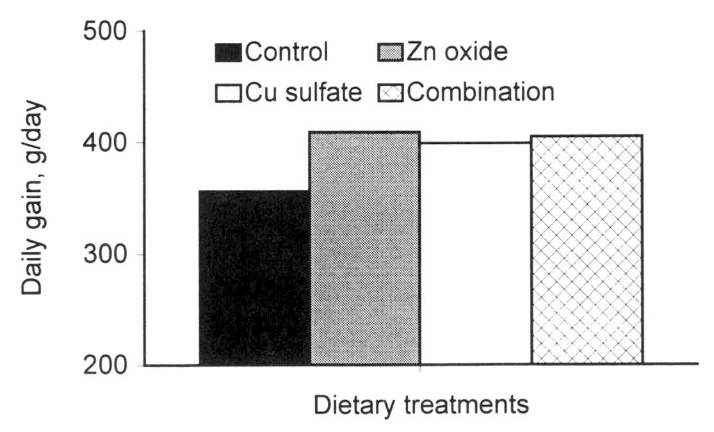

Figure 2. Effects of high levels of zinc oxide (3000 ppm Zn) or copper sulfate (250 ppm Cu) or their combination when fed to 3 week old weaning pigs (Hill *et al.*, 1996).

Summary

The pig secretes HCl, but the quantity secreted is low at birth and limited while the animal is nursing the sow. Upon weaning, HCl production remains low during the initial weeks post-weaning resulting in a lower apparent protein digestion and post-weaning growth of the pig even when pig starter diets contain feeds (Ultimate Protein 1672, dried whey, plasma protein) which have naturally high sodium and chloride contents. The addition of salt to the weanling pig diet at 0.40 to 0.60% has resulted in improved performance responses. The addition of high levels of copper sulfate (250 ppm Cu), or zinc oxide (3000 ppm Zn) have been shown to increase post-weaning pig growth rates by 12 to 15%. Copper appears to reduce urease microbial activity in the small intestine, while zinc oxide kills the bacteria. Environmental concerns about the high use of these elements in animal feeds may limit their application in many countries.

References

Cranwell, P.D. 1995. Development of the neonatal gut and enzyme systems. In: The Neonatal Pig. (Ed. M.A. Varley), C.A.B. International, pp. 99–154.

Flemström, G. 1987. Gastric and duodenal mucosal bicarbonate secretion: In: (Johnson, L.R., Ed.), Physiology of the Gastrointestinal Tract. Vol. II, 2nd ed. Raven Press, New York, pp. 1011–1029.

Hill, G.M., G.L. Cromwell, T.D. Crenshaw, R.C. Ewan, D.A. Knabe, A.J. Lewis, D.C. Mahan, G.C. Shurson, L.L. Southern and T.L. Veum. 1996. Impact of pharmacological intakes of zinc and(or) copper on performance of weanling pigs. J. Anim. Sci. 74(Suppl. 1):181 (Abstr.).

Ito, S. 1987. Functional gastric morphology. In: (Johnson, L.R., Ed.), Physiology of the Gastrointestinal Tract, Vol. II, 2nd ed. Raven Press, New York, pp. 817–851.

Johnston, R. 1991. Acidification of pig diets to improve health and growth of weaners. Pork Talk 2000. pp. 19–34.

Kidder, D.E. and M.J. Manners. 1978. Digestion in the Pig. Kingston Press, Bath, Canada.

Newman, K.E., B.E. Moore and V.E. Chandler. 1995. Minimum inhibitory concentrations (MIC) of different sources of copper and zinc on pure strains of gastrointestinal bacteria. J. Dairy Sci. 78(Suppl. 1):237.

NRC. 1988. Nutrient Requirement of Swine. (9th Ed.) National Academy Press, Washington, D.C.

Varley, M.A. 1995. The Neonatal Pig – Development and Survival. C.A.B. International.

Varel, V.H., I.M. Robinson and W.G. Pond. 1987. Effect of dietary copper sulfate, Aureo SP250, or clinoptilolite on ureolytic bacteria found in the pig large intestine. Appl. Environ. Microbiol. 53:2009–2012.

EARLY STAGE TURKEY NUTRITION AND IDEAL PROTEIN: IMPLICATIONS FOR ALL ASPECTS OF POULTRY PRODUCTION

JEFFRE D. FIRMAN

Department of Animal Sciences, University of Missouri, Columbia, Missouri, USA

Introduction

Protein/amino acids is one of the major cost components of the diets of turkeys. It is also one of the components of the diet that affect performance in a practical sense. On a wholesale price basis, the turkey industry has an output of about \$3 billion in the US. About 40%, or \$1.2 billion, of this value goes into the feed for turkeys. Considering the major importance of feed to the turkey industry, there is, relatively speaking, little research on the feeding of turkeys. This has led to wide variation in feeding programs found in the US and worldwide. Performance on diets with similar nutrient contents varies as well. One of the reasons for this is the variation in amino acid digestibility in different feeds currently used for feeding turkeys. Some of the concepts and data that relate to amino acid nutrition, low protein diets, ideal proteins, digestible amino acid requirements, and where we can go in the future will be covered below.

Background

FUNDAMENTALS OF AMINO ACID NUTRITION

Protein is the foundation of the lean meat that the industry is trying to produce. Protein is made up of long chains of amino acids, but may contain other non-amino acid portions. Amino acids are small molecules which contain an amino group (NH_2) and a carboxylic acid group (COOH), thus deriving the name. Amino acids can be placed in several different categories based on their structural properties and are measured with a relatively complex assay system using high pressure liquid chromatography. Examples of amino acids include lysine, methionine, and threonine. All amino acids can be produced commercially, although the cost of production can vary dramatically and this, combined with demand, has determined which amino acids are cost effective. As an animal consumes feed, it takes in protein and amino acids. The protein is broken down into its constituent amino acids and these are eventually absorbed into the bloodstream. They then are used by the body for a variety of functions which include enzymes, transmitters such as epinephrine and for building muscle tissue. In the growing animal, the largest share of the amino acids consumed becomes muscle tissue and other primary body constituents. Each animal species uses different levels of each amino acid to incorporate into tissue proteins resulting in somewhat different requirements for amino acids for each species. With this background, we will examine some of the more practical aspects of protein/amino acid nutrition of the turkey.

As the poult grows it builds muscle tissue, adds feathers and various other body components that require amino acids. The needs of the poult are somewhat pre-programmed by the bird's genetic makeup. A poult with the genetic potential for fast growth requires more amino acids on a daily basis than does one genetically determined to grow more slowly. The nutritionist then attempts to match dietary amino acids with the needs of the bird for its most rapid growth. If we provide too little, the poult lacks amino acids it needs for growth. Muscle growth occurs when the body takes the amino acids consumed and produced in the body and places them together in long chains of protein. When the substrate amino acid is not available, protein building ceases and the protein deposition does not occur, thus limiting growth. If we provide too much, the poult probably achieves maximal growth, but the excess will be excreted at considerable expense. Feeding the poult its exact needs is obviously the goal, but one that has yet to be achieved.

Needless to say, there are a number of factors affecting requirements of the poult. Some of these include: genetic potential for growth, age, sex, temperature, intake, quality of the protein, disease state, feedstuff digestibility and management conditions. All of these components will interact to affect the actual amino acid needs of the bird. Poults actually have no need for protein *per se*, but have needs for dietary essential and non-essential amino acids. Essential amino acids must be provided in the diet since they cannot be synthesized rapidly enough in the body to meet needs. The non-essential amino acids can be made in the body, but do require some form of utilizable nitrogen for bodily synthesis. Obtaining the proper level of amino acids in the diet will be the focus of this article and much research in the future.

LOW PROTEIN DIETS

A number of attempts have been made over the years to utilize low protein diets and to determine the limiting amino acids as we reduce total protein. At least in theory, as we reduce protein in the diet we can save money by replacing the amino acids that become limiting as each increment of protein is removed. Unfortunately, much of what we term 'limiting amino acids' is really a function of dietary components used for formulation. In other words, if we have a strict corn/soy diet, we have a certain profile of amino acids. If we change the diet by adding other ingredients or reducing the crude protein content, we change where the amino acids come from and thus the potential order of limitation. Thus this type of work has a limited benefit. Following is a review of the major literature over the years that has dealt with low protein diets in turkeys.

PROTEIN REQUIREMENTS

The current National Research Council (NRC) (1994) protein requirement is set at 28% of the diet with a 2800 kcal/kg energy concentration. A variety of factors affect the requirement for protein. These include: amino acid balance, protein quality and digestibility, intake and dietary energy content. The current requirement was set over 40 years ago when it was found that the turkey showed improved growth rate when fed 26–28% versus 24% crude protein (Fritz *et al.*, 1947; Scott *et al.*, 1948). Since this time there has been little work specifically on the protein requirement of the turkey, although a variety of other factors have been found to affect protein requirements. Baldini and coworkers (1954) found the protein requirement to be less than 28% if lysine was supplemented. Ferguson and coworkers (1956) found that additions of supplemental methionine also improved growth with the best results obtained at 28% crude protein and 0.05% added methionine. Although there have been reports of requirements higher than the current NRC (1994) requirement (Summers and Moran, 1971), it would appear that the current requirement is adequate for maximal growth at moderate energy levels if one allows for a safety margin (Bowyer and Waldroup, 1986). A moderate drop in the protein fed can be made up with additions of limiting amino acids (Bowyer and Waldroup, 1986; Harms and Ruiz, 1986) and this appears to be a relatively common practice in times of high protein prices. Certainly a number of diets are formulated on an amino acid basis where total protein is left to float to achieve the cheapest diet. Protein requirements beyond the starter period have not been well researched. Although the data do not always correspond to the requirement time periods as set forth in the NRC publication, it would appear that the NRC protein requirements are accurate for the energy levels noted (Potter *et al.*, 1980; Auckland, 1971; Eberst *et al.*, 1972; Potter and Shelton, 1980).

ENERGY RELATIONSHIPS TO PROTEIN REQUIREMENTS

It is well known that increasing energy will increase growth rate of turkeys although at some point in the energy addition the growth becomes merely an increase in lipid deposition (Auckland and Morris, 1971; Potter and McCarthy, 1985; Hurwitz *et al.*, 1988). For years there has been discussion over whether there is an optimum protein:energy ratio for formulation of turkey diets. Several reports have indicated that the effects of protein and energy are separate and thus no ratio is needed for diet formulation (Sell *et al.*, 1985; Firman; 1990). While no interaction effects of protein and energy are found in factorial experiments, it appears obvious that there must be an optimum protein/amino acid ratio required at the tissue level for a given dietary energy content. Other factors such as digestibility may also affect ratios and in fact may make it difficult to find optimum ratios using factorial experiments where one may or may not find significant interaction terms. Changes in protein and energy have obvious effects on growth rate and feed efficiency. Protein has also been found to influence breast meat yield, while increased energy has been related to increased lipid deposition. For a review of this subject please see Lilburn (1992).

TEMPERATURE EFFECTS ON PROTEIN REQUIREMENTS

While there do not appear to be any adverse effects on nitrogen absorption by high environmental temperature (Wolfenson *et al.*, 1987), it does appear that increasing temperature does have an effect on protein/amino acid requirements. Additions of lysine and methionine in the summer grow-out period provided a beneficial growth response while similar additions in cooler weather did not (Firman, 1988). Increasing temperature increased the requirement for protein and essential amino acids and improved feed efficiency at moderate temperatures (Bowyer and Waldroup, 1988; Rose and Michie, 1987; Oju *et al.*, 1987; Oju *et al.*, 1984). Most of the effects of temperature can be explained by the changes in intake that occur with increased temperature. Noll and Waibel (1989) showed that any variability in gain due to temperature was primarily due to the reduced intake that occurred at higher temperatures.

LIMITING AMINO ACIDS IN TURKEY DIETS

Early research performed with turkeys indicated that lysine and methionine were the two limiting amino acids in a corn/soybean meal ration of sufficient protein (Baldini *et al.*, 1954; Waibel, 1959). Attempts to formulate rations without the availability of synthetic amino acids result in overfeeding of protein. Significant cost savings result through the addition of methionine and lysine to the formulation matrix. Formulation on an amino acid basis without the constraint of a protein requirement may reduce the cost of the ration as well as result in a formula that can be substantially lower in protein. This is highly dependent on the costs associated with both the protein source and the synthetic amino acids used for addition.

Early work by Klain *et al.* (1954) and Ferguson *et al.* (1956) showed excellent responses to dietary additions of lysine and methionine to low protein diets, but performance was not similar to that of the higher protein control diets. Similar experiments by Fitzsimmons and Waibel (1962) confirmed that methionine and lysine are the first limiting amino acids in conventional diets for young poults, but growth equal to that of the control diet did not occur. Lysine was considered first limiting in low protein diets in work by Bixler *et al.* (1969) with growth being depressed and body fat being increased. Older birds appeared to be more responsive to reductions in protein with a three percentage point drop in protein with added methionine and lysine producing similar results to higher protein rations (Carter *et al.*, 1962). Balloun (1962) reported that additions of methionine and lysine will replace some protein in older birds as well. More recent data have shown that a 25 or 27% crude protein diet with 0.11% methionine added will provide similar growth to that of a 30% crude protein diet although it is difficult to determine if the methionine requirement was being reached even in the 30% crude protein diet used in these experiments (Ting and Balloun, 1972). A variety of studies testing levels of methionine, lysine and crude protein have been completed with results similar to those found previously (Atkinson *et al.*, 1976; Potter and Shelton, 1979, 1984; Kashani *et al.*, 1986; Tuttle and Balloun, 1974; Jensen *et al.*, 1976; Murillo and Jensen, 1976).

While the detrimental effects of low protein diets on performance of poults appear to be well researched, there have been relatively few attempts to determine what is limiting growth other than lysine and methionine. Adequate levels of sulfur amino acids (Waibel *et al.*, 1991; Buresh *et al.*, 1986) and lysine are certainly necessary for proper growth and adequate body composition (Rogers *et al.*, 1991), but for realistic reductions in protein content of turkey diets to occur, it is essential to know the order of limitation or the true requirements for each of the essential amino acids as well as the actual non-essential amino nitrogen requirements. These may differ among strains (Jackson and Potter, 1984) and may be related to the age of the bird.

There are two basic mechanisms that have been used to determine the amino acids limiting growth in poultry rations. These include the addition method and the deletion method. In the deletion method, a low protein diet is supplemented with a group of essential amino acids and then individual or groups of amino acids are removed to determine order of limitation. In the addition method, a low protein diet is supplemented with additions of essential amino acids either singly or in combinations to determine which are necessary for maximal growth. The deletion method is a faster method of determining limitation order, but may be confounded by the relative excesses of the other amino acids. The addition method eliminates this problem, but may require a large number of trials to determine the order of limitation since there are many combinations of essential amino acids which may be used (Edmonds *et al.*, 1985). The order of limitation is also highly dependent on the feedstuffs being used and ideally should be calculated based on a digestible amino acid basis. While there is some disagreement on the first two limiting amino acids, most commercial diets will focus on lysine and the sulfur amino acids as the limiting amino acids in a corn-soy ration. If these levels are maintained at or above the NRC (1994) recommendations and protein is reduced incrementally, we see that significant differences in growth and feed efficiency occur when protein is less than 24% plus lysine and methionine (Firman, unpublished).

Deletion studies

A great deal of work on the order of amino acid limitation in turkey diets was done by Potter's group at Virginia Polytech. Stas and Potter (1982) performed a series of experiments in which a 30% crude protein diet was compared to a 22% protein diet with an amino acid mixture added to the diet that would bring all essential amino acids up to the level of that found in the 30% diet. Growth, intake and feed efficiency were all similar when the mixture was added to the low protein diet. Removal of arginine, glycine, phenylalanine, tyrosine, histidine, leucine and tryptophan from the mixture did not affect performance. However, removal of valine, lysine, threonine or isoleucine resulted in decreased performance although none of the deletions resulted in performance as poor as that found in the unsupplemented 22% diet. Thus it would appear that if the other essential amino acids can provide the necessary substrate for the synthesis of the non-essential amino acids, valine, isoleucine, threonine and lysine are limiting in a 22% protein corn-soybean meal ration. In a series of follow up experiments (Jackson *et al.*, 1983), similar diets with deletions of lysine, valine or threonine were used with the addition of 4% glutamic acid to determine if nitrogen was limiting in the low protein diets. Similar results were obtained. The possible effect of amino acid interactions was also investigated (Jackson and Potter, 1984). Removal of lysine from the essential amino acid mixture resulted in the expected decrease in performance. Additions of arginine in an attempt to induce an imbalance did not affect performance of the diet relative to lysine level. However, increasing dietary leucine above the requirement adversely affected performance unless valine was also increased. In a final experiment, amino acids that were found to be limiting in previous experiments were added to the 22% diet. Methionine alone improved growth when added to the low protein diet. When a non-essential amino acid mixture was added to this diet, a growth response was elicited, indicating a nitrogen deficiency. Addition of an essential amino acid mixture including methionine, lysine, threonine and valine improved growth rate, but did not provide similar feed efficiency when compared to the high protein control diet. Addition of nitrogen did not affect growth rate, but did improve feed utilization, although not to the level of the high protein diet. Interestingly, isoleucine, which was found to be limiting in the initial deletion trials, was not used in the final essential amino acid additions. Although the data are incomplete, it would appear that a 22% crude protein diet is deficient in methionine, lysine, threonine, valine and possibly isoleucine based on the deletion method.

Addition studies

In a series of experiments run in our laboratory, the addition method was used to determine order of amino acid limitation. Amino acids were added to a 22% protein diet with both lysine and methionine levels maintained at those

Table 1. Average 13-day body weight gain and feed:gain of poults fed different levels of protein with lysine and methionine added*.

	Dietary protein (%)				
	20	22	24	26	28
Body weight gain, g	296.1[a]	335.2[b]	366.1[c]	384.9[c]	364.2[c]
Feed:gain	1.772[a]	1.652[b]	1.509[c]	1.450[c]	1.472[c]

* Means + pooled standard error. Numbers with differing superscripts are significantly different ($p<0.05$).
Adapted from Firman (unpublished).

Table 2. Protein and amino acid composition of a 22% versus 28% protein diet (%)*.

Ingredient	22% Crude protein	28% Crude protein	Difference
Protein	22.00	28.00	6.00
Arginine	1.55	2.02	0.47
Glycine-serine	2.27	2.94	0.67
Histidine	0.58	0.72	0.14
Isoleucine	1.13	1.48	0.35
Leucine	1.98	2.36	0.38
Lysine	1.27	1.74	0.47
Methionine	0.36	0.61	0.25
Methionine-cystine	0.72	1.05	0.33
Phenylalanine	1.14	1.44	0.30
Phenylalanine-tyrosine	1.89	2.31	0.42
Threonine	0.92	1.15	0.23
Tryptophan	0.26	0.38	0.12
Valine	1.20	1.50	0.30

*All values are percentages and were calculated from analyzed corn and soybean meal values.

Table 3. Performance of hen poults from one to three weeks when fed additions of individual amino acids to a 22% protein diet. All additions were made to bring the individual amino acids to the level found in the 28% protein control ration*.

	Treatment	Control 22% basal	+ Arginine	+ Histidine	+ Leucine	+ Isoleucine	+ Phenylalanine	+ Threonine	+ Tryptophan	+ Valine
Weight gain, g	293	238	258	254	247	243	218	235	233	256
Feed:gain	1.44	1.74	1.64	1.70	1.64	1.75	1.77	1.74	1.75	1.72

*Data combined from several trials.

Table 4. Performance of hen poults from one to three weeks when fed combinations of essential amino acids to a 22% protein corn-soybean meal ration. All additions were made to bring the individual amino acids to the level found in the 28% protein control ration*.

	Treatment control	Basal	+ Threonine, valine	+ Threonine, valine, isoleucine	Basal + all essential amino acids
Weight gain	419	360	398	407	414
Feed:gain	1.54	1.73	1.64	1.64	1.61

*Data were combined from several trials.

found in the standard 28% crude protein ration. The 22% crude protein ration was selected as it would not support maximal growth and feed efficiency while the 24 and 26% rations fed were similar in performance to the 28% control (Tables 1 and 2). Based on the previous work using the deletion method, it could be assumed that the best chance for performance improvements lay in addition of lysine and methionine (already added), threonine, valine or isoleucine or some combination of these. Additions of nitrogen or these amino acids individually showed no improvement in performance (Table 3). When added in combinations, a mix of all essential amino acids provided similar growth to the control group while additions of valine, threonine and isoleucine also provided growth similar to the control diet with a slight reduction in efficiency (Table 4). Although a variety of other dietary additions have been tried, only the additions of valine and threonine or a combination of threonine, valine and isoleucine appear to support growth similar to that of the control diet (Firman, 1994).

LATER PERIOD PROTEIN REDUCTION

Needless to say, the reduction of protein later in the growth cycle has the potential for far greater cost savings as the cost of the starter diet is comparatively minor due to the small amounts fed. Each reduction in dietary protein will result in a reduction in feed cost. The addition of crystalline amino acids to maintain growth will have a cost associated with it that may eat up most of those cost savings. It also becomes much easier to utilize high energy diets without substantial fat additions as we decrease protein. Changes in protein *per se* have relatively little impact on overall performance (Table 5, data from late 1980s). If we were to summarize these data, the main impact on performance with these types of diets comes at the point where protein gets below 90% of the NRC recommendation with the limiting amino acids at this level as well. If we feed similar diets with lysine and sulfur amino acids held at 100% of NRC for the low protein diets and at 105–115% of NRC for the higher protein diets, the effects of protein do not differ significantly. Thus it would appear that we can reduce protein by 10–15% if we maintain our first two limiting amino acids. We have demonstrated this at Missouri and Dr Sell at Iowa State has done similar work. Several questions come to mind. How accurate are our amino acid requirement values? Can we reduce protein further with a concomitant cost savings? Should we express requirements on a daily intake basis rather than a percentage of the diet? The answers to these questions are probably not very accurate; maybe we can further reduce protein; and yes, we should express requirements on a daily intake basis. Our current requirements are based on very little hard data. Reductions in protein are possible, but the cost savings may prove elusive. Finally, in the author's opinion, we do need to modify the way we treat requirements. Changes in temperature or energy level of the diet can certainly affect the intake of protein/amino acid and the current expression does not take this into account.

Table 5. Effects of different levels of protein on performance of turkeys.

	Protein (% of NRC)	Body weight (lbs)	Feed/gain
Hens	85	14.4	2.75
	95	14.6	2.72
	105	14.8	2.69
	115	15.2	2.59
Toms	85	26.1	2.83
	95	27.5	2.90
	105	28.1	2.91
	115	27.8	2.81

Diets fed were corn-soy with lysine and sulfur amino acids maintained at the same percentage relative to NRC as protein. (15-week hens, 19-week toms).

PRACTICAL LOW PROTEIN DIETS

The reduction of protein in turkey diets has potential to reduce dietary costs. Although in periods of low protein prices it is more difficult to see the benefits of reducing protein, in general some reduction in protein should prove cost effective. Given current protein prices, reducing protein may be necessary to maintain profitability. As mentioned earlier, reductions of 10–15% of the crude protein with maintenance of lysine and sulfur amino acids has been used with some success. As we continue to reduce protein, several items are of concern. Which amino acids become limiting next and is breast meat yield compromised? In our studies with turkeys to market age, using 85% crude protein with sulfur amino acids and lysine at 100% of NRC (4-week feed intervals), a slight reduction in breast meat yield may be found although this protein point seems to be somewhat borderline. Based on these studies with limiting amino acids, several studies were designed to look at adding amino acids back in an attempt to maintain breast meat yield with the reduced protein diets. The first study was conducted with 1600 toms and began at 12 weeks of age. The addition of threonine or threonine, isoleucine and valine improved gain at 15 weeks, but was not statistically different by 18 weeks. These data would indicate that additions of threonine or the combination of threonine, isoleucine and valine may be useful in maintenance of growth rate. Breast meat yield did not differ in this trial. A second trial was conducted with birds being placed on diets at 6 weeks with dietary changes made at 3-week intervals based on the NRC report. Threonine was the only amino acid added in this trial. Based on these data it would appear that there is a good response to threonine from 9–12 weeks, but little response prior to that period. By 15 weeks there is no significant difference. At 18 weeks, threonine additions provided a response, but growth was not as great as the control diet. Overall it appears that we can practically reduce dietary protein, but will need a bit more work on later period nutrition to get our amino acid additions to function reliably.

Current data

DIGESTIBLE AMINO ACIDS

Certainly as we move to the future of nutrition, the concepts of digestible amino acid nutrition, formulation and ideal proteins come to the forefront. Relatively speaking, there is little work on either digestible amino acid contents of feedstuffs for turkeys or the requirements for amino acids. The closest work is that of Hurwitz and coworkers (1983) who did work modeling growth of turkeys and determining maintenance needs, needs for feather growth and the requirements based on a total digestibility which appeared to be set arbitrarily at 85%. Other than the recent work here at Missouri (Firman, 1992; Firman and Remus, 1993, 1994), there are only a few older papers noting information on amino acid digestibility in turkeys (Chu and Potter, 1969; Pierson *et al.*, 1980; Parsons and Potter, 1981). Most of the digestibility work to date has been performed in the adult Leghorn rooster. The methodology used in the rooster has been beautifully described by Sibbald (1986) and the data recently summarized by Parsons (1991). While the methodology for determining digestible amino acids in feedstuffs continues to be debated in academia, the method using cecectomized birds appears to be relatively accurate and has been more or less accepted for several years.

Recent data on amino acid requirements of turkeys are somewhat sketchy as well with only one publication referenced in the new NRC (1994) report in the past decade. Most of the current recommendations are based on a publication from over 20 years ago (Warnick and Anderson, 1973) using the bronze turkey as the model. The last serious attempt to determine amino acid requirements of turkeys was outlined by Hurwitz and coworkers (1983). In these experiments, carcass composition, maintenance costs and feather growth were all used to determine amino acid requirements. A coefficient of absorption was used to account for digestibility (85%) and did not take into account different feedstuffs. It is not clear how the amino acid requirements were calculated and Nixey (1989), when comparing several modeling approaches, found the data to be very unsatisfactory when used to determine practical formulations. More data on intake patterns and digestibility values will be needed before a usable model can be made.

IDEAL PROTEIN

The concept of an ideal protein is not new. As early as 1946, Mitchell (Mitchell and Block, 1946) had discussed the concept of a perfect balance of amino acids. Ideal protein may be thought of as a protein that provides for the exact needs of the animal with respect to amino acids without any excess or deficiencies, but with maximum growth potential of the animal reached. A variety of attempts have been made to define an ideal amino acid pattern through carcass composition (Price *et al.*, 1953; Williams *et al.*, 1954; Summers and Fisher, 1961; Robel and Menge, 1973). While carcass composition may be used as a starting point for an ideal ratio, it does not take into account the dynamics of the live animal such as maintenance costs. Dean and Scott (1965) were the first to use dietary testing to begin determination of ideal proteins. Since that time, a number of studies have sought to determine the ideal ratio of both the chick and the pig (Fuller *et al.*, 1989; Wang and Fuller, 1989; Chung and Baker, 1992; Baker and Chung, 1993; Baker and Han, 1994; Brown, 1994). The current objective of our program at Missouri is to conduct work to define digestible requirements and the ideal protein in turkeys.

As noted previously, very little recent data exist on amino acid nutrition of turkeys, let alone the digestible requirements and ideal proteins. We have done a good bit of work with cecectomized turkeys to calculate digestibilities of a variety of feedstuffs (Firman and Remus, 1993) and have some data on starter turkey diets, but really there is little to go on at present. Given the paucity of data, I believe one must take a close look at how we can best utilize what is currently available and where we need to go in the future to make some progress.

Over the years a number of us have argued about what are the correct protein/amino acid requirements, how low can we go on protein, which amino acids are limiting, whether or not there is an energy relationship, whether poults can handle fat, etc. For the most part, I believe these arguments to be moot. Lets back up and take a look at some of the practices that we currently follow, where they come from and how we might change in the future.

The NRC requirements are listed below for the starter period (probably the most accurate period):

Amino acid	Percent of diet
Protein	28.00
Lysine	1.60
Methionine + cystine	1.05
Arginine	1.60
Histidine	0.58
Isoleucine	1.10
Leucine	1.90
Phenylalanine + tyrosine	1.80
Threonine	1.00
Tryptophan	0.26
Valine	1.20
Glycine + serine	1.00
Total amino acid requirement	13.09

Based on these numbers we need to feed a little over 13% crude protein plus some nitrogen to allow for construction of the amino acids that are not dietary essentials but are indispensible. Certainly we do not need 15% crude protein to build these other amino acids. Our research has indicated that we can reduce total crude protein to about 22% with similar performance as on a 28% diet. It may be lower than that level in actuality, but formulation using practical components becomes quite difficult. Why the apparent overfeeding of protein? Based on costs of dietary amino acids, it is cheaper to overfeed protein and make sure we get all the amino acid requirements met than to feed closer to the exact requirements, especially when we probably do not have the best data. This is where we have been in the past, but let's take a look at the future and some approaches that I believe will eventually bear fruit.

APPROACHES TO PROTEIN/AMINO ACIDS FOR THE FUTURE

A. Reduce dietary protein in a cost effective fashion.

B. Formulate on a digestible amino acid basis:

1. Determine amino acid digestibility of feedstuffs used;
2. Determine digestible amino acid requirements;
3. Determine ideal protein.

C. Determine energy effects on amino acid requirements:
  1. Determine the effects of energy on intake;
  2. Determine an energy/lysine ratio (ideal basis).

D. Determine amino acid partitioning needs:
  1. Efficiency of deposition;
  2. Maintenance costs;
  3. Effects on body composition.

E. Model the above data.

F. Determine the metabolic fate and attempt modifications.

The pig industry incorporates much of this into a useful equation shown below:

$$\frac{\text{Protein (g) deposited for period} \times \text{amino acid composition (\%)}}{\text{Efficiency of deposition} \times \text{amino acid digestibility (period)}} = \text{Amino acids (g)}$$

Looking at protein/amino acid nutrition from this standpoint is why I believe we have been focused on the wrong areas. If we continue to do work based on gross feed values we will continue to have difficulty making sense of the data. In other words, if we have company X feeding strictly corn/soy and company Y feeding a complex diet, we can have a 10% difference in what the turkey actually can use from the feed. It is therefore not surprising that we see differences in terms of response to the various diets we feed. Let's take a look at where our research has led us in the past few years at Missouri.

Reduce dietary protein and determine energy effects
Although I have discussed the effects of protein reduction in a previous section, some of our data on protein and energy relationships may be useful here. Below are several tables that are typical of some of the protein/energy work we finished several years ago. For the sake of brevity only data for the starter period and final data are included. Table 6 and 7 show the effects of changes in protein and energy relative to the NRC guides. Amino acids were kept relative to protein and the diets were corn/soy with meat meal added. Fat content varied dramatically as one might expect, with the highest energy/protein diets having fat levels commercially difficult to mix. This is one of four identical trials with toms (1600 birds each). As is obvious from the data, the lowest level of protein is too low. However, if we add back lysine and methionine, growth comes back almost to the previous level. Somewhere in that 85–90% of NRC level it appears that we have another limiting amino acid. The energy portion of the trial is quite interesting. Contrary to what I had been told, the starting poult handled high levels of added fat (10%+) by growing as fast as lower energy diets and with much better efficiency. We have used these levels in at least eight large floor pen trials as well as in field trials with only beneficial effects. Feed efficiency was dramatically improved and this held throughout the bird's lifetime. The economics of the high energy diets change constantly,

Table 6. Body weight and feed efficiency of turkey toms (4 weeks) in relation to protein and energy.

	Energy (% of NRC)				
	88	96	104	112	Mean
Protein, % of NRC					
85	1.38	1.55	1.81	1.74	1.62
95	1.48	1.64	1.95	2.01	1.77
105	1.49	1.70	2.01	1.98	1.79
115	1.36	1.84	2.02	2.11	1.83
Mean	1.43	1.68	1.95	1.96	1.75
Feed/gain					
85	1.98	1.87	1.63	1.37	1.71
95	2.03	1.73	1.46	1.34	1.64
105	1.85	1.75	1.52	1.42	1.63
115	2.01	1.43	1.52	1.42	1.60
Mean	1.97	1.69	1.53	1.39	1.65

Table 7. Body weight and feed efficiency of turkey toms (19 weeks) in relation to protein and energy.

	Energy (% of NRC)				
	88	96	104	112	Mean
Protein, % of NRC					
85	24.4	24.8	26.5	26.2	25.5
95	25.4	26.5	28.5	28.6	27.2
105	26.2	28.3	30.2	31.0	28.9
115	25.3	27.3	29.5	31.7	28.5
Mean	25.3	26.7	28.7	29.4	27.5
Feed/gain					
85	3.59	3.04	2.75	2.56	2.99
95	3.52	3.01	2.66	2.50	2.92
105	3.23	3.08	2.77	2.20	2.82
115	3.31	3.14	2.65	2.36	2.86
Mean	3.41	3.07	2.71	2.41	2.90

but if we go to lower protein diets, energy can more easily come from the grains fed. The relationship between protein and energy is difficult to determine in a study such as this. No interactions were noted in most cases.

Formulation on a digestible amino acid basis

Several years ago I made the decision that we needed a more defined set of requirements that could be used to take into account all the different factors affecting how much of an amino acid the turkey will use. Some of these include dietary energy content, sex, strain, temperature, management, etc. Two goals came to mind: the first of these was to formulate more accurately for our current bird and second to find a more rapid method of updating requirements. Our first step on this path was to determine the digestibility of a number of feedstuffs in cecectomized turkeys and determine that there were few differences with age or sex. These data have been presented previously. Based on this information we have moved toward formulation based on digestible amino acid and attempted to determine the digestible amino acid requirements in the starting period and the ideal protein. Before reviewing these data, let's take a look at the value of digestible amino acid formulations.

Given that the amino acid digestibility data set for formulation is not yet available, is there any point worrying about this until the data set is complete? Some benefits can be found with existing information. Table 8 shows some data collected on several other trials over the past few years where one group of birds had diets formulated on a digestible amino acid basis (toms) versus total amino acid basis (hens). These diets used a wide variety of by-product meals included at what would be considered high levels, with similar amino acid requirements used in each study and between diets. As can be seen from the data, formulation with digestible amino acids resulted in similar response to the control even with very high levels of by-products; while differences were noted when the diets were formulated on a total amino acid basis.

Thus it appears that there is some benefit to digestible formulation even without the benefit of a complete knowledge of digestible amino acid requirements. On the surface the benefits to digestible amino acid formulation appear obvious. If the exact amino acid needs of the turkey are known in theory at least, any feedstuffs could be used as long as the amino acid requirements could be met with the ingredients used. While there would still be some constraint on certain feedstuffs due to other potential problems, amino acids should not be one of them and the ability to use by-products in rations will be enhanced. Since we would know the exact requirements for each amino acid, we should be able to reduce the overfeeding of protein and utilize lower protein diets. This could lead to reduced nitrogen excretion and less potential waste product produced. If we fed less excess nitrogen there would also be less energy expended for nitrogen excretion, thus improving feed efficiency. This is difficult to show experimentally, but is probably a real effect. Overall, this has the potential to reduce feed costs for the industry. We have had more difficulty showing the same effects as the birds get older and consume more feed. Thus, we must tediously collect a substantial amount of data to see all of the benefits.

We are currently working at Missouri on the digestible amino acid requirements of turkeys. This has been one of the major focuses of my research for the past five years or more. Several pieces of information need to be in place before one can actually run an experiment on the digestible requirements. The first of these was the data on digestible amino acids which have been

Table 8. Digestible formulation versus non-digestible*.

Soybean meal (%)	Hen gain	Tom gain
30% by-product	1.61[a]	1.80[a]
20% by-product	1.74[b]	1.77[a]
10% by-product	1.82[c]	1.77[a]
Control	1.86[c]	1.84[a]

*Weight gain of hens (total amino acid) and toms (digestible amino acid) during the starter period.

mentioned previously. The second bit of information is a low protein diet that can be used to titrate the amino acid on a digestible basis. To the best of my knowledge, Potter at VPI had the lowest protein diet that supported maximal growth at 22% crude protein plus amino acids. We have used a similar diet with good results, but needed something with even lower protein for our titrations. This led us to formulate a very low protein starter diet (17.8% + amino acids) with corn and soy that would support maximal growth (Table 9). To get a diet at this level, we also were led into the ideal protein area and now have an estimate of the ideal protein (Table 10).

Once these data had been collected we could then determine requirements for essential amino acid on a digestible basis. As one might expect, lysine was the first amino acid studied. Based on a number of experiments, 1.32% lysine appears to support maximal growth (Table 11) and 1.34% supports maximum efficiency. We have also worked on the sulfur amino acid requirement with 0.78% of the diet supporting maximal growth (Table 12). While these numbers appear low, keep in mind that this is on a digestible basis, with the other amino acids balanced based on our ideal protein ratio. This also does not take into account any safety factor and is only at one energy level. We are currently looking at different energy levels to see how this will affect requirements.

Let's take a moment to define ideal protein and examine how it should and should not be used. Although several definitions are possible, the ideal protein is the theoretically exact balance of amino acids that meet the animal's needs. There should be no excess, no deficiency and use of the amino acid for energy must be minimized. Nitrogen excretion would be minimized in

Table 9. Reduced protein starter diet.

Ingredient	Percent in diet
Corn	53.39
Soy 48	27.48
Fat	4.00
Dical-P	2.64
Limestone	1.43
Sugar, premixes + amino acids to 100%*	
Analysis (digestible basis)	
Crude protein	17.86
ME (kcal/kg)	3171
Calcium	1.29
Avail-P	0.60
Lysine	0.86
SAA	0.50
Threonine	0.52
Valine	0.71
Arginine	1.09
Histidine	0.43
Isoleucine	0.66
Leucine	1.32
Phe+Tyr	1.26
Tryptophan	0.18

*Amino acids were added to ideal ratio.

Table 10. Estimated ideal protein ratio for starting turkeys (chick and pig ratio for comparison)*.

Amino Acid	Turkey	Chick	Pig
Lysine	100	100	100
SAA	59	72	60
Threonine	55	67	65
Valine	76	77	68
Arginine	105	105	–
Histidine	36	31	32
Isoleucine	69	67	60
Leucine	124	100	111
Phe+Tyr	105	105	95
Tryptophan	16	16	18

*Expressed as a percentage of the lysine requirement.

Table 11. Digestible lysine requirement of hens during the starter period.

Lysine level (%)	Gain (g)
1.26	267
1.29	269
1.32	293
1.35	288
1.38	280
1.41	289

Table 12. Digestible methionine-cystine requirement of hens during the starter period based on weight gain of hens fed different levels of digestible sulfur amino acids.

Methionine+cystine level	Weight gain
0.59	222
0.62	261
0.65	305
0.68	318
0.71	303
0.74	333
0.77	347
0.80	355
0.83	350

this situation and all dispensable amino acids should be provided in another form (i.e. not as indispensable amino acid). The concept is that all of the amino acids can be related to lysine (by choice) and that if the lysine needs of the animal increase due to genetics, etc., then the amino acid needs remain the same relative to the lysine requirement. The ratio probably changes throughout the growth cycle of the animal as it moves to periods of feather growth versus breast meat accretion. The ideal ratio is useful from several standpoints. It allows for determination of digestible amino acid requirements which can then be easily modified through the changes in the requirement for lysine. This allows for rapid research response to improving genetics and the new amino acid requirements that accompany it. It has allowed us to formulate very low protein diets for determination of amino acid requirements. It forces us to move to formulation based on digestibility, which is useful in order to more accurately meet the turkey's needs, reduce overfeeding and value feedstuffs based on the capacity of the animal to use them. Pricing of ingredients based on the actual usable nutrient content may be the most useful portion of the switch to digestible requirements and formulation.

All of this sounds very good, but there are also some inherent difficulties when we try to use the ideal values for practical formulation. Use of the ideal protein has taken some criticism based on the lack of practicality in formulations. This has probably occurred for several reasons. First, nutritionists need to be prepared for digestible formulation or the ratios have little meaning. Some information on changing to formulation based on digestible amino acid is presented below. Second, the numbers should be used as minimum requirement values for each amino acid as they will not meet the exact requirements. Meeting the exact requirements means extensive reduction in total protein and adding back substantial amounts of crystalline amino acid (extremely expensive). Thus the ideal ratio is probably most useful in determination of digestible requirements and for quickly adjusting the requirement to changing genetics.

Determine amino acid partitioning needs, etc.

We are just starting collection of data that will yield efficiency and maintenance requirement numbers (Figure 1). This is done with a similar titration followed by grinding of the carcasses and amino acid analysis to determine how much deposition has occurred relative to digestible amino acid intake. Obviously a good deal of work with grinding, drying, etc. on several hundred carcasses for each experiment must be done. Modeling of these data and the post-absorptive work will certainly have to wait until more data are collected.

MOVING TO DIGESTIBLE FORMULATION

The first step in formulating based on digestible amino acid is to set up or modify the formulation database. Basically one can add new ingredients (i.e. digest/corn) or add new nutrients to your current ingredients (i.e. digest/lysine). The numbers can be derived by multiplying published digestibility values by the percentage of amino acid in each feedstuff. This is the digestible amino acid content of the feedstuff. Obviously this will take several hours of work. Setting up new ingredients with new standard amino acid values is probably the easiest approach but lacks the benefit of comparison to total amino acid content.

Once this is complete, the second step is to put in digestible requirements. While this appears to be the most critical aspect, it is probably less important than the feedstuff data. Obviously having the correct requirement is valuable, but consistency across formulation may be more valuable. Given this,

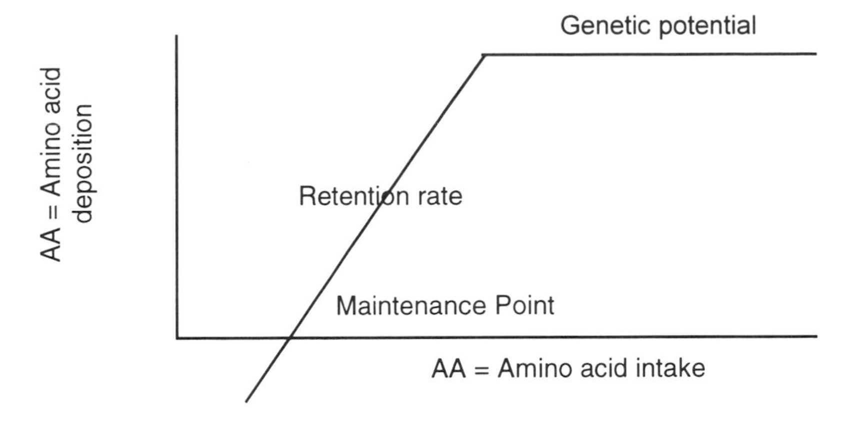

Figure 1. Effects of increasing amino acid intake on amino acid deposition.

one can estimate the requirement in several ways. The easiest is to just reduce the requirement by 15% as a general guideline. While this is better than nothing it does not give a very good estimate. The better approach is to take a formulation that has worked well for you and is relatively straightforward (ingredients that tend to have low variability) and calculate the requirement from this. This can be done by putting the ingredient profile into the computer that uses the digestible database and letting the computer do the calculations. Given the newly calculated requirements, you can now proceed to formulate with more exotic ingredients.

How can you make this system more useful? If we could spend several hundred thousand dollars and several years of effort, we could have the needed data. Short of this, regular analysis of your feedstuffs for total amino acid content is valuable. A baseline data set needs to be established for each ingredient and supplier so that changes from harvests or time of the year can be noted. Having digestibility assays run on cecectomized turkeys will also be beneficial after the data set is developed. Ultimately we will need to determine the digestible amino acid needs of the turkey on a daily basis to gain the most efficiency from the nutritional system.

Summary

Low protein diets, digestible formulation and ideal proteins all have a place in the nutrition and formulation of turkey diets. A renewed effort to delineate some of the necessary information must be undertaken to move the industry toward knowledge of the actual nutrient needs of the bird.

References

Atkinson, R.L., K.K. Krueger, J.W. Bradley and W.F. Krueger. 1976. Amino acid supplementation of low protein turkey starting rations. Poultry Science. 55:1572.

Auckland, J.N. 1971. A note on the protein requirements for early finishing of male turkeys. British Poultry Science. 12:283.

Auckland, J.N. and T.R. Morris. 1971. The effect of dietary nutrient concentration and calorie to protein ratio on growth and body composition of male and female turkey poults. British Poultry Science. 12:305.

Baker, D.H. and T.K. Chung. 1993. Ideal protein for swine and poultry. Proceedings of the Arkansas Nutrition Conference.

Baker, D.H. and Y. Han. 1994. Ideal amino acid profile for chicks during the first three weeks posthatching. Poultry Science. 73:1441–1447.

Brown, R.H. 1994. Worldwide movement aimed at lowering protein in rations. Feedstuffs, June 6, 24.

Baldini, J.T., H.R. Rosenberg and J. Waddell. 1954. The protein requirement of the turkey poult. Poultry Science. 33:539.

Balloun, S.L. 1962. Lysine, arginine and methionine balance of diets for turkeys to 24 weeks of age. Poultry Science. 41:417.

Bixler, E.G., G.F. Combs and C.S. Shaffner. 1969. Effect of protein level on carcass composition of turkeys. Poultry Science. 48:261.

Bowyer, B.L. and P.W. Waldroup. 1986. Evaluation of minimum protein levels for growing turkeys and development of diets for estimating lysine requirements. Poultry Science. 65(Suppl. 1):16 (Abstr.).

Bowyer, B.L. and P.W. Waldroup. 1988. Comparison of protein and energy feeding programs for large white turkeys during the growing period. Poultry Science. 67(Suppl. 1):4 (Abstr.).

Buresh, R.E., R.H. Harms and R.D. Miles. 1986. A differential response in turkey poults to various antibiotics in diets designed to be deficient or adequate in certain essential nutrients. Poultry Science. 65:2314.

Carter, R.D., E.C. Naber, S.P. Touchburn, J.W. Wyne, V.D. Chamberlain and M.G. McCartney. 1962. Amino acid supplementation of low protein turkey growing rations. Poultry Science. 41:305.

Chu, A.B. and L.M. Potter. 1969. ME and protein and fat digestibility evaluations of fish solubles in diets of young turkeys. Poultry Science. 48:1169–1174.

Chung, T.K. and D.H. Baker. 1992. Ideal amino acid pattern for 10-kilogram pigs. J. An. Sci. 70:3102–3111.

Dean, W.F. and H.M. Scott. 1965. The development of an amino acid reference diet for the early growth of chicks. Poultry Science. 44:803–808.

Eberst, D.P., B.L. Damron and R.H. Harms. 1972. Protein requirements of growing turkeys. Poultry Science. 51:1309.

Edmonds, M.S., C.M. Parsons and D.H. Baker. 1985. Limiting amino acids in low-protein corn-soybean meal diets fed to growing chicks. Poultry Science. 64:1519.

Ferguson, T.M., H.P. Vaught, B.L. Reid and J.R. Couch. 1956. The effect of amino acid supplements to the diet of broad breasted bronze turkey poults fed various levels of protein and productive energy. Poultry Science. 35:1069.

Firman, J.D. 1988. Effects of reduced dietary amino acid content followed by addition of lysine and methionine on performance of large white toms. Poultry Sci. Assn. meetings, July 1988.

Firman, J.D. 1990. Effect of differing protein and energy levels during different seasons on performance and body composition of large white toms. Poultry Science. 69(Suppl. 1):50 (Abstr.).

Firman, J.D. 1992. Amino acid digestibilities of soybean meal and meat meal in male and female turkeys of different ages. J. Applied Poultry Research 1:350–354.

Firman, J.D. 1994. Utilization of low protein diets for turkeys (Review). Biokyowa Technical Review Series.

Firman, J.D. and J.C. Remus. 1993. Amino acid digestibilities of feedstuffs in female turkeys. J. Applied Poultry Research. 2:171–176.

Firman, J.D. and J.C. Remus. 1994. Fat additions increase digestibility of meat and bone meal. J. Applied Poultry Research 3:80–83.

Fitzsimmons, R.C. and P.E. Waibel. 1962. Determination of the limiting amino acids in corn-soybean oil meal diets for young turkeys. Poultry Science. 41:260.

Fritz, J.C., J.L. Halpin and J.H. Hooper. 1947. Studies on the nutritional requirements of poults. Poultry Science. 26:78.

Fuller, M.F., R. McWilliam, T.C. Wang and L.R. Giles. 1989. The optimum dietary amino acid pattern for growing pigs. Brit. J. Nutr. 62:255–267.

Harms, R.H. and N. Ruiz. 1986. Evidence for a straight line response to supplemental methionine when the diet contains adequate cystine. Poultry Science. 65(Suppl. 1):54. (Abstr.).

Hurwitz, S., I. Plavnick, I. Bengal, H. Talpaz and I. Bartov. 1983. The amino acid requirements of growing turkeys. 1. Model construction and parameter estimation. Poultry Science. 62:2208–2217.

Hurwitz, S., I. Plavnick, I. Bengal and I. Bartov. 1988. Response of growing turkeys to dietary fat. Poultry Science. 67:420.

Jackson, S. and L.M. Potter. 1984. Influence of basic and branched chain amino acid interactions on the lysine and valine requirements of young turkeys. Poultry Science. 63:2391.

Jackson, S., R.J. Stas and L.M. Potter. 1983. Relative deficiencies of amino acids and nitrogen per se in low protein diets for young turkeys. Poultry Science. 62:1117.

Jensen, L.S., B. Manning, L. Falen and J. McGinnis. 1976. Lysine needs of rapidly growing turkeys from 12–22 weeks of age. Poultry Science. 55:1394.

Kashani, A.B., H. Samie, R.J. Emerick and C.W. Carlson. 1986. Effect of copper with three levels of sulfur containing amino acids in diets for turkeys. Poultry Science. 65:1754.

Klain, G.J., D.C. Hill and S.J. Slinger. 1954. Supplementation of poult diets with lysine. Poultry Science. 33:1280–1298.

Lilburn, M.S. 1992. Influence of dietary protein and amino acid intake on carcass composition and parts yields of turkeys. Biokyowa Technical Review. 3:1–12.

Mitchell, H.H. and R.J. Block. 1946. Some relationships between the amino acid contents of proteins and their nutritive values for the rat. J. Biol. Chem. 163:599–620.

Murillo, M.G. and L.S. Jensen. 1976. Methionine requirement of developing turkeys from 8–12 weeks of age. Poultry Science. 55:1414.
National Research Council. 1994. Nutrient Requirements of Poultry, Eighth Revised Edition. National Academic Press, Washington D.C.
Nixey, C. 1989. Nutritional responses of growing turkeys. Recent Advances in Turkey Science. Butterworths.
Noll, S.L. and P.E. Waibel. 1989. Lysine requirements of growing turkeys in various temperature environments. Poultry Science. 68:781.
Oju, E.M., P.E. Waibel, S.L. Noll and J.C. Halvorson. 1984. Protein requirements of growing female turkeys under various environmental temperatures. Poultry Science 63(Suppl. 1):157 (Abstr.).
Oju, E.M., P.E. Waibel and S.L. Noll. 1987. Protein, methionine, and lysine requirements of growing hen turkeys under various environmental temperatures. Poultry Science. 66:1675.
Parsons, C.M. 1991. Amino acid digestibilities for poultry: Feedstuff evaluation and requirements. Biokyowa Technical Review.
Parsons, C.M. and L.M. Potter. 1981. TME and AA digestibility of dehulled soybean meal. Poultry Science. 60:2687–2696.
Pierson, E., L.M. Potter and R.D. Brown. 1980. AA digestibility of dehulled soybean meal by adult turkeys. Poultry Science. 59:845–848.
Potter, L.M. and J.P. McCarthy. 1985. Varying fat and protein in diets of growing large white turkeys. Poultry Science. 64:1941.
Potter, L.M. and J.R. Shelton. 1979. Methionine and protein requirements of young turkeys. Poultry Science. 58:609.
Potter, L.M. and J.R. Shelton. 1980. Methionine and protein requirements of turkeys 8 to 16 weeks of age. Poultry Science. 59:1268.
Potter, L.M. and J.R. Shelton. 1984. Methionine, cystine, sodium sulfate, and Fermacto-500 supplementation of practical-type diets for young turkeys. Poultry Science. 63:987.
Potter, L.M., J.R. Shelton and J.P. McCarthy. 1980. Lysine and protein requirements of growing turkeys. Poultry Science. 59:1652.
Price, W.H., W.M. Taylor and W.C. Russell. 1953. The retention of essential amino acids by the growing chick. J. Nutr. 51:413–422.
Robel, E. and H. Menge. 1973. Performance of chicks fed an amino acid profile based on carcass composition. Poultry Science. 52:1219–1221.
Rogers, S.R., C.C. Miller and M.E. Cook. 1991. Meat yield of male turkeys fed various protein and methionine levels. Poultry Science 70(Suppl. 1):100 (Abstr.).
Rose, S.P. and W. Michie. 1987. Environmental temperature and dietary protein concentrations for growing turkeys. Brit. Poultry Science. 28:213.
Scott, H.M., G.F. Heuser and L.C. Norris. 1948. Studies in turkey nutrition using a purified diet. Poultry Science. 27:770.
Sell, J.L., R.J. Hasiak and W.J. Owings. 1985. Independent effects of dietary metabolizable energy and protein concentrations on performance and carcass characteristics of tom turkeys. Poultry Science. 64:1527.
Sibbald, I. R. 1986. The TME system of feed evaluation: Methodology, feed composition data and bibliography. Tech. Bull. 1986–4e Agric. Canada.

Stas, R.J. and L.M. Potter. 1982. Deficient amino acids in a 22% protein corn-soybean meal diet for young turkeys. Poultry Science. 61:933.

Summers, J.D. and H. Fisher. 1961. Net protein values for the growing chicken as determined by carcass analysis: exploration of the method. J. Nutr. 75 (4):435–442.

Summers, J.D. and E.T. Moran, Jr. 1971. Protein requirement of poults to 8 weeks of age. Poultry Science. 50:858.

Ting, Keh-chuh and S.L. Balloun. 1972. Effect of protein level and methionine supplementation on several metabolic responses in turkey poults. J. Nutr. 102:681.

Tuttle, W.L. and S.L. Balloun. 1974. Lysine requirements of starting and growing turkeys. Poultry Science. 53:1698.

Warnick, R.E. and J.O. Anderson. 1973. Essential amino acid levels for starting turkey. Poultry Science. 52:445.

Waibel, P.E. 1959. Methionine and lysine in rations for turkey poults under various dietary conditions. Poultry Science. 38:712.

Waibel, P.E., J.K. Liu, J. Brannon and S.L. Noll. 1991. Amino acid supplementation of low protein market turkey diets. Poultry Science. 70(Suppl. 1):126 (Abstr.).

Wang, T.C. and M.F. Fuller. 1989. The optimum dietary amino acid pattern for growing pigs I. Experiments by amino acid deletion. Brit. J. Nutr. 62:77–89.

Williams, H.H., L.V. Curtin, J. Abraham, J.K. Loosli and L.A. Maynard. 1954. Estimation of growth requirements for amino acid by assay of the carcass. Journal of Biological Chemistry 208 (1):277–286w.

Wolfenson, D., D. Skylan, Y. Graber, O. Kedar, I. Bengal and S. Hurwitz. 1987. Absorption of protein, fatty acids and minerals in young turkeys under heat and cold stress. Brit. Poultry Science. 28:739.

TRACE MINERAL NUTRITION: APPLICATION FOR MINERAL PROTEINATES AND SELENIUM YEAST

EGGSHELL QUALITY AND ECONOMIC LOSSES: THE POTENTIAL FOR IMPROVEMENT WITH DIETARY TRACE MINERAL PROTEINATES

JUAN GOMEZ-BASAURI

Alltech Inc., Ithaca, New York, USA

Introduction

Eggshell quality is a primary concern of the layer industry due to the economic losses associated with the incidence of inferior shell quality (Hamilton *et al.*, 1979; Hamilton, 1982; Roland, 1988). Estimates of reduced shell quality range from 6 to 8% (Figure 1). However, these figures do not take into account a 7.77% additional loss due to shell-less eggs or eggs which fall through the cage into the manure (Roland, 1976, 1977a,b). In other words, for 100 collected eggs 7.7 eggs remain uncollected at the barn.

Recently, a survey involving a large portion of the US layer industry was conducted to determine the percentage of eggs lost or cracked prior to consumer use. The study revealed that 4% of all eggs produced were cracked (Roland, 1988). Similarly, a study in the Pennsylvania layer industry found that 5.5% of all eggs were cracked and 1.7% were undergrades (Mast, 1987, Table 1).

Worldwide, estimates of 6.7% cracked eggs were found in the UK (Anderson

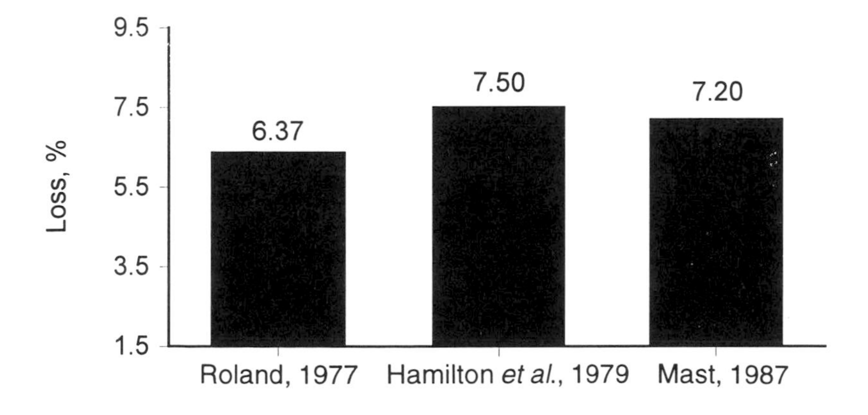

Figure 1. Estimates of egg loss associated with poor shell quality.

Table 1. Estimates of cracker and 'leaker' eggs.

Source	Year	Cracked and leaker eggs	Downgrades
Mast, (1987)	1982	4.4	1.7
Roland, (1988)	1988	4.0	–

Table 2. Financial losses associated with poor shell quality.

Parameter	Average, (%)	Value of loss ($/dozen)	Loss per annum (million $US)
Cracked eggs	4.4	0.75	222.8
Downgrades	1.7	0.25	28.7
Total	–	–	251.8

and Carter, 1976) and 8% in Germany (Folkerts, 1976). Studies in Australia have found that 47% of all eggs purchased have defective shell faults (Balnave and Yoselewitz, 1988). Based on the above estimates, the financial loss translates into almost $300 million in the US alone (Table 2), not including losses for uncollected eggs and those eggs with decreased value which go to the breaking plant (Washburn, 1982; Roland, 1988).

Since egg producers derive their income from the sale of eggs and look for the highest possible margin per unit of capacity over a given time interval (Wills, 1975), nutritional and management factors influencing egg quality are of critical importance. This chapter will review various aspects of eggshell quality.

Eggshell organic matrix

The eggshell consists primarily of mineral matter (Table 3), most of which is calcium carbonate deposited on an organic matrix. This matrix is made up of a combination of protein and mucopolysaccharides better known as the shell membranes (Austic and Nesheim, 1990). These membranes (an inner and an outer) are made up of many interlacing fibers which are known to consist of a protein core enveloped by a glycoprotein mantle (Board and Tranter, 1995). The proteins are thought to be classed as keratins due to the high concentration of sulfur containing amino acids (70–75%) and minor amounts of collagen (10%) (Li-Chan *et al.*, 1995; Leach, 1982). The organic matrix is an important determinant of shell quality. Shell membranes undergo changes in amino acid composition as the bird ages, with subsequent effects on shell quality (Britton and Hale, 1977; Blake *et al.*, 1985). Changes in the synthesis and secretion of the shell membranes as part of the organic matrix of the eggshell can also impair eggshell formation and affect eggshell quality (Leach, 1982).

Table 3. Approximate chemical composition of the eggshell.

Constituent	Percentage
Water	2
Dry matter	98
Protein	6
Fat	–
Carbohydrates	–
Ash	92

MINERAL NUTRITION AND THE ORGANIC MATRIX

Trace mineral nutrition has a significant role in development of the shell organic matrix. Copper deficiency (Baumgartner *et al.*, 1978) and manganese deficiency (Lyons, 1939; Leach and Gross, 1983) affect membrane formation, shell morphology, shell thickness and egg production. The above is probably due to the role that manganese plays in the synthesis of mucopolysaccharides (Leach *et al.*, 1969), since this important glycoprotein mantle as well as other organic components in the shell matrix appear to influence initiation of shell calcification (Garlich, 1982). Additionally, zinc is essential for proper formation of keratin.

Minerals and eggshell quality

CALCIUM

Calcium metabolism is a contributing factor to shell formation. Calcium represents approximately 38–40% of the eggshell by weight (2.2–2.4 g) and extensive studies have been conducted to determine dietary calcium levels required for optimum shell formation (Keshavarz, 1995). Calcium is derived either from the diet or from the body reserves, but increasing the level of calcium and phosphorous in the diet cannot improve shell quality (Keshavarz and Nakajima, 1993). It was concluded that physiological changes should be used as criteria for changing calcium content from growing to laying rations. A calcium intake of about 3.75–4.00 g daily was found to be required for optimum shell formation along with the 'resident time' of the calcium in the digestive system (Keshavarz, 1995).

ZINC-DEPENDENT CARBONIC ANHYDRASE

The eggshell is formed in the shell gland and therefore an adequate supply of calcium ions is required by the shell gland. In addition, the presence of carbonate ions in the shell gland fluid in sufficient quantities to form calcium carbonate is necessary for eggshell formation (Austic and Nesheim, 1990). Carbonate, another major component of the eggshell, is usually overlooked

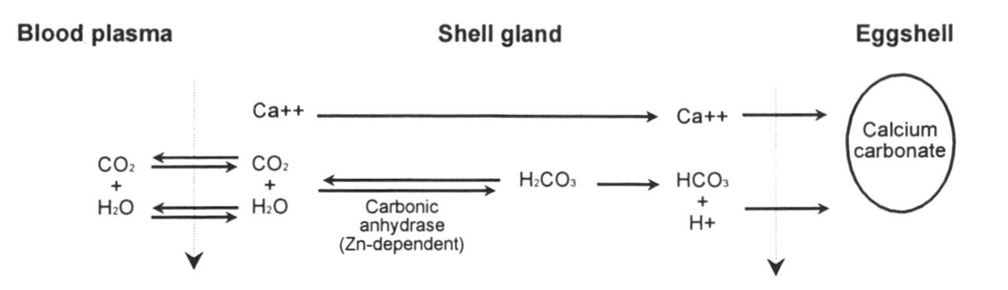

Figure 2. Carbonate production for eggshell formulation (adapted from Austic and Nesheim, 1990).

as a possible factor in shell quality problems (Balnave, 1996). One of the major sources of carbonate ions for shell formation is carbon dioxide produced during normal metabolism of the cells in the shell gland or from the blood. Carbonic anhydrase, a zinc-dependent enzyme, catalyzes formation of bicarbonate ions from carbon dioxide and water (Figure 2). Decreased activity of carbonic anhydrase has been observed in hens rested from lay compared to that of control hens remaining in lay (Balnave *et al*., 1992), which suggests that carbonate supply for eggshell formation is related to the activity of this enzyme. Saline water intake has also been associated with depressed activity of carbonic anhydrase, and in turn an increase in eggshell defects (Balnave and Yoselewitz, 1988; Balnave, 1996).

The above observations indicate that carbonic anhydrase may play an important role in shell formation. The fact that zinc is a cofactor for this enzyme makes enzme activation and proper function potentially susceptible to trace mineral interaction and bioavailability. It is well known that calcium competes for the same sites of absorption as zinc, copper and manganese; and in fact, diets high in calcium may depress carbonic anhydrase activity. Recently zinc from inorganic sources has been shown to be antagonistic to the effect of calcium and interfere with zinc absorption (Lowe, 1996).

TRACE MINERAL PROTEINATES: A POSSIBLE SOLUTION TO EGGSHELL PROBLEMS

Bioavailability of minerals other than calcium and phosphorous may have been overlooked in studies of factors affecting eggshell formation. Further consideration of zinc and manganese nutrition may well be warranted based on the role these minerals have as cofactors and (or) structural components of enzyme systems responsible for carbonate formation and mucopolysaccharide synthesis.

The use of more bioavailable and bioactive forms of these trace minerals may help reduce some of the problems associated with poor shell quality. For example, it has been shown that zinc proteinate was better absorbed than an inorganic form of zinc in the presence of high calcium concentrations (Lowe, 1996). Recent data (Klecker *et al*., 1997) suggest an overall improvement in

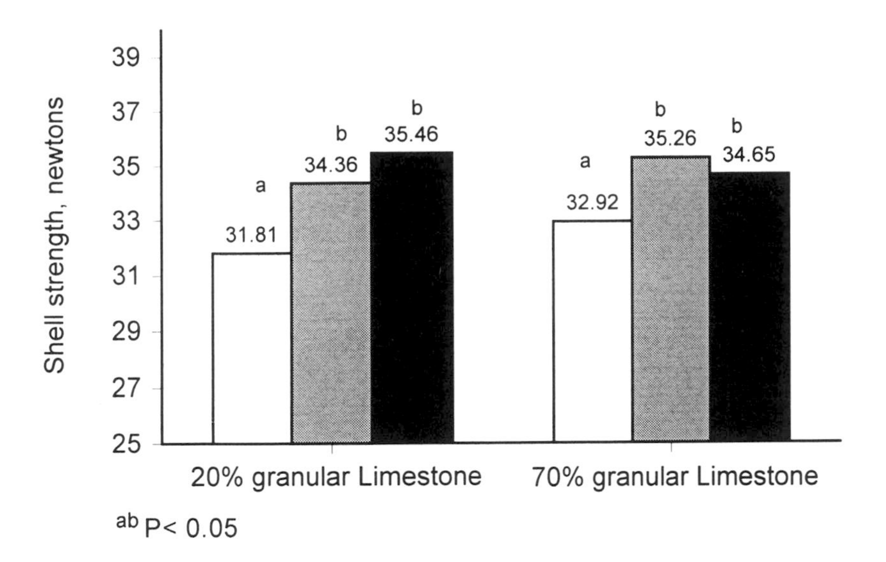

Figure 3. Effect of limestone level and mineral proteinate (Eggshell 49) on shell strength.

eggshell quality measured as shell strength, specific gravity, and shell thickness, when zinc and manganese proteinates are substituted for inorganic forms (Figure 3). These results indicate that the increased bioavailability and bioactivity of these organic minerals in carbonic anhydrase activation and glycoprotein formation must be taken into account in order to improve eggshell quality. From this a new commercial product, Eggshell 49, was developed.

Commercial usage of trace mineral proteinates has also shown exciting results in terms of reduction of 'bucket' eggs and checks. Improvement in egg grading results has also been observed when the new product is added at 1 kg/T from week 57 on. (Tables 4 and 5). What this means is that significant

Table 4. Effect of Eggshell 49 supplementation on egg grade at a northeastern US commercial farm*.

	Control period (weeks 52–56)	Eggshell 49 period (weeks 57–61)	Difference (%)
Number of birds	25,000	25,000	
Grade A eggs, %	85.08	87.67	+ 3.04
Checks‡	8.48	6.60	–22.20
Loss, %	6.46	5.74	–11.10

* The average percentage of grade A eggs, checks and losses from grading are presented for five weeks prior to inclusion of the mineral proteinate supplement (control period) and for the five-week period after beginning supplementation (mineral proteinate period)

‡ Includes downgrades.

Table 5. Effect of Eggshell 49 supplementation on egg grade.

	Control*	Eggshell 49	Difference (%)
Number of birds	120,000	72,000	
Grade A eggs, %	81.90	84.60	+ 3.30
Checks[†]	14.00	12.03	–14.00
Loss	4.10	3.37	–17.80

* Birds were 61 weeks of age at the beginning of the experiment.
[†] Includes downgrades.
Values are the means of a 10-week period and adjusted for the total number of birds for comparison purposes.

savings are possible and increased profitability can be achieved since cracked and broken eggs have already been paid for in production costs.

Summary

Eggshell quality problems represent a major financial loss to the layer industry. Based on available estimates and current number of layers, the dollar value associated with reduced shell quality amounts to $300 million in the US alone. In addition to factors contributing to shell defects, minerals other than calcium and phosphorous may have been overlooked, particularly zinc and manganese which function both in enzymes which provide components for shell synthesis and in formation of the shell membranes. In both research and commercial trials addition of a new product containing zinc and manganese proteinates along with activators has been shown to improve shell quality as measured by shell strength, specific gravity, and shell thickness. The commercial result is an improvement in the number and grade of eggs sold.

References

Anderson, C.B. and T.C. Carter. 1976. The hen's egg: shell cracking at impact on a heavy, stiff body and factors that affect it. Br. Poultry Sci. 17:613–626 (as cited by Washburn, 1982).

Austic, R.E. and M.C. Nesheim. 1990. In: Poultry Production. 13th Edition. Lea & Febiger. Philadelphia, London. pp 47–55.

Balnave, D. 1996. Carbonate limitation as a cause of poor egg shell quality. In: Arkansas Nutrition Conference Proceedings of the Meeting. September 10–12.

Balnave, D. and I. Yoselewitz. 1988. Eggshell quality is affected by salinity in water. Poultry-Misset 4(6):16–17.

Balnave, D., N. Usayran El-Khatib and D. Zhang. 1992. Poultry Sci. 71:2035–2040 (as cited by Balnave, 1996).

Baumgartner, S.G., D.J. Brown, E. Salevsky, Jr. and R.M. Leach, Jr. 1978, Copper deficiency in the laying hens. J. Nutr. 108:804–811 (as cited by Leach, 1982).

Blake, J.P., L.J. Kling and W.A. Halteman. 1985. The relationship of the amino acid composition of a portion of the outer eggshell membrane to eggshell quality. Poultry Sci. 64:176–182.

Board R.G. and H.S. Tranter. 1995. The microbiology of eggs. In: Egg Science and Technology. Fourth Edition. Eds. W.J. Staldeman and O.J. Cotteril. The Haworth Press Inc., New York, London. pp. 81–104.

Britton, W.M. and K.K. Hale, Jr. 1977. Amino acid analysis of shell membranes of eggs from young and old hens varying in shell quality. Poultry Sci. 56:865–871.

Folkerts, J. 1976. Influence of feeding and husbandry on eggshell quality. Page 580 in 5th Europ. Poultry Conf. Vol.1. (as cited by Washburn, 1982).

Garlich, J.D. 1982. Symposium: Egg shell quality. Poultry Sci. 61:2004.

Hamilton, R.M.G. 1982. Methods and factors that affect the measurement of eggshell quality. Poultry Sci. 61:2022–2039.

Hamilton, R.M.G., K.G. Hollands, P.W. Voisey and A.A. Grunder. 1979. Relationship between eggshell quality and shell breakage and factors that affect shell breakage in the field. A review. World's Poul. Sci. J. 35:177–190 (as cited by Hamilton, 1982).

Keshavarz, K. 1995. An overview of calcium and phosphorous nutrition of growing pullets and laying hens. In: Proceedings of the Cornell Nutrition Conference for Feed manufacturers. October 24–26, Rochester N.Y. pp. 161–170.

Keshavarz, K. and S. Nakajima, 1993. Re-evaluation of calcium and phosphorus requirements of laying hens for optimum performance and eggshell quality. Poultry Sci. 72:144–153.

Klecker, D., L. Zeman and J. Gomez-Basauri. 1997. Influence of trace mineral proteinate supplementation on eggshell quality. Presented at Southern Poultry Science, January 20, Atlanta, GA.

Leach, R.M. Jr. 1982. Biochemistry of the organic matrix of the eggshell. Poultry Sci. 61:2040–2047.

Leach, R.M., Jr. and R. Gross. 1983. The effect of manganese deficiency upon the ultrastructure of the eggshell. Poultry Sci. 62:499–504.

Leach, R.M. Jr., A.M. Muenstert and E.M. Wein. 1969. Studies of the role in manganese in bone formation. II. Effect upon chondroitin sulfate synthesis in chick epiphyseal cartilage. Arch. Biochem. Biophys. 133:22–28 (as cited by Leach, 1982).

Li-Chan, W.D. Powrie and S. Nakai. 1995. The chemistry of eggs and egg products. In: Egg Science and Technology. Fourth Edition. Eds. W.J. Staldeman and O.J. Cotteril. The Haworth Press Inc., New York, London. pp. 105–175.

Lowe, J.A. 1996. An investigation into the metabolism of supplemental protected zinc with reference to the use of isotopes. In: Biotechnology in the Feed Industry. Proceedings of Alltech's Twelfth Annual Symposium. (T.P. Lyons and K.A. Jacques, Eds). Nottingham University Press, Loughborough, Leics. UK. pp. 195–216.

Lyons, M. 1939. Some effects of manganese on eggshell quality. Bull 374, Arkansas Agric. Exp. Stn. (as cited by Leach, 1982).

Mast, M.G. 1987. 'Loss' eggs should be more fully utilized. Poultry-Misset.3(4):11–13.
Roland, D.A., Sr. 1976. The extent of uncollected eggs due to inadequate shell. Poultry Sci. 55:2085. (as cited by Roland, 1977b).
Roland, D.A., Sr. 1977a. The extent of uncollectable eggs due to inadequate shell. Poultry Sci. 56:1517–1521.
Roland, D.A., Sr. 1977b. Incidence of uncollectable eggs. Poultry Sci. 56:1327–1328.
Roland, D.A., Sr. 1988. Research note: Egg shell problems: Estimates of incidence and economic impact. Poultry Sci. 67:1801–1803.
Washburn, K.W. 1982. Incidence, cause, and prevention of egg shell breakage in commercial production. Poultry Sci. 61:2005–2012.
Wills, R.J.R. 1975. Economic components of egg production. In: Economic Factors Affecting Egg Production. roceedings of the Tenth Poultry Science Symposium. Eds. B.M. Freeman and K.N. Boorman. British Poultry Science Ltd. pp. 105–120.

TRACE ELEMENT SUPPLEMENTATION IN LATIN AMERICA AND THE POTENTIAL FOR ORGANIC SELENIUM

LEE R. MCDOWELL

Animal Science Department, University of Florida, Gainesville, Florida, USA

Introduction

Undernutrition is commonly accepted to be the most important limitation to livestock production in Latin America. Lack of sufficient energy and protein is often responsible for suboptimum production. Numerous investigators, however, have observed that ruminants sometimes deteriorate in spite of an abundant feed supply (McDowell *et al.*, 1993). Mineral deficiencies or imbalances in soils and plants have long been held responsible for low production and reproduction problems among tropical livestock.

Mineral deficiencies and imbalances for herbivores are reported from almost all tropical regions of the world. Phosphorus (P) deficiency has been reported in 25 Latin American countries and deficiencies of calcium (Ca) in 11, sodium (Na) in 15, magnesium (Mg) in 14, cobalt (Co) in 13, copper (Cu) in 21, selenium (Se) in 17 and zinc (Zn) in 16 (McDowell *et al.*, 1993). Iodine (I) deficiency is reported worldwide.

Mineral supplementation is required to correct mineral deficiencies in livestock and poultry diets. Prior to the turn of the 20th Century, the only supplemental minerals that were generally recognized as of value and provided to livestock and (or) humans, were common salt (NaCl), iron (Fe), or I, but administered on a very infrequent basis. In the early part of this century, pasture, other forages, distiller's solubles or grains, brewer's grains, fermentation products, meat and bone scraps, bone meal and fish by-products were used in large quantities in livestock diets. In addition to providing energy and proteins, these feeds likewise provided sources of minerals and vitamins. As animal feeding became more sophisticated, fewer feed sources were used, faster-growing and higher-producing animals were developed and the industry tended toward more intensified operations, and it became necessary to add an increasing number of minerals to properly fortify animal diets.

For many classes of livestock including pigs, poultry, feedlot cattle, and dairy cows, mineral supplements are incorporated into concentrate diets, which generally ensures that animals are receiving required minerals. However, for grazing livestock to which concentrate feeds cannot be economically fed, it is necessary to rely on both indirect and direct methods of providing minerals. Self-feeding of 'free-choice' mineral supplements is widely used for grazing livestock.

In recent years, the importance of Se deficiency in Latin America has been realized (McDowell, 1992; McDowell *et al.*, 1993). Prior to 1970, there were few Se analyses completed in Latin America and clinical signs of Se deficiency were unrecognized and confused with other disease conditions. Techniques for determining low concentrations of Se are relatively new and few tissue and feed samples had been analyzed for this element in Latin America prior to the 1980s. Illustrative of this fact is that of 3390 feeds included in the 1974 Latin American Tables of Feed Composition, there were Se values for only 0.06% of the samples (McDowell *et al.*, 1977). With more recent improvements in analytical methods for Se and an increased awareness by researchers and veterinarians of the likelihood of deficiencies, a number of reports in the 1980s and 1990s have established that Se is deficient in vast regions of Latin America.

There are two purposes of this paper: (1) to discuss methods and considerations for mineral supplementation, with emphasis on providing free-choice minerals for grazing livestock; and (2) to review Se deficiency in Latin America and to evaluate the potential for use of organic Se as a supplemental source for livestock.

Factors influencing mineral requirements

Many factors affect mineral requirements including kind and level of production, age, level and chemical form of elements, interrelationships with other nutrients, mineral intake, breed and animal adaptation. Mineral requirements are highly dependent on the level of productivity. The criterion of adequacy is important as illustrated by the fact that minimum Zn requirements for spermatogenesis and testicular development in male sheep are higher than for growth, and Mn requirement is similarly lower for growth than for fertility (Underwood, 1981).

Improved practices that lead to improved egg or milk production and growth rates will necessitate more attention to mineral nutrition. Mineral deficiencies, often marginal under low levels of production, are likely to become important and previously unsuspected nutritional deficiency signs may occur as production level increases.

Specific mineral requirements are difficult to pinpoint since exact needs depend on chemical form and numerous mineral interrelationships. The chemical form of mineral elements varies greatly in amount of dietary mineral supplied and in biological availability. As an example, elemental Se is largely unavailable for chicks but is quite effective in protecting against Se deficiency in sheep and cattle (Underwood, 1981).

Mineral supplementation is much less important for livestock if energy-protein requirements are inadequate. However, when energy and protein supplies are adequate, livestock gain weight rapidly resulting in high mineral requirements. In some countries of the world during the dry winter season unsupplemented cattle and sheep grazing extensive grassland may lose 25–30% or more of their maximum summer body mass. The concept that the most limiting nutrient dictates productivity is illustrated in Figure 1. In this example, water is lost from the lowest slat in the barrel (P) and the effect of other limiting nutrients (e.g. energy and protein) would not be realized until dietary P is increased. It is therefore uneconomical, and often of no benefit, to provide mineral supplements to grazing livestock if the main nutrients which they are lacking are energy and (or) protein.

Important differences in mineral metabolism can be attributed to breed and adaptation. The effect of breed differences on mineral requirements has often been observed in ruminants. Marked ruminant animal variation within breeds in the efficiency of mineral absorption from the diet is reported to be 5–35% for Mg, 40–80% for P and 2–10% for Cu (Field, 1981). It is not unusual for cattle introduced into an area to show deficiency signs while the indigenous breeds that are slow-growing and late-maturing do not exhibit the deficiencies to the same degree. Unacclimatized cattle of temperate types which sweat profusely and lose saliva and mucus from the mouth may lose significant quantities of minerals, particularly in the arid tropics.

Factors affecting the mineral content of plants

Of the mineral elements in soils, only a fraction is taken up by plants. Plant minerals are dependent upon a number of factors, including soil, plant species, stage of maturity, yield, pasture management and climate (Reid and Horvath, 1980; McDowell, 1985; McDowell *et al.*, 1993). Most naturally occurring mineral deficiencies in herbivores are associated with specific

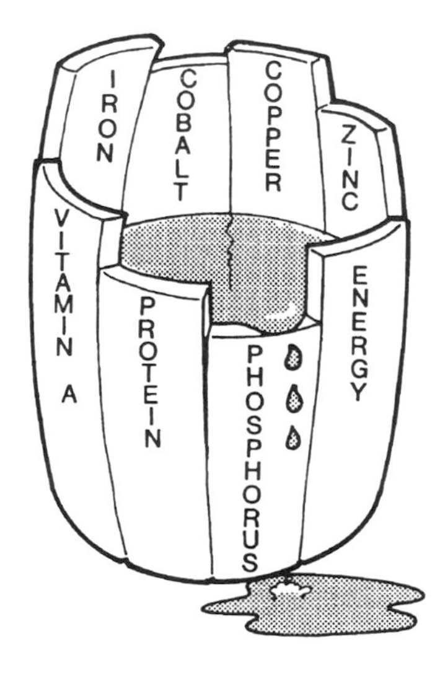

Figure 1. A barrel losing water at the shortest stave illustrates that the most limiting nutrient dictates productivity. In the example, phosphorus is the most limiting element. Since this is the shortest stave, the other nutrients would not really be limited until more phosphorus is provided.

regions and are directly related to soil characteristics. Young and alkaline geological formations are more abundant in most trace elements than the older, more acid, coarse, sandy formations. Trace element fertilization is quite effective in elevating mineral concentrations in crops. However, trace mineral deficiencies in grass, in particular Cu, Co, Se, Mn and I, are becoming more common as increased non-trace element fertilizer (N, P, K) usage results in increasing amounts of grass being utilized per hectare (Leaver, 1988). Poor drainage conditions often increase extractable trace elements (e.g. Mn and Co), thereby resulting in a corresponding increase in plant uptake. Almost all soils that produce plants containing sufficient molybdenum (Mo) to cause molybdenosis in animals are poorly drained. As plants mature, mineral contents decline due to a natural dilution process and translocation of nutrients to the root system. In most circumstances, Cu, Co, Fe, Se, Zn and Mo decline as the plant matures (Reid and Horvath, 1980).

Pasture management, forage yield, and climate influence the species of forage predominating and also change the leaf-stem ratio radically, thereby having a direct bearing on the mineral content of the sward. Increasing crop yields remove minerals from the soil at a faster rate so deficiencies are frequently found on the most progressive farms. As an example, corn produced from soils fertilized to produce 200 bushels per acre would have less Se than lower yielding corn from the same soils. Over liming can accentuate a Se or Mo toxicity in livestock by increasing plant concentrations of these elements and at the same time favor Co and Mn deficiencies due to lowered plant uptake.

Indirect and direct mineral supplementation methods

The most efficient method of providing supplemental minerals is through use of mineral supplements combined with concentrates. This assures an adequate intake of mineral elements by each animal as it consumes other nutrients. This procedure represents an ideal system for providing supplemental minerals but it cannot be used with grazing livestock which receive little concentrate and depend on forages.

Many pig and poultry operations will purchase complete concentrate feeds which will include mineral supplements. However, other farmers may be using their own grains and (or) protein supplements and will purchase mineral and other supplement additives to be mixed with their feeds on the farm. These commercially prepared additives would be added at the level recommended by the manufacturer to meet suggested requirements. Usually the quantities of both mineral and vitamin supplements required are less than 3% of the total diet. The majority of this section will emphasize providing supplemental minerals by either indirect or direct methods to grazing livestock.

INDIRECT METHODS OF PROVIDING MINERALS

The significance of indirect provision of minerals to livestock would mainly involve grazing animals. Indirect provision of minerals to grazing livestock includes use of mineral-containing fertilizers, altering soil pH, and encouraging growth of specific pasture species. A number of reports have indicated that increasing soil pH influences forage mineral uptake, thereby potentially causing deficiencies of Cu and Co and excesses of Se and Mo. Large variations in mineral content of different plant species growing on the same soil can be used to promote or discourage availability of specific minerals to grazing livestock. Underwood (1981) reports that the indirect approach as a means of controlling mineral deficiencies is not without its problems arising from the great complexity of soil-plant-mineral interrelationships and difficulties related to erratic climate and cost.

Where economic and climatic considerations are favorable, fertilizer treatment of the soil is an effective means of improving both yield and mineral composition of herbage. Where animals graze in small and well-defined pastures with just a few plant species present, the element may be added to the soil if it is one that is readily taken up by plants. However, if the element is not readily taken up by plants, addition of the deficient element to the soil will be ineffective. Phosphorus and Se fertilization of highly acid soils would have a marginal value due to the low uptake of these minerals by plants under these conditions. In more favored areas, where regular fertilizer applications are made to increase pasture or crop yields, inclusion of a proportion of the deficient element, notably Cu, Co and Zn, is added to the fertilizer (Underwood, 1981). Valle *et al.* (1993) suggested that spraying with 24 g/ha Se can be an adequate means to meet requirements of the mineral for grazing cattle. Under sparse grazing or range conditions, trace element fertilization is usually uneconomical and unreliable because of low forage productivity per unit area, variable uptake of the element, and high application and transport costs.

Increased mineral content of forages through fertilization has an advantage of assuring a more uniform mineral consumption, since all cattle would be consuming higher quantities of minerals in the forage. The major problem with free-choice mineral supplements is that not all cattle in a herd will consume adequate quantities. However, unless there are definite forage yield increases that can be utilized effectively by grazing herbivores, use of mineral-containing fertilizers is economically prohibitive.

DIRECT METHODS OF MINERAL SUPPLEMENTATION

Direct administration of minerals to livestock in water, mineral licks, mixtures, drenches, rumen preparations, and injections is generally the most economic method of supplementation. Benefits and disadvantages of mineral supplementation methods have been presented (Underwood, 1981; McDowell *et al.*, 1993).

Supplemental minerals from water, drenches, injections and rumen preparations

Systems have been considered for metering soluble forms of microelements into the drinking water, but apparently there is limited use of this method. However, this is a very effective method of assuring adequate mineral intakes, assuming there is only the one source of water available.

As with provided minerals in water, oral dosing or drenching of animals with mineral solutions or pastes has the advantage that all animals receive known amounts of the required mineral at known intervals. This type of treatment is unsatisfactory where labor costs are high and animals have to be driven long distances and handled frequently and specifically for treatment. With minerals such as Cu and Se that are readily stored in the liver to provide reserves against periods of inadequate intake, drenching with large doses several months apart has proved quite satisfactory. Sometimes oral dosing of these minerals can be combined with administration of anthelmintics at the same time. Cobalt deficiency, however, cannot be prevented fully if the oral doses of Co salts are more than one or two weeks apart.

Intramuscular injections (i.m.) of trace minerals have been highly successful in preventing or curing deficiencies of Cu, Se, I and Zn. Likewise, vitamin B_{12} is effective in preventing this vitamin deficiency in animals deficient in Co. Injectable organic complexes of these minerals when injected intramuscularly are absorbed slowly into the tissues and provide protection against deficiency for lengthy periods. As an example, Cu deficiency can be prevented when injections (e.g. glycinate or EDTA forms) are given at 3–6 month intervals.

Rumen preparations to supply minerals are based on the principle that ingested heavy particles are retained in the gastrointestinal tract, where they allow a sustained release of one or more specific minerals. For example, heavy pellets which remain in the reticulo-rumen and which contain Co, Se or Zn have been developed. Heavy pellets containing an alloy of Mg have been used with very limited success in the prevention of grass tetany.

Cobalt pellets (developed in Australia) have been used satisfactorily under practical conditions for more than 30 years while Se, Mg and Zn pellets represent more recent developments. Although heavy pellets are used widely and successfully, they are not without problems arising from regurgitation and from coating of the pellets with an impervious layer of calcium phosphate. Australian experience indicates that this latter problem can be minimized by using two pellets (or one cleaning screw) which keep each other clean by abrasion. A steady supply of Co to the rumen by these means for more than five years has been demonstrated (Lee and Marston, 1969).

A recent rumen mineral preparation is the use of soluble glass pellets as a means of providing trace elements for ruminants. As with other pellets, they remain in the reticulo-rumen and degree of solubility of the glass determines element release rate. The trace elements are an inherent part of the glass structure. One product which supplies Co, Se and Cu has been tested commercially and has been found to be effective as a source of the three elements. Pellets which were tested remained in the reticulo-rumen about 18 months. There appeared to be less problems with insoluble coatings forming on the glass pellets than experience has shown with either Co or Se pellets. In addition to Co, Se and Cu, glass pellets are adaptable to also contain Zn and Mn or could be formulated to contain only single elements. A disadvantage of glass pellets is that they remain in the rumen and may affect the quality of slaughterhouse by-products used for animal feed. With single- or limited-element supplementation, each animal receives the needed element(s), compared to a free-choice consumption method in which intakes of individual animals cannot be well controlled.

Single element or limited element administration to grazing animals appears to be most effective where there is a very severe deficiency of the particular element(s) being administered. For example, Co deficiency in parts of Australia and Se deficiency in New Zealand have been so severe that the administration of these elements alone can result in a positive response. The major disadvantage of single or limited element administration to grazing livestock in most parts of the world is that there are other minerals (e.g. P and Na) that cannot be administered in this way. In most world regions there also will be deficiencies of Na and P which must be corrected, and this indicates the use of a mixture of salt, Ca, P and trace minerals offered free-choice to be the probable system of choice. An additional disadvantage of the pellets or injections is that the elements in a free-choice mixture are generally less expensive.

Free-choice (free access) mineral supplementation

Voluntary consumption of individual minerals or mineral mixtures by animals is referred to as free-choice or free access feeding. This practice of feeding minerals free-choice to ruminants has been used for many years to supply needed minerals, but is often based on the erroneous assumption that the animal knows which minerals are needed and how much of each mineral is required.

Arnold (1964) cited evidence that most mammals exhibit little nutritional wisdom and that animals will select a palatable but poor quality diet in preference to an unpalatable, nutritious diet, even to the point of death. Gordon *et al.* (1954) had earlier measured the preferences of P-deficient cattle and sheep for supplemental Ca carbonate alone or combined with an equal part of dicalcium phosphate. The animals failed to consume enough of the P-containing supplement to prevent aphosphorosis. Under conditions of low Ca or P intake, lactating dairy cows did not consume enough dicalcium phosphate free-choice to meet requirements or to correct the deficiencies (Coppock *et al.*, 1972). Therefore, it was concluded that lactating dairy cows had no, or only limited, appetite for Ca or P.

Another approach to providing free-choice minerals is the use of a 'cafeteria style' mineral feeder, which offers the animal a choice of as many as 10 or more minerals. Dairy cows did not consume sufficient amounts of 'cafeteria-style' minerals to meet requirements and acceptability rather than appetite or craving for minerals influenced free-choice consumption (Hutjens and Young, 1976; Muller *et al.*, 1977). Maller (1967) presented evidence that domestication has produced an animal that is more responsive to the sensory qualities of feed than to nutritive value. Thus, the ability to select needed nutrients may have been lost through domestication.

Providing supplemental minerals for livestock

RECEIVING MINERALS AS PART OF CONCENTRATE MIXTURES

Livestock which receive supplemental minerals mixed with palatable concentrate mixtures, including pigs, poultry, dairy cows and feedlot cattle are more assured of receiving adequate mineral intakes compared to grazing livestock. Nevertheless, when feed intake is reduced, mineral allowances should be adjusted to assure adequate mineral intake for optimum performance. Restricting feed intake practices and (or) improved feed conversion will decrease dietary intake of all nutrients, including minerals. Restricted feeding of broiler breeders, turkey breeder hens, and gestating sows and gilts may result in marginal mineral intake if diets are not adequately fortified. Reduced feed intake may also result from stress and disease.

Use of high-energy feeds (i.e. fats) to provide diets with greater energy density requires a higher mineral concentration in feeds. Nonruminant species provided diets *ad libitum* consume quantities sufficient to meet energy requirements. Thus, mineral fortification must be increased for higher energy diets as animals will consume less total feed. Feed consumption comparisons for broilers receiving metabolizable energy ranging from 2800 to 3550 kcal/kg were made (Friesecke, 1975). Feed consumption (including minerals) was 19.1% lower for the diet with the greater energy density.

Ambient temperature also has an important influence on diet consumption, as animals consume greater quantities during cold temperatures and reduced amounts as a result of heat stress. Minerals as well as other nutrients must, therefore, be adjusted to reflect changing dietary consumptions.

Cunha (1987) reviewed some problems for animals fed complete mixed diets. Some diets with supposedly proper mineral levels may not always have a proper level or balance of minerals since factors such as level of production, age, feed intake, environmental stress, etc. may not be accounted for. Also, the producer may not feed the diet as recommended by the manufacturer. Farmers may be using a purchased feed well balanced in minerals, but this is fed in combination with farm-produced feeds (i.e. pasture, hay, silage and grains) which often results in inadequate mineral intake. Under these conditions it would be recommended to also provide a free-choice mineral mixture.

GRAZING LIVESTOCK

Animals not consuming concentrates are less likely to receive an adequate mineral supply. Free-choice minerals are much less palatable than concentrates and are often consumed irregularly. Intakes of free-choice mineral mixtures by grazing cattle are highly variable and not related to mineral requirements (McDowell, 1985). Coppock *et al.* (1972) measured individual daily consumption of dicalcium phosphate by lactating dairy cows and found individual variation to be large, ranging from 0 to more than 1000 g/head daily.

Little information is available on the individual intake of free-choice mineral supplements. Tait *et al.* (1992) using a computer system developed innovative research to electronically monitor individual mineral consumption for grazing cattle. The system is based on electronic animal identification and one or more weatherproof electronic scales located at feed stations connected to a computer. The software program identifies and records individual animals, time and duration of visits and quantity of supplement consumed. Using this equipment with grazing Holstein steers (averaging 350 kg) in a 3.25 ha pasture, Tait *et al.* (1992) reported consumption of a mineral mixture to range between 60 to 330 g per day. The average number of daily visits to the station was three per animal, and it was most interesting that a high proportion of visits were in the late evening between 20:00 and 23:00 hours. These mineral consumptions were quite high compared to cattle intakes on extensive grazing systems where animals travel further to obtain supplements. Factors that affect the consumption of mineral mixtures have been listed by Cunha *et al.* (1964) and McDowell (1992).

Soil fertility and forage type consumed
Usually, the higher the level of soil fertility, the lower the consumption of minerals. Barrows (1977) reported that for cattle, salt, Ca, P, and Mg each appears to be consumed in relation to the content of the particular element in the grass. A number of reports have shown that cattle on native range consume more mineral supplement than those cattle on improved pastures. Cattle on low quality or overgrazed pastures consume more mineral supplements.

Season of year
Season of the year affects mineral intake (Cunha, 1987). Mineral intake is often greatest during the winter or dry season when forages stop growing, lose green color and become high in fiber and lignin and low in digestibility and mineral availability. As plants mature the contents of most minerals decline (McDowell, 1985). Mineral supplement intake is lower during the period of the year when forage quality and quantity is optimum. Under drought conditions, mineral supplement intake is increased to counteract the low mineral availability in the forage and the low level of forage intake due to its reduced palatability.

Available energy-protein supplements
The kind and level of protein-energy supplementation will influence mineral supplement intake. Protein and energy supplements that likewise provide minerals will decrease both the need and desire for free-choice minerals. Weber *et al.* (1992) reported a wide day-to-day variability in free-choice mineralized salt and protein block consumption by British-bred beef cows. Variation was much greater for salt-type blocks than for the softer, protein-type blocks, with several cows consuming none of the salt-type blocks for periods of several weeks.

Individual requirements
Growth rate, percentage of calf crop, and milk production influence mineral needs. Added requirements of gestation and lactation increase mineral needs

and, thereby, consumption. The higher the level of productivity, the more important an adequate level of mineral intake. Barrows (1977) reported that mineral consumption tended to decline as cows increased in age.

Salt content of drinking water

Naturally high salt concentration of drinking water decreases mineral supplement intake. Livestock have a natural craving for salt. However, if that desire is fulfilled from drinking water high in salt, grazing livestock will consume less or none at all of a free-choice mineral mixture based on salt. Where naturally occurring salt content of water is high, mineral supplements cannot be based on salt and should be reformulated with other palatability stimulators such as cottonseed meal and molasses.

Palatability of mineral mixture

As previously mentioned, research has shown that livestock have no particular desire for the majority of minerals, with the exception of common salt. In a review on salt appetite, Denton (1967) noted that all mammals have the ability to taste salt, and there is a universal liking for it. Becker *et al.* (1944) noted that the attitude of cattle toward salt in a mineral supplement is inversely related to the amount of salt present in feeds and water. Common salt, because of its palatability, is a valuable 'carrier' of other minerals. If mixtures contain 30–40% salt or more, they are generally consumed on a free-choice basis in sufficient quantities to supply supplementary needs of other minerals.

Dew *et al.* (1954) allowed dairy cows free access to combinations of salt and steamed bonemeal. When salt was furnished in a separate container, bonemeal consumption dropped. However, when bonemeal and salt were mixed together, bonemeal consumption increased eight-fold. Many reports testify to the beneficial effects of bonemeal in free-choice supplements. Processing methods for bonemeal and other supplements affect both the nutritive value of the products and also palatability and, consequently, consumption. Improperly processed bonemeals can emit an unpleasant odor, which reduces consumption. Also, the danger of botulism and other disease conditions, can be transmitted from inadequately processed bonemeal.

Palatability and appetite stimulators such as cottonseed meal, dried molasses, dried yeast culture, and fat help achieve more uniform, herdwide consumption. Some of these products not only give the supplement a dust-free, moist, and free-flowing character, but also provide energy, protein and other benefits. Ingredients that increase palatability must be used in moderation or they will cause overconsumption.

Availability of fresh mineral supplies

Previous diet or access to mineral supplements is a factor affecting short-term consumption of minerals. When animals are not allowed access to minerals for long periods of time, they may become so voracious that they often injure each other in attempting to reach salt. Under these conditions, they will consume two to ten times the normal daily quantities of minerals until appetite is satisfied.

Rainproof mineral feeders help increase mineral intake by preventing caking, molding, and blowing away during windy weather. The choice of palatability or appetite stimulators is important when considering the keeping value of a supplement. Cornmeal is a good appetite stimulator when included in a mineral mixture but is more easily fermentable than a proteinaceous product such as cottonseed meal. The use of 20–40% salt prevents molding and blowing.

Mineral feeders will be used more frequently by livestock if they are located near water tanks, shaded loafing areas, back rubbers, and areas of best grazing. Mineral feeders should be constructed low enough so that calves can also consume minerals. They should be located on dry ground accessible to trucks for checking and servicing throughout the year. Mineral boxes should be filled frequently and not allowed to get empty. Keeping the mineral supply fresh increases its consumption. Feeders should be spaced at intervals of less than one-half mile and be adequate in number for the stocking capacity of the pasture. One suggestion is to have approximately one mineral feeder per 50 head of livestock. Less minerals are consumed if grazing livestock must travel long distances to the mineral box.

In some regions with vast grazing areas, there are great difficulties in locating feeders so that animals have constant access to minerals. This is a particular problem where animals graze over large areas with no central location for drinking water. Also, in regions that seasonally flood, locating mineral feeders above the water level is sometimes a problem.

Physical form of minerals

Mineral consumption is often 10% less when provided in block versus loose form. Mineral blocks can be developed on the basis of degree of hardness to take into consideration rainfall, humidity and other environmental conditions. Rain will dissolve too soft a block causing mineral losses, and yet livestock experience difficulty consuming enough of a hard block to fulfill mineral requirements. If the animals remain only a limited time in the vicinity of mineral blocks, then excessive block hardness will result in reduced mineral consumption.

Biological availability of mineral sources

GENERAL CONSIDERATIONS

There is considerable difference in the availability of a mineral element provided from different sources. Chemical analysis of a mineral element in a feed or mineral supplement does not provide information on availability of an element for animals (Ammerman *et al.*, 1995). Biological availability may be defined as that portion of the mineral which can be used by the animal to meet its bodily needs. The bioavailability and percentage of mineral elements in some inorganic sources commonly used in mineral supplements are shown in Table 1. These variations in bioavailability of sources must be taken into consideration when evaluating or formulating a mineral supplement.

Calculations are required to account for both the amount of element in mineral

Table 1. Percentage of mineral elements and relative bioavailability.*

Element	Source compound	Element in compound (%)	Bioavailability
Calcium	Steamed bonemeal	29.0 (23–37)	High
	Defluorinated rock phosphate	29.2 (19.9–35.7)	Intermediate
	Calcium carbonate	40.0	Intermediate
	Soft phosphate	18.0	Low
	Ground limestone	38.5	Intermediate
	Dolomitic limestone	22.3	Intermediate
	Monocalcium phosphate	16.2	High
	Tricalcium phosphate	31.0–34.0	—
	Dicalcium phosphate	23.2	High
	Hay sources		Low
Cobalt	Cobalt carbonate	46.0–55.0	—[†]
	Cobalt sulfate	21.0	—[†]
	Cobalt chloride	24.7	—[†]
Copper	Cupric sulfate	25.0	High
	Cupric carbonate	53.0	Intermediate
	Cupric chloride	37.2	High
	Cupric oxide	80.0	Low
	Cupric nitrate	33.9	Intermediate
Iodine	Calcium iodate	63.5	High
	Ethylenediamine dihydroiodide	80.0	High[‡]
	Potassium iodide, stabilized	69.0	High
	Cuprous iodide	66.6	High
Iron	Iron oxide	46.0–60.0	Unavailable
	Ferrous carbonate	36.0–42.0	Low[§]
	Ferrous sulfate	20.0–30.0	High
Magnesium	Magnesium carbonate	21.0–28.0	High
	Magnesium chloride	12.0	High
	Magnesium oxide	54.0–60.0	High
	Magnesium sulfate	9.8–17.0	High
	Potassium and magnesium sulfate	11.0	High
Manganese	Manganous sulfate	27.0	High
	Manganous oxide	52.0–62.0	High
Phosphorus	Defluorinated rock phosphate	13.1 (8.7–21.0)	Intermediate
	Calcium phosphate	18.6–21.0	High
	Dicalcium phosphate	18.5	Intermediate
	Tricalcium phosphate	18.0	—
	Phosphoric acid	23.0–25.0	High
	Sodium phosphate	21.0–25.0	High
	Potassium phosphate	22.8	—
	Soft phosphate	9.0	Low
	Steamed bonemeal	12.6 (8–18)	High
Potassium	Potassium chloride	50.0	High
	Potassium sulfate	41.0	High
	Potassium and magnesium sulfate	18.0	High
Selenium	Sodium selenate	40.0	High
	Sodium selenite	45.6	High

Table 1. Continued

Element	Source compound	Element in compound (%)	Bioavailability
Sulfur	Calcium sulfate (gypsum)	12.0–20.1	Low
	Potassium sulfate	28.0	High
	Potassium and magnesium sulfate	22.0	High
	Sodium sulfate	10.0	Intermediate
	Anhydrous sodium sulfate	22.0	—
	Sulfur, flowers of	96.0	Low
Zinc	Zinc carbonate	52.0	High
	Zinc chloride	48.0	Intermediate
	Zinc sulfate	22.0–36.0	High
	Zinc oxide	46.0–73.0	High

* From Ellis *et al.* (1988).
† Critical tests not done, but source effective.
‡ Some liberation of free iodine when mixed with trace minerals.
§ Some samples are fairly high in availability – but not as available as ferrous sulfate.

salts as well as bioavailability. For example, copper sulfate contains about 25% Cu, and copper carbonate contains about 53% of the mineral element. Therefore, it takes about twice as much copper sulfate to provide the same amount of the elemental Cu as copper carbonate. Since there are differences in the bioavailability of these two sources of Cu for the animal involved, then a correction factor should also be used so that the same amount of available Cu would be supplied to the animal.

MINERAL CHELATES AND COMPLEXES OF Cu, Co, Mn AND Zn

This section will discuss chelates and organic complexes of the trace minerals Cu, Co, Mn and Zn. The beneficial responses of a supplemental organic Se derived from yeast will be discussed in the last section of this paper.

Several mineral chelates are available for Cu, Co, Mn and Zn. Excellent reviews on the significance of chelates and complexes for the feed industry have been prepared (Nelson, 1988; Kincaid, 1989; Patton, 1990; Spears *et al.*, 1991). Spears *et al.* (1991) concluded that the use of certain organic trace mineral complexes or chelates in ruminant diets has increased performance (growth and milk production), carcass quality and immune responses and decreased somatic cell counts in milk compared with animals fed inorganic forms of the mineral.

Once in the circulatory system, true minerals are bound to proteins for transport through the blood to various tissues. Binding trace minerals to amino acids prior to feeding may facilitate transport-protein binding. Perhaps one reason for the improved mineral status of animals fed organic trace minerals is a greater solubility of the minerals in the digestive tract. Wedekind *et al.* (1992) reported that Zn was considerably more bioavailable in Zn methionine than in Zn sulfate form for chicks fed corn/soybean meal diets.

Organically complexed mineral forms also have a lack of interaction with vitamins and other ions and are effective at low levels. In cases where there is high dietary Mo, Cu in chelated form would have an advantage over an inorganic form as it may escape the complexing that occurs in the digestive system among Mo-Cu-S (Nelson, 1988).

Some studies have shown no benefit from chelated and complexed minerals, but most have shown positive responses when compared to inorganic sources. Zinc and Cu complexed with proteins or amino acids tended to have an advantage over inorganic forms of trace elements when given to stressed cattle. Ward *et al.* (1992) suggested that a mixture of Zn, Mn, Cu and Co in amino acid complexed forms may stimulate feed intake and growth during the initial stress period of feedlot steers compared to the oxide or sulfate forms.

Zinc methionine has been reported to have reduced the incidence of foot rot 55% in grazing steers (Muirhead, 1992). The Zn methionine fed steers gained 0.21 kg of body weight per day more than control steers.

Weaning weights were higher for Zn and Mn methionine-supplemented calves compared to control or oxide-supplemented calves (Spears and Kegley, 1991). Spears (1989) showed that Zn-deficient lambs retained Zn from Zn methionine better than from Zn oxide. Herrick (1989) reviewed Zn methionine feeding in four dairy trials and concluded that the Zn complex-treated animals had lower somatic cell counts and higher milk yields than control cows.

For chicks Mn chelates provided a 58% more effective value for egg production and a 99% more available value for bone deposition when compared to Mn oxide (Savage, 1973). In tests with broiler chicks, relative bioavailability of test compounds were as follows: Mn oxide, 90%; Mn sulfate, 100%; and Mn methionine, 120% (Henry *et al.*, 1989). In addition, Mirando *et al.* (1993) reported that the dietary supplementation of proteinated trace minerals improved embryo and fetal survival in sows.

Kincaid *et al.* (1986) compared Cu proteinate and Cu sulfate in terms of ability to increase Cu status in calves fed a diet naturally high in Mo (3.1 ppm) and low in Cu (2.8 ppm). Calves fed Cu proteinate had higher plasma (0.87 vs 0.75 mg/l) and liver (325 vs 220 ppm) Cu concentrations than calves supplemented with a similar level of Cu from the sulfate form after 84 days. Zinc in the form of Zn lysine resulted in the highest levels of metallothionein in liver, pancreas and kidney compared to other Zn sources; thus indicating a more bioavailable source of Zn (Rojas *et al.*, 1995).

Much more needs to be learned about the selectivity of chelating agents toward minerals, the kind and quantity most effective, their mode of action, and their behavior with different species of animals and with varying diets. Dietary requirements for minerals may be greatly reduced by the addition of chelating agents to animal diets, but cost to benefit relationships need to be established.

Selecting a free-choice mineral supplement

Even though grazing livestock have been found not to balance their mineral needs perfectly when consuming a free-choice mixture, there is usually no other practical way of supplying mineral needs under grazing conditions. As a low cost insurance to provide adequate mineral nutrition, 'complete' mineral supplements should be available free-choice to grazing livestock (Cunha *et al.*, 1964). A 'complete' mineral mixture usually includes salt, a low fluoride-phosphorus source, Ca, Co, Cu, Mn, I, Fe and Zn. Except where selenosis is a problem, most free-choice supplements should contain Se. Magnesium, K, S, or additional elements can also be incorporated into a mineral supplement or can be included at a later date as new information suggests a need.

Calcium, Cu, or Se, when in excess, can be more detrimental to ruminant production than any benefit derived by providing a mineral supplement. In regions where high forage Mo predominates, three to five times the Cu content in mineral mixtures is needed to counteract Mo toxicity (Cunha *et al.*, 1964). As little as 3 ppm Mo have been shown to decrease Cu availability by 50%. Sulfur at 500 ppm can have the same effect. Thus, the exact level of Cu to use in counteracting Mo or S antagonism is a complex problem and should be worked out for each area.

A number of so called 'authorities' feel there is no justification for the use of 'shotgun' (complete) free-choice mineral mixtures that are designed to cover a wide range of environments and feeding regimens and that contain a margin of safety as an insurance against deficiency. These people feel that 'shotgun' mixtures are economically wasteful and can also be harmful. This author is in disagreement with this viewpoint regarding 'shotgun' mixtures for grazing ruminants. There is little danger of toxicity or excessive cost in relation

Table 2. Characteristics of a 'good' complete free-choice cattle mineral supplement.

- Contains a minimum of 6–8% total P. In areas where forages are consistently lower than 0.20% P, mineral supplements in the 8–10% phosphorus range are preferred.
- Has a calcium-phosphorus ratio not substantially over 2:1.
- Provides a significant proportion (e.g. about 50%) of the trace mineral requirements for Co, Cu, I, Mn, and Zn*. In known trace-mineral-deficient regions, 100% of specific trace minerals should be provided.
- Includes high-quality mineral salts that provide the best biologically available forms of each mineral element, and avoidance of minimal inclusion of mineral salts containing toxic elements. As an example, phosphates containing high F should be either avoided or formulated so that breeding cattle would receive no more than 30–50 ppm F in the total diet. Fertilizer or untreated phosphates could be used to a limited extent for feedlot cattle.
- Is sufficiently palatable to allow close to adequate consumption in relation to requirements.
- Is backed by a reputable manufacturer with quality control guarantees as to accuracy of mineral-supplement label.
- Has an acceptable particle size that will allow adequate mixing without smaller size particles settling out.
- Is formulated for the area involved, the level of animal productivity, the environment (temperature, humidity, etc.) in which it will be fed, and is as economical as possible in providing the mineral elements used.

* For most regions it would be appropriate to include Se, unless toxicity problems have been observed. Iron should be included in temperate region mixtures but often both Fe and Mn can be eliminated for acid soil regions. In certain areas where parasitism is a problem Fe supplementation may be beneficial.

to the high probability of increased production rates for cattle from administering a complete 'shotgun' free-choice mineral mixture following the guidelines in Table 2 (McDowell, 1992). Copper and Se added at recommended levels would be the minerals of most concern for toxicity. However, cattle, contrary to sheep, are much less sensitive to Cu toxicity, and inorganic forms of Se (e.g. sodium selenite) are less well utilized by livestock when administered in excess of the requirements. In conclusion, it is best to formulate free-choice mixtures on the basis of analyses or other available data. However, when no information on mineral status is known for a given region, a free-choice complete ('shotgun') mineral supplement is definitely warranted.

Information required for mineral supplement formulation

Mineral ratios and interrelationships are always important, but not as important as adequate concentrations of individual minerals in the mixtures. As an example, the Ca to P ratio is of minimal importance to ruminants provided the P level is adequate. However, a narrower ratio of Ca to P is much more important for monogastric species, such as growing poultry and pigs. Likewise, high levels of Ca are extremely detrimental in relation to Zn requirements for monogastric species, but apparently have much less effect in ruminant diets.

To evaluate a free-choice mineral supplement, it is necessary to have an approximation of: (1) grazing livestock requirements for the essential nutrients, which includes the age of the animals involved, stage of current production or reproduction cycle, and intended purpose for which the animals are being fed; (2) relative biological availability of the minerals in the sources from which they will be provided; (3) approximate daily intake per head of the mineral mixture and total dry matter that is anticipated for the target animals; and (4) concentration of the essential minerals in the free-choice mixture.

REQUIREMENTS

The needs of individual animals under varying conditions may differ from averages. It is recognized that with the introduction of crossbreeding and of exotic breeds of livestock that growth rates have been increased with a consequent increase in mineral requirements. In spite of these shortcomings, most researchers agree that this is the best information we have and that these requirements should be used as a guide.

BIOLOGICAL AVAILABILITY

Biological availability of a mineral element implies the availability of the element. For mineral supplement formulation, bioavailability of sources must be considered.

INTAKES OF MINERAL SUPPLEMENT AND DRY MATTER

Palatability of the supplement affects intake more than do physiological needs. In formulating mineral mixes, estimating the possible need must coincide with adequate intake. A number of reports conclude that grazing cattle do not always consume mineral mixtures well. Cunha *et al.* (1964) in Florida, Weber *et al.* (1992) in Oregon and Rode and Beauchemin (1993) in Lethbridge, Canada have presented data showing a wide monthly variation in the consumption in specific regions. When evaluating mineral supplements where consumption is not known, researchers commonly start with an intake figure of 50 g/day and adjust this figure according to local conditions. For sheep a starting intake figure would be 15 g/day (McDowell, 1996).

It is virtually impossible to measure accurately the total dry matter consumption of livestock on pasture. However, this is essential since requirements are based on intakes of dry matter. The quality of a pasture will determine intake to a great degree. Although 2% body weight is considered a rough estimate of forage dry matter intake by cattle, they may eat much less if the forage is of poor quality. Actual dry matter consumption often becomes a matter of judgment on the part of the researcher or rancher. For grazing mature cattle, often daily dry matter consumption is between 7 and 10 kg, for sheep 1.8 kg could be an estimated value (McDowell, 1996).

ELEMENT CONCENTRATION IN MINERAL MIXTURE

After an evaluation has been made of the biological availability of elements to be supplied by mineral mixture, and a judgment has been made as to approximate daily intake of mineral mixture and of total dry matter, the concentration of each element in the mineral mixture can be used to calculate the amount of each element that will be furnished per animal, expressed as a percentage or parts per million (ppm) of total dry matter intake. The concentration of each element furnished by the mineral mixture can be compared to total requirements for that element to determine if a significant amount is being furnished by the supplement. It is difficult to determine what constitutes a significant portion of the requirement for each mineral that should be supplied by the mineral mixture, but it is generally believed the figure should be 25–50% for the trace elements. In zones known to have trace mineral deficiencies, 100% of the requirements for these elements should be provided.

Calculations for free-choice mineral supplement formulation

The calculations are as follows:

$$\frac{\text{\% element in mineral mixture} \times \text{daily intake of mineral mix (g)}}{\text{total daily dry matter intake (g)}}$$

$$\times\ 100 = \text{\% element in total from mineral mixture}$$

If, for example,

Copper in mineral mixture (%) = 0.12

Daily intake of mineral mixture (g) = 50

Total daily intake of dry matter (g) = 10,000

then:

$$\frac{0.0012 \times 50g \times 100}{10{,}000\ g} = 0.0006\% \text{ or } 6 \text{ ppm}$$

Note that to convert percentage to ppm, the decimal is moved four places to the right. If approximately 10 ppm is considered the allowance for Cu, then 60% of the Cu requirement would thus be supplied by this particular mixture. For sheep use estimated intakes of mineral mixtures and total dry matter as 15 and 1800 g, respectively (McDowell, 1996).

Table 3 illustrates the estimated trace mineral requirements and percentages of each element required in a beef cattle mineral mixture to meet 25, 50 or 100% of the requirement. These figures are based on an estimated daily mineral consumption of 50 g. With less consumption, the mineral supplement should contain a higher percentage of each mineral. Likewise, a lower intake of dry matter would reduce the percentage of minerals required in the mixture. Each producer should determine mineral consumption for his herd and change products if higher consumption rates are required (e.g. increase the cottonseed meal, as a palatability factor, in the mixture from 5 to 10%).

Table 3. Trace minerals in an adequate supplement*†.

Element	Estimated maximum requirement (ppm)†	Percentage of mineral need in mixture to supply 25, 50 or 100% of requirement*		
		25%	50%	100%
Cobalt	0.1	0.0005	0.001	0.002
Copper	10	0.05	0.10	0.20
Iodine	0.8	0.004	0.008	0.016
Manganese	25	0.125	0.25	0.50
Zinc	50	0.25	0.50	1.0
Iron	50	0.25	0.50	1.0
Selenium	0.2	0.001	0.002	0.004

* From McDowell *et al.* (1993).

† This assumes an average consumption of 50 g/day of mineral mixture for cattle and 10 kg of total dry feed per animal daily.

Free-choice mineral supplement evaluation

Problems concerned with mineral supplementation programs in diverse world regions have been summarized (McDowell *et al.*, 1993) and include: (1) insufficient chemical analyses and biological data to determine which minerals are required and in what quantities; (2) lack of mineral consumption data needed for formulating supplements; (3) inaccurate and (or) unreliable information on mineral ingredient labels; (4) supplements that contain inadequate amounts or imbalances; (5) standardized mineral mixtures that are inflexible for diverse ecological regions (e.g. supplements containing Se distributed in a Se-toxic region); (6) farmers not supplying mixtures as recommended by the manufacturer (e.g. mineral mixtures diluted 10:1 and 100:1 with additional salt; (7) farmers not keeping minerals available to animals continually; and (8) difficulties involved with transportation, storage, and cost of mineral supplements. Many of these problems are more related to tropical vs temperate regions as in temperate regions (more developed countries) there is better quality control of products produced. However, some of the problems with temperate mineral mixes are related to inadequate quantities of Cu and Zn in mixtures, with some products low in P while others still not providing Se.

Responsible firms that manufacture and sell high-quality mineral supplements provide a great service to individual farmers. However, there are companies that are responsible for exaggerated claims of advertising, and some that produce inferior products that are of little value, or worse, those likely to be of detriment to animal production. Table 4 provides an example of an inferior mineral mixture available in Latin America. This particular mineral supplement is recommended for cattle, sheep, pigs, and chickens. It is impossible to adequately meet requirements of both ruminants and monogastric animals with the same mixture. This unbalanced mineral mixture, which is extremely high in Ca (29.4%) and low in P (1.8%), would likely be more detrimental to grazing cattle than having no access to supplemental minerals, and may actually contribute to a P deficiency.

Table 4. An inferior mineral mixture available in Latin America*†‡.

Element	Dietary allowance	Mineral mixture (%)	Allowance provided from mineral mixture	Allowance for mineral mixture (%)
Sodium chloride	0.50%	20.00	0.10%	20.0
Calcium	0 30%	29.44	0.147%	49.1
Phosphorus	0.25%	1.80	0.009%	3.6
Magnesium	2000 ppm	3.2	0.016%	8.0
Iron	100 ppm	0.88	44.0 ppm	44.0
Zinc	50 ppm	0.02	1.0 ppm	2.0
Cobalt	0.1 ppm	0.002	0.1 ppm	100.0
Iodine	0.80 ppm	0.001	0.05 ppm	6.25
Copper	10 ppm	0.015	0.75 ppm	7.5
Manganese	25 ppm	0.075	3.75 ppm	15.0
Selenium	0.1 ppm	0.0005	0.025 ppm	25.0

* From McDowell *et al.* (1993).

† Mineral mixture is recommended for cattle, sheep, pigs, and chickens. It is assumed that mineral consumption will average approximately 0.5% of the total dietary intake. This is based on an estimated intake of 50 g of mineral mixture for cattle and 10 kg of total dry feed per head daily.

‡ Criticisms of mineral mixture are as follows: (1) Mixture extremely low in P and exceptionally high in Ca. The Ca:P ratio is 16.4:1. (2) The supplement does not provide a significant proportion (i.e., 50%) of the trace mineral requirements of Cu, I, Mn and Zn. (3) The majority of the Fe is from ferric oxide, an unavailable form of this element. (4) Since this diet contains 29.4% Ca and only 20% salt (NaCl), it is likely to be of low palatability.

Manufacture of mineral mixes

Mineral elements exist in many chemical forms including sulfates, carbonates, chlorides, oxides and organic forms (e.g. amino acid complexes and proteinates). The form chosen for use should depend on its biological value, cost, availability in the use area, its stability and effect in the type of diet used and other functions. It is important to know what combinations of mineral salts to mix together to avoid their reacting with each other and causing adverse effects in a mineral mixture (Cunha, 1987). Moreover, mineral mixes need to travel in trucks, railroad cars and by other means under hot, cold, humid, dry or other weather conditions. Likewise, both loose and block mineral forms also need to be stored and fed under a wide range of weather conditions and under varying levels of rainfall or snow. Therefore, proper formulation of mineral supplements requires considerable knowledge. A particular expertise is required to produce mineral blocks; if the block is too hard, consumption will be reduced, if not hard enough, the product will crumble.

If a mineral mix is to be fed free-choice, the characteristics of the finished product are likely to be different than if it is to be used as a base mix in a finished feed. The mineral mix formulator does need some rather specific information concerning the various ingredients such as: (1) biological availability; (2) compatibility; (3) toxicity; (4) solubility; (5) hydroscopicity; (6) relative particle size; (7) density; (8) chemical stability; and (9) moisture and nutrient content. The final mineral product should not be dusty and should have the proper degree of palatability so the animal will consume, as close as possible, the correct amount to meet its nutritional needs.

The user of mineral supplements must rely on the reputation and integrity of the mineral feed manufacturer, who provides properly balanced essential mineral fortification for the particular target species, level of production, and for the correct season of the year to complement the available minerals in the forages and concentrates, considering individual animal variation in requirements. Safe, biologically available and palatable forms of the minerals, at a fair price, allow both the user and manufacturer to realize a profit from its use.

Selenium nutrition in Latin America

SELENIUM DEFICIENCY

Selenium has a dual role in the nutrition of livestock, as a toxic element (feeds containing over 2–5 ppm) and as a deficient nutrient (feeds containing less than 0.1 to 0.3 ppm). Latin American countries where selenosis is encountered include Argentina, Chile, Colombia, Ecuador, Honduras, Mexico, Peru and Venezuela (McDowell, 1985). In each of these countries there are regions that are also deficient in Se. Research has now established that the total areas of the world affected by Se deficiency are far greater and the consequences are more economically important than those afflicted with Se excess (Underwood, 1981).

Selenium deficiency is confirmed or suspected in 17 Latin American countries (Argentina, Bahamas, Bolivia, Brazil, Colombia, Costa Rica, Dominican Republic, Ecuador, Guatemala, Guyana, Honduras, Mexico, Nicaragua, Paraguay, Peru, Uruguay and Venezuela) (McDowell *et al.*, 1993). A number of Latin American countries, including Brazil, Ecuador, Mexico, and Uruguay, have reported clinical cases of calves exhibiting white muscle disease. From Peru, Terry (1964) reported a disease condition in ruminants that resembled Se-vitamin E deficiency. When treated with Se-vitamin E, 70% of the animals recovered from the condition. In Paraguay, Se administration increased weight gains and increased pregnancy rate in heifers over controls (Boggino *et al.*, 1973). Low Se levels in feeds from Mexico (Gutierrez *et al.*, 1974), the Bahamas (A. Dorsett, personal communication), Brazil (Moxon, 1971), and Honduras (A.L. Moxon, personal communication) have been reported. Lang *et al.* (1975) reported borderline to deficient Se concentrations in 33% of the liver samples analyzed from beef cattle slaughtered in Guanacaste, Costa Rica. Hall (1977) reports undocumented evidence of Se deficiency in southern Brazil and northern Uruguay on the basis of typical symptoms (i.e. muscular dystrophy and incoordination) and low forage Se levels (0.045 ppm). On the basis of forage and serum Se concentrations, Se was found to be borderline to deficient in the provinces of Chaco and Formosa, Argentina (Balbuena *et al.*, 1990).

A total of 81% of forage and 48% of serum samples were deficient in Se. In subclinical Se deficiency performance may be reduced with slower gains and lowered reproduction involving an increased number of services needed per conception. Poor reproductive performance has been shown to include retained placenta, with high incidence of retained placentas in cattle greatly reduced by administration of adequate dietary levels of Se, as shown by research in Brazil, the United States and Scotland. High incidences of retained placentas have been reported when blood serum levels were below 0.04 ppm of Se. Selenium levels below 0.04 ppm were reported in 75% of serum samples from 974 dairy cattle in 12 regions of the state of São Paulo, Brazil (Lucci *et al.*, 1983b). Lucci and Moxon (1982) have analyzed serum Se from 43 dairy herds in São Paulo, Brazil, with only five herds having an average concentration greater than 0.04 ppm; likewise, confirmed white muscle disease in calves was also encountered in this region.

Many regions of Latin America are deficient in Se on the basis of Se feed analyses. Forage analyses from 80 farms in São Paulo, Brazil, revealed low Se concentrations, with a mean of 0.072 for the rainy season and 0.054 ppm for the dry season, with corn silage averaging 0.034 ppm of Se (Lucci *et al.*, 1983a). Other than Brazil, Table 5 summarizes forage Se concentrations from 12 research investigations by season from both Central and South America locations as well as Florida in the United States. In all of these studies, a significant number of total forages were deficient. Of the 21 seasonal mean values, more than half of the total forages analyzed were based on Se concentrations of less than 0.1 ppm. The estimated Se requirement for livestock and poultry species range between 0.1 to 0.3 ppm. The majority of soils in Latin America are acid (Cox, 1973); uptake of soil Se is reduced in acid soils as an unavailable ferric iron and selenite complex is formed.

Table 5. Forage selenium concentrations (ppm) for selected warm climate regions in Latin America and Florida.

Location	Season	Number	Mean	Percent of samples below 0.1 ppm
Argentina[a]	Wet	57	0.07	81
Bolivia[b]	Dry	8	0.11	33
	Wet	16	0.14	22
Bolivia[c]	Dry	17	0.07	88
	Dry	42	0.10	47
Colombia[d]	Dry	34	0.11	74
	Wet	35	0.12	38
Colombia[e]	Dry	64	0.09	69
	Wet	67	0.11	43
Costa Rica[f]	Both wet and dry	409	0.12	63
Dominican Republic[g]	Dry	33	0.14	48
Florida[h]	Dry	10	0.05	90
	Wet	19	0.07	84
Guatemala[i]	Dry	84	0.14	63
	Wet	84	0.40	49
Guatemala[j]	Both wet and dry	81	0.07	77
Guatemala[k]	Both wet and dry	72	0.16	29
Nicaragua[l]	Dry	112	0.21	27
	Wet	192	0.25	14
Venezuela[m]	Dry	98	0.09	71
	Wet	100	0.08	76

[a] (Balbuena *et al.*, 1990)
[b] (McDowell *et al.*, 1982a)
[c] (Peducassé *et al.*, 1983)
[d] (Vargas *et al.*, 1984)
[e] (Pastrana *et al.*, 1991)
[f] (Vargas *et al.*, 1992)
[g] (Jerez *et al.*, 1984)
[h] (McDowell *et al.*, 1982b)
[i] (Tejada *et al.*, 1987)
[j] (Knebusch *et al.*, 1988)
[k] (Valdes *et al.*, 1988)
[l] (Velasquez-Pereira *et al.*, 1996)
[m] (Rojas *et al.*, 1993)

Grains grown in low Se soils will produce Se deficiency diseases in swine and poultry. Moxon (1971) analyzed corn samples for Se from 19 locations in Brazil and found average values of 0.060 and 0.036 ppm for 1969 and 1971, respectively. Corn from locations within Minas Gerais, Rio Grande do Sul, and São Paulo had particularly low Se concentrations. A report from Campinas, Brazil indicated Se/vitamin E deficiency problems in poultry when corn containing 0.035 ppm Se was fed and fish meal (a good source of Se) was omitted from the diet. This problem was corrected when 0.2 ppm of Se was added to the ration (Moxon, personal communication). The Se deficiency disease, exudative diathesis, in poultry has even been reported in a region with adequate Se in soils and feed crops. The reason for this is that the poultry were being fed diets composed mainly of corn and soybean meal imported from Se-poor regions of the United States (midwest) vs consuming feeds of local origin.

POTENTIAL USE OF SUPPLEMENTAL ORGANIC SE IN LATIN AMERICA

The supplemental form of Se most widely used is the inorganic form, sodium selenite. An alternative organic Se source derived from yeast has been developed whereby Se is incorporated into the protein structure of growing yeast cells (Mahan, 1996). The yeast organism used in this process has a high sulfur requirement, but when fed a sulfur-deficient medium with a concurrent high Se content, the resulting yeast product contains a high concentration of Se (1000 ppm Se). The Se-fed yeast incorporates Se into its body protein structure effectively replacing the sulfur component. This process produces an array of Se amino acid analogs in the yeast protein, the most notable one being selenomethionine (Kelly and Power, 1995). The supplemental Se-yeast product, produced by Alltech Inc., is called Sel-Plex 50. The product is so named because approximately 50% of Se is present as selenomethionine and the rest as other selenoproteins and amino acids.

Supplemental organic Se has a greater potential to meet the needs of livestock and poultry than does sodium selenite. Selenium deficiencies seem to be more prevalent from regions where sodium selenite serves as the major contributor of dietary Se, which implies that selenite may not be as biologically effective as the Se indigenous in grains or grain by-products. Grains incorporate Se in an organic form largely as selenomethionine (Levander, 1986) with those grains from Se-deficient areas having a lower organic Se content. Organic forms of Se result in higher tissue Se levels than inorganic forms (Cary *et al.*, 1973). Selenium from grains (0.5 ppm) resulted in four to five times as much Se in muscle as the same amount of sodium selenite. For pigs, milk and tissue Se content of the reproducing female were increased when a 0.3 ppm Se level was provided from a Se-enriched yeast compared with sodium selenite (Mahan and Kim, 1996). The tissue Se content of the pig can be increased when a Se-enriched yeast source is fed to gestating and lactating dams at 0.3 ppm.

Although organic Se has shown definite benefits for monogastric animals, the potential is even greater for ruminants. The absorption of selenite is high in nonruminants, but lower in ruminants, whereas both ruminants and nonruminants can effectively absorb the organic seleno-amino acid compounds from the small intestine. Wright and Bell (1966) found Se absorbed less in ruminants than in monogastrics, with oral selenite Se retained 77% in pigs but only 29% in sheep. Lower absorption for sheep was thought due to selenite being reduced to insoluble compounds in the rumen. More unabsorbed Se is found in the feces of dairy cattle (Cast, 1994) compared to pigs (Mahan and Parrett, 1996) when these species are fed sodium selenite. The provision of either dietary inorganic or organic Se has been shown to result in equal effectiveness in glutathione peroxidase activity in both grower pigs and reproducing pigs (Mahan, 1996). In contrast, dairy cattle have exhibited a higher glutathione peroxidase activity when the organic form of selenium is fed compared to when selenite is provided (Pehrson *et al.*, 1989). These responses suggest a more equal bioavailability of both Se sources for nonruminants, whereas, the inorganic form has a lower absorption and bioavailability value for ruminants.

Supplemental Se is provided to pigs and poultry as part of the concentrate mixture. The benefits of the Se yeast product for both pigs and poultry have been reviewed (Torrent, 1996; Edens, 1996). As an example, Se^- yeast (Sel-Plex 50) has a significant effect in slowing the process of water loss from pork and poultry products and in improving poultry feathering and performance. Edens (1996) and Mahan (1996) have suggested that organic Se may replace sodium selenite within five years.

For grazing livestock, it is more difficult to provide supplemental Se. The principal methods of increasing Se intake by grazing livestock include: (1) a free-choice Se mineral supplement; (2) Se fertilization; (3) injections of Se; (4) as a drench; and (5) Se ruminal pellets (heavy boluses). Due to the extensive grazing areas in Latin America, Se fertilization would not be economically feasible. Likewise, Se injections, drenches and ruminal pellets can be more labor intensive and expensive. Free-choice Se mineral supplementation is the most practical method of providing supplemental Se to grazing livestock in Latin America. In the future, organic Se will likely be an extremely important source of supplementation of this element.

An understanding of the way the different forms of Se are absorbed and metabolized by both ruminants and nonruminants is essential in understanding why the deficiency is still occurring and why the Se yeast product may potentially be a better supplemental source for all species, particularly in those areas of the world where Se contents of grains and forages are low. This would include the majority of Latin American locations as well as Latin American regions which import large quantities of Se-deficient corn and soybean meal produced in the Se-deficient midwestern states of the United States.

References

Ammerman, C.B., D.H. Baker and A.J. Lewis. 1995. Bioavailability of Nutrients for Animals. Academic Press, San Diego.

Arnold, G.W. 1964. Some principles in the investigation of selective grazing. Proc. Aust. Soc. Anim. Prod. 5:285.

Balbuena, O., L.R. McDowell, C.A. Luciani, J.H. Conrad, N. Wilkinson and F.G. Martin. 1990. Estudios de la Nutricion mineral de los bovinos para carne del este de las provincias de Chaco y Formosa (Argentina) 5. Cobalto y Selenio. Vet. Arg. VII No 61:25.

Barrows, G.T. 1977. Research efforts have lagged in free-choice feeding. Anim. Nutr. Hlth. May: 12–14.

Becker, R.B., G.K. Davis, W.G. Kirk, R.S. Glasscock, A.P.T. Dix and J.E. Pace. 1944. Defluorinated superphosphate for livestock. Bull. 401, Fla. Agric. Exp. Stn., Gainesville.

Boggino, E., V.J. Romero and M.A. Cano. 1973. Effecto de la administración de cobre, cobalto, y selenio sobre el crecimiento y fertilidad de vaquilas. Informe Anual 1973. Programa Nacional de Investigación y Extensión (Ganadera). República del Paraguay.

Cary, E.E., W.H. Allaway and M. Miller. 1973. Utilization of different forms of dietary selenium. J. Anim. Sci. 36:285.

CAST. 1994. Risks and benefits of selenium in agriculture. Issue paper no. 3 supplement.

Coppock, C.E., R.W. Everett and W.G. Merrill. 1972. Effect of ration on free-choice consumption of calcium-phosphorus supplements by dairy cattle. J. Dairy Sci. 55:245.

Cox, F.R. 1973. Micronutrients. In: A Review of Soils Research in Tropical Latin America. North Carolina Agriculture Experiment Station, Raleigh, N.C.

Cunha, T.J. 1987. Salt and Trace Minerals. Salt Institute, Alexandria, Virginia.

Cunha, T.J., R.L. Shirley, H.L. Chapman, Jr., C.B. Ammerman, G.K. Davis, W.G. Kirk and J.F. Hentges. 1964. Minerals for beef cattle in Florida. Fla. Agr. Exp. Stn. Bull. 683, Gainesville.

Denton, D.A. 1967. Salt appetite. Handbook of Physiology I, Am. Physiological Society, Washington, DC..

Dew, M.I., G.E. Stoddard and G.O. Bateman. 1954. Phosphorus supplements made more palatable with salt. Utah Farm and Home Sci. 15:36.

Ellis, G.L., L.R. McDowell and J.H. Conrad. 1988. Evaluation of mineral supplements for grazing cattle in tropical areas. Nutr Rpts. Int. 38(6):1137.

Edens, F. 1996. Organic selenium: From feathers to muscle integrity to drip loss. Five years onward: No more selenite. In: Proceedings of the 12th Annual Symposium on Biotechnology in the Feed Industry (T.P. Lyons and K.A. Jacques, eds.), Nottingham Univ. Press. Loughborough, Leics. UK. p. 165.

Field, A.C. 1981. Information needs in mineral nutrition of ruminants. Florida Nutrition Conference p. 233, University of Florida, Gainesville.

Friesecke, H. 1975. Pantothenic acid. No. 1533, Hoffmann-LaRoche, Basel, Switzerland.

Gordon, J.G., D.E. Tribe and T.C. Graham. 1954. The feeding behavior of phosphorus-deficient cattle and sheep. Br. J. Anim. Behav. 2:72.

Gutierrez, J.L., G.S. Smith, J.D. Wallace and A.B. Nelson. 1974. Selenium in plants, water and blood. New Mexico and Chihuahua. J. Anim. Sci. 38:1331 (Abstr).

Hall, G.A. 1977. Phosphorus and trace mineral nutrition of grazing livestock in southern Brazil. Symposium on Feed Composition, Animal Nutrient Requirements and Computerization of Diets. Utah State Univ., Logan. p. 369.

Henry, P.H., C.B. Ammerman and R.D. Miles. 1989. Relative bioavailability of Mn in a Mn-methionine complex for broiler chicks. Poultry Sci. 68:107.

Herrick, J. 1989. Zinc methionine: Feedlot and dairy indications. Large Anim. Vet. 44:35.

Hutjens, M.F. and C.W. Young. 1976. Evaluation of cafeteria-style free-choice minerals by dairy cows. Proc. 71st Amer. Dairy Sci. Assoc. Ann. Meeting, p. 30 (abstr.).

Jerez, M., L.R. McDowell, F.G. Martin, W.A. Hargus and J.H. Conrad. 1984.

Mineral status of beef cattle in Eastern Dominican Republic. Trop. Anim. Prod. 9:12.

Kelly, M.P. and R.F. Power. 1995. Fractionation and identification of the major selenium compounds in selenized yeast. J. Dairy Sci. 78(Suppl. 1):237 (Abstr.).

Kincaid, R.L. 1989. Availability, biology of chelated, sequestered minerals explored. Feedstuffs 61(11):22.

Kincaid, R.L., R.M. Blauwiekel and J.D. Cronrath. 1986. Supplementation of copper as copper sulfate or copper proteinate for growing calves fed forages containing molybdenum. J. Dairy Sci. 69: 160.

Knebusch, C.F., J.L. Valdes, L.R. McDowell and J.H. Conrad. 1988. Seasonal effect of mineral supplementation on micro-element status and performance of grazing steers. Nutr. Repts. Inter. 38:399.

Lang, C.E., L.R. McDowell, J.H. Conrad and H. Fonseca. 1975. Estado mineral del ganado en Guanacaste, Costa Rica. 5a Reunión Latinoamericana de Producción Animal (ALPA). p. R-2.

Leaver, J.D. 1988. The Contribution of Grass and Conserved Forages to the Nutrient Requirements for Milk Production. (W. Haresign and D.J.A. Cole, eds.), p. 213, Butterworths, London.

Lee, H.J. and H.R. Marston. 1969. The requirement for cobalt of sheep grazed on cobalt-deficient pastures. Aust. J. Agric. Res. 20:905.

Levander, O.A. 1986. Selenium. In: Trace Elements in Human and Animal Nutrition. 5th ed. By W. Mertz. New York, NY, Academic Press, Inc. p. 209.

Lucci, C.S. and A.L. Moxon. 1982. Teor de selênio no soro de bovinos leiteiros do estado de São Paulo. In: Anais da Sociedade Brasileria de Zootecnia. p.104 (abstr.). Sociedade Brasileira de Zootecnia. Piracicaba, São Paulo, Brazil.

Lucci, C.S. A.L. Moxon, M.A. Zanetti, E. Schalch, R.L. Pettinati, R.S. Fukushima, R. Franzolin Neto and D.G. Marcomini. 1983a. Selênio em rebanhos leiteiros do estado de São Paulo. I analises de forragenş – nota prévia. In: Anais da Sociedade Brasileira de Zootecnia. Sociedade Brasileira de Zootecnia. Pelotas, Brazil. p. 192 (abstr.).

Lucci, C.S., A.L. Moxon, M.A. Zanetti, E. Schalch, R.L. Pettinati, R.S. Fukushima, R. Franzolin Neto and D.G. Marcomini. 1983b. Selênio em Rebanhos leiteiros do Estado de São Paulo. I. Niveis de Selênio em soros sanquíneos – Nota Prévia In: Anais da Sociedade Brasileira de Zootecnia. p. 193 (abst.) Sociedade Brasileira de Zootecnia. Pelotas, Brazil.

Mahan, D. 1996. Organic selenium for the livestock industry: conclusions and potential from research studies. North American University Tour, Alltech Biotechnology Center, Nicholasville, Kentucky.

Mahan, D.C. and Y.Y. Kim. 1996. Effect of inorganic or organic selenium at two dietary levels on reproductive performance and tissue selenium concentrations in first parity gilts and their progeny. J. Anim. Sci. 74:2711.

Mahan, D.C. and N.A. Parrett. 1996. Evaluating the efficacy of Se-enriched yeast and inorganic selenite on tissue Se retention and serum glutathione peroxidase activity in grower and finisher swine. J. Anim. Sci. (submitted).

Maller, O. 1967. Specific appetite. The Chemical Senses and Nutrition (M.R. Kare and O. Maller, eds.), Johns Hopkins Press, Baltimore, Maryland p. 201.

McDowell, L.R. 1985. Nutrition of Grazing Ruminants in Warm Climates. Academic Press New York.

McDowell, L.R. 1992. Minerals in Animal and Human Nutrition. p. 524 Academic Press, San Diego, CA.

McDowell, L.R. 1996. Free-choice mineral supplements for grazing sheep in developing countries. In: Detection and Treatment of Mineral Nutrition Problems in Grazing Sheep (D.G. Masters and C.L. White, eds.). ACIAR, Canberra, Australia.

McDowell, L.R., J.H. Conrad, J.E. Thomas, L.E., Harris and K.R. Fick. 1977. Nutritional composition of Latin American forages. Trop. Anim. Prod. 2:273.

McDowell, L.R., B. Bauer, E. Galdo, M. Koger, J.K. Loosli and J.H. Conrad. 1982a. Mineral supplementation of beef cattle in the Bolivian tropics. J. Anim. Sci. 55, 964.

McDowell, L.R., M. Kiatoko, J.E. Bertrand, H.L. Chapman, F.M. Pate, F.G. Martin and J.H. Conrad. 1982b. Evaluating the nutritional status of beef cattle herds from four soil order regions of Florida II. Trace Minerals. J. Anim. Sci. 55:38.

McDowell, L.R., J.H. Conrad and F.G. Hembry. 1993. Minerals for Grazing Ruminants in Tropical Regions 2nd Ed., Univ. of Florida, Gainesville.

Mirando, M.A., D.N. Peters, C.E. Hostetler, W.C. Becher, S.S. Whiteaaker and R.E. Rompala. 1993. Dietary supplementation of proteinated trace minerals influences reproductive performance of sows. J. Anim. Sci. 7(Suppl.1):180.

Moxon, A.L. 1971. Deficiencia Nutricional de Selenio. Suplemento Agrícola 855. Estado de S. Paulo, p. 10.

Muirhead, S. 1992. Zinc methionine appears to improve hoof condition of grazing beef cattle. Feedstuffs 64:Oct. 12.

Muller, L.D., L.V. Schaffer, L.C. Ham and M.J. Owens. 1977. Cafeteria-style free-choice mineral feeder for lactating dairy cows. J. Dairy Sci. 60:1574.

Nelson, J. 1988. Review of trace mineral chelates and complexes available to the feed industry. Western Nutrition Conference, Winnipeg, Manitoba, Canada.

Pastrana, R., L.R. McDowell, J.H. Conrad and N.S. Wilkinson. 1991. Mineral status of sheep in the Paramo region of Colombia II. Trace minerals. Small Ruminant Research 5:23.

Patton, R.S. 1990. Chelated minerals: What are they, do they work? Feedstuffs 62(9):14.

Peducassé, C.A., L.R. McDowell, L.A. Parra, J.V. Wilkins, F.G. Martin, J.K. Loosli and J.H. Conrad. 1983. Mineral status of grazing beef cattle in the tropics of Bolivia. Trop. Anim. Prod. 8: 118.

Pehrson, B., M. Knutsson and M. Gyllenswoad. 1989. Glutathione peroxidase activity in heifers fed diets supplemented with organic and inorganic selenium compounds. Swedish J. Agr. Res. 19:53.

Reid, R.L. and D.J. Horvath. 1980. Soil chemistry and mineral problems in farm livestock. A review. Anim. Feed Sci. Technol. 5:95.

Rode, L.M. and K.A. Beauchemin. 1993. Cattle are poor judges of their mineral needs. Weekly Letter Agric. Canada No. 3082. Lethbridge, Canada.

Rojas, L.X., L.R. McDowell, N.S. Wilkinson and F.G. Martin. 1993. Mineral status of soils forages and beef cattle in southeastern Venezuela. II. Microminerals. Int. J. Anim. Sci. 8:183.

Rojas, L.X., L.R. McDowell, R.J. Cousins, F.G. Martin, N.S. Wilkinson, A.B. Johnson and J.P. Velasquez. 1995. Relative bioavailability of two organic and two inorganic zinc sources fed to sheep. J. Anim. Sci. 73: 1202.

Savage, S.T. 1973. Chelated trace minerals in chick diets. Poultry Sci. 52:2082(abstr.).

Spears, J.W. 1989. Zinc methionine for ruminants: Relative bioavailability of zinc in lambs and effects on growth and performance of growing heifers. J. Anim. Sci. 67:835.

Spears, J.W. and E.B. Kegley. 1991. Effect of zinc and manganese methionine on performance of beef cows and calves. J. Anim. Sci. 69(Suppl. 1):59 (Abstr.).

Spears, J.W., E.B. Kegley and J.D. Ward. 1991. Bioavailability of organic, inorganic trace minerals explored. Feedstuffs 27(47):12.

Tait, R.M., L.J. Fisher and J. Upright. 1992. Free-choice mineral consumption by grazing Holstein steers. Can. J. Anim. Sci. 72:1001.

Tejada, R., L.R. McDowell, F.G. Martin and J.H. Conrad. 1987. Evaluation of cattle trace mineral status in specific regions of Guatemala. Trop. Agric. 64:55.

Terry, T. 1964. Tratamiento de la coquera en vacunos mediante la inyección intramuscular de selenio y vitamina E. Annales del II Congreso Nacional de Medicina Veterinaria y Zootecnia, Lima, Perú. p. 163.

Torrent, J. 1996. Selenium yeast and pork quality. In: Proceedings of the 12th Annual Symposium on Biotechnology in the Feed Industry (T.P. Lyons and K.A. Jacques, eds.), Nottingham Univ. Press. Loughborough, Leics. UK. p. 165.

Underwood, E.J. 1981. The Mineral Nutrition of Livestock. Commonwealth Agricultural Bureaux, London.

Valdes, J. L., L.R. McDowell and M. Koger. 1988. Mineral status and supplementation of grazing beef cattle under tropical conditions in Guatemala. II. Microelements and Animal Performance. J. Prod. Agr. 1(4): 351.

Valle, G., L.R. McDowell and N.S. Wilkinson. 1993. Selenium concentration of bermudagrass after spraying with sodium selenate. Commun. Soil Sci. Plant Anal. 24(13 and 14):1763.

Vargas, R., L.R. McDowell, J.H. Conrad, F.G. Martin, C. Buergelt and G.L. Ellis. 1984. The mineral status of cattle in Colombia as related to a wasting disease ('secadera'). Tropical Animal Prod. 9:103.

Vargas, E., R. Solís, M. Torres and L. McDowell. 1992. Selenio y cobalto en algunos forrages de Costa Rica: Efecto de la epoca climatica y el estado vegetativo. Agronomía Costarricense 16: 171.

Velasquez-Pereira, J., L. McDowell, J. Conrad, N. Wilkinson and F. Martin. 1996. Nivel mineral existente en suelos, forrages y ganado bovino en Nicaragua: Microminerales. Revista de las Facultal de Agronomía de la Universidad del Zulia. (In press).

Ward, J.D., J.W. Spears and E.B. Kegley. 1992. Effect of trace mineral source on mineral metabolism, performance and immune response in stressed cattle. J. Anim. Sci. 70(suppl. 1):300 (abstr.).

Weber, D.W., T.O. Dill, J.E. Oldfield, R. Frobish, K. Vandebergh, and W. Zollinger. 1992. Daily intake of free-choice salt and protein blocks by beef cows was highly variable. Prof. Anim. Sci. 8(2):15.

Wedekind, K.S., A.E. Hortin and D.H. Baker. 1992. Methodology for assessing zinc bioavailability: Efficacy estimates for zinc-methionine, zinc sulfate and zinc oxide. J. Anim. Sci. 70:178.

Wright, P.L. and M.C. Bell. 1966. Comparative metabolism of selenium and tellurium in sheep and swine. Am. J. Physiol. 211:6.

ECONOMIC IMPLICATIONS OF COPPER AND ZINC PROTEINATES: ROLE IN MASTITIS CONTROL

R. J. HARMON and P. M. TORRE

Department of Animal Science, University of Kentucky, Lexington, Kentucky, USA

Introduction

Considerable emphasis in mastitis prevention has focused on management routines; however, progress in enhancing resistance of the dairy cow to mastitis has been slow in comparison. The increased interest and research effort in the area of nutritional relationships to host defense has encouraged new and potentially beneficial approaches for enhancing resistance of the dairy cow to intramammary infection by major mastitis pathogens or limiting the severity of response to invasion of the mammary gland when it does occur. The positive impact of adequate dietary Se and vitamin E on host defense and mastitis resistance has been documented extensively (Erskine, 1993; Erskine *et al.*, 1989; Grasso *et al.*, 1990; Smith *et al.*, 1984, 1985). Another area of micronutrient nutrition that shows potential in this regard is the influence of Cu and Zn status on host defense and mastitis. Further, supplementation with Cu and Zn proteinates shows promise in improving udder health and somatic cell counts (SCC). The reduction in mastitis and lowering SCC in a dairy herd will result in significant economic benefits as well as improved welfare of the cattle.

Economic losses due to bovine mastitis

It is crucial to understand the economic impact of mastitis on the dairy industry if the value of reduction in mastitis and SCC is to be understood. Recent estimates suggest that mastitis results in losses of approximately $185 per cow annually in the US or a total cost of approximately $1.8 billion each year (Bramley *et al.*, 1996). About 66% of this loss is due to decreased milk production because of infections.

Eberhart *et al.* (1982) estimated a 6% milk production loss in herds with a bulk tank SCC of 500,000 cells per ml and 18% loss in milk yield at bulk tank SCC of 1,000,000 (Table 1). Normal SCC from uninfected udders are generally considered to be below 200,000 cells per ml. The Dairy

Table 1. Estimated infection prevalence and losses in milk production associated with elevated bulk tank somatic cell counts.

BTSCC/ml*	Infected quarters in herd (%)	Production loss† (%)
200,000	6	0
500,000	16	6
1,000,000	32	18
1,500,000	48	29

* BTSCC = bulk tank somatic cell count.
† Production loss calculated as a percent of production expected at 200,000 cells/ml.

Table 2. Estimated differences in lactation milk yield associated with an increase in the SCC score.

Lactation average SCC Score (SCCS)	Lactation average SCC (1000s/ml)	Difference in milk yield (lbs/305 Days)	
		Lactation 1	Lactation ≥ 2
0	12.5	—	—
1	25	—	—
2	50	—	—
3	100	-200	-400
4	200	-400	-800
5	400	-600	-1200
6	800	-800	-1600
7	1600	-1000	-2000

Herd Improvement Association (DHIA) in the US has adopted an SCC scoring system which divides the SCC of composite milk of individual cows into 10 categories from 0 to 9 known as somatic cell count score (SCCS). The DHIA programs determine the SCC of each milking cow each month and report either the SCC or SCCS. Table 2 shows the relationship between SCCS, SCC, and estimated milk yield losses, but the average SCCS for the lactation most accurately reflects reduced milk yield (Bramley *et al.*, 1996).

The advantage of the SCCS is that there is a linear relationship between the SCC score and the loss in milk production. Each increase of one SCCS corresponds to a doubling of the SCC. Also each increase of one SCCS (above the 50,000 level or SCCS of 2) results in an additional 400 lb decrease in milk production for the lactation. For example, an increase in the lactation average SCCS from 4 to 5 (corresponding SCC of 200,000 to 400,000) will result in an additional 400 lb loss in milk production from 800 to 1200 lb per cow in that lactation. Conversely, reducing the average SCC by 50% will increase milk production by an estimated 400 lb per cow over a lactation. This provides a rather simple way of estimating the milk production losses in a dairy herd due to mastitis.

Copper and host defense

The importance of Cu as an essential trace element has been recognized for nearly 60 years, with the early discovery that Cu was necessary for normal hemoglobin synthesis in young rabbits and rats (Hart *et al.*, 1928). Since that time, the importance of Cu for normal growth, production and reproductive performance has been established. The biological role of Cu is exerted through a number of Cu-containing proteins including ceruloplasmin and superoxide dismutase (SOD) (Prohaska and Lukasewycz, 1990). When Cu is inadequate in animals, physiological and metabolic functions related to the Cu enzymes may be impaired and clinical deficiency symptoms will appear. In addition, inadequate Cu status may influence the magnitude of tissue injury that occurs during inflammation. Although low Cu content of feedstuffs is a common cause of Cu inadequacy, reduced bioavailability of Cu in ruminants may occur when dietary S, Mo, Zn, or Fe is high (Hemken *et al.*, 1993). Recent evidence in cattle suggests that Cu proteinate (Bioplex Cu) is stored more readily than inorganic Cu (Hemken *et al.*, 1993) and Cu proteinate may be absorbed in the organic form (Du *et al.*, 1995).

Ceruloplasmin is the main Cu-binding protein in blood, binding approximately 90% of the total circulating Cu (Holmberg and Laurell, 1948). Ceruloplasmin is recognized as an acute phase protein in cattle. Plasma from cattle with clinical mastitis have elevated ceruloplasmin compared with that from normal cows (Conner *et al.*, 1986). Gutteridge (1983) has suggested that ceruloplasmin may function as an antioxidant, scavenging oxygen free radicals, and could serve as an endogenous modulator of the inflammatory response in addition to its function of transporting Cu to tissue sites. Other work has suggested that ceruloplasmin may play a role in the modulation of tissue injury associated with the production of oxygen metabolites by phagocytes during inflammation (Broadley and Hoover, 1989; Galdston *et al.*, 1984). Clinical Cu deficiency results in a decline in ceruloplasmin levels in the blood (Mulhern and Koller, 1988; Prohaska and Lukasewycz, 1981) and could potentially compromise host defense.

The enzyme activity of SOD, a Cu-containing enzyme found in various tissues, was first described by McCord and Fridovich (1969) in bovine erythrocytes. The enzyme catalyzes the dismutation of toxic superoxide radicals to hydrogen peroxide and oxygen. This reaction may be essential in protecting host tissues from membrane oxidation by oxygen-derived free radicals from neutrophils and macrophages during the inflammatory response. SOD levels in tissues and erythrocytes of pigs (Williams *et al.*, 1975), rats (Paynter *et al.*, 1979), sheep (Andrewartha and Caple, 1980), and cattle (Jones and Suttle, 1981) may be decreased during dietary Cu deficiency potentially contributing to decreased protection from oxidation.

Deficiencies in protein, energy, vitamins, and minerals are known to compromise immune function (Beisel, 1982). A relationship between Cu and immune function has been shown by decreased resistance to infection in animals that were Cu deficient. Jones and Suttle (1983) reported that subclinically Cu-deficient mice had an increased susceptibility to experimental infections with *Pasteurella haemolytica*. A study with lambs that were genetically selected for low and high Cu status showed that the lambs with low Cu status were highly vulnerable to bacterial infections (Woolliams *et al*., 1986). The susceptible genotypes were protected by Cu supplementation. Reports by Jones and Suttle (1988) support a relationship between Cu status and disease susceptibility. Low liver Cu levels have been linked circumstantially with the occurrence of abomasal ulceration and pathogenic bacterial infections in beef calves in a Wyoming survey (Libbey *et al*., 1985).

The phagocytosis and killing of microorganisms are important functions of nonspecific cellular immune effectors such as the macrophage and neutrophil. This is especially important in the protection of the bovine mammary gland against infection by bacterial pathogens (Craven and Williams, 1985). The ability of peripheral blood granulocytes to kill ingested *Candida albicans* was reduced in Cu-deficient ewes (Jones and Suttle, 1981) and in Cu-deficient steers (Boyne and Arthur, 1981). Furthermore, leukocyte and erythrocyte SOD activities were significantly decreased in the hypocupremic ewes and cattle (Arthur and Boyne, 1985). Xin *et al*. (1991) showed that neutrophils from dairy steers made Cu deficient by feeding 10 ppm molybdenum (as ammonium molybdate) had significantly lower capacity to kill *S. aureus* than neutrophils from Cu-supplemented (20 ppm) animals. The low Cu status resulted in a significant decrease in SOD activities in whole blood, neutrophils, and erythrocytes and significantly lower Cu levels in the liver and neutrophils.

Liver Cu concentrations can be influenced by numerous factors including species, age of animal, disease conditions, and dietary composition. Dietary Cu is accumulated in the liver with a nearly linear response in liver Cu concentration to dietary Cu supplementation. Studies by Xin *et al*. (1991) at Kentucky indicated that liver Cu concentration, but not total plasma Cu, was a good indicator of Cu status of animals with subclinical Cu deficiency. Although liver Cu concentrations in these studies approached deficiency levels (18.1 ppm at trial termination), no influence on growth was observed. This is a significant observation, because it suggests that a subclinical Cu deficiency may compromise host defense mechanisms in the absence of clinical symptoms or an effect on growth. Although Puls (1988) reported that 87.5 to 350 ppm Cu in the liver reflects an adequate Cu status, the level at which host defenses or antioxidant activities may be compromised is not clear.

Copper and mastitis

Recent studies (Harmon *et al*., 1994) at the University of Kentucky have focused on the role of Cu status on inflammation and infection of the mammary gland. Eighteen Holstein heifers were assigned in pairs (by date of expected calving) to two dietary treatments from 84 days prepartum through 105 days of lactation. No supplemental Cu had been fed since about 5 months of age. The diets were basal diet (6 to 7 ppm Cu) with no supplemental Cu (-CU) and basal diet plus 20 ppm supplemental Cu (+CU) as copper sulfate. The NRC recommendation for dairy cattle is 10 ppm. Liver biopsies were performed pre- and postpartum to monitor Cu status, and duplicate quarter foremilk samples were taken for bacteriological analysis after calving. Two quarters of each cow were challenged with 25 µg *E. coli* endotoxin at 35 days of lactation and with 200 to 300 CFU *S. aureus* at 70 days of lactation. Responses were measured for 6 and 15 days.

Analysis of the data showed that liver Cu in the +CU and -CU cows were 209 and 14 ppm at calving and 474 and 29 ppm at 105 days postpartum. However, plasma Cu levels in both groups of heifers were normal. The +CU cows had more uninfected quarters (60%) than the -CU cows (36%) at calving ($P<0.05$). The +CU and -CU cows had 6 and 28% of quarters with intramammary infections (IMI) by major pathogens ($P<0.01$); however, there was no difference in coagulase-negative staphylococci IMI. Major pathogen IMI were 0 and 14% on day 7.

The -CU cows tended to have higher SCC than +CU on day 1 ($P<0.10$) and day 3 ($P<0.07$) after endotoxin challenge (3954 vs 2624 × 10^3/ml; 1213 vs 877 × 10^3/ml). Mean clinical severity scores ($P<0.05$) and the number of abnormal quarters ($P<0.01$) were lower in +CU cows on day 1. No differences in milk production or other measures of inflammation were observed. In contrast, no differences in SCC, clinical scores, or other measures were found between groups in response to *S. aureus* challenge. Numbers of *S. aureus* in milk tended to be lower in +CU cows than in -CU cows. Torre *et al.* (1996) further demonstrated that neutrophils from the -CU cows had reduced ability to kill *S. aureus in vitro*.

This study demonstrates that inadequate Cu status is possible in the absence of Cu supplementation of normal diets of heifers. The failure to provide Cu supplementation in heifer diets is likely a common occurrence in some dairy herds. These data suggest that inadequate Cu status may result in increased infection prevalence at calving and an increase in clinical severity following challenge, perhaps accompanied by increased SCC, compared with that observed in cows with an adequate Cu status. Studies are currently in progress at the University of Kentucky to evaluate effects of Bioplex Cu on copper status, infections of heifers at calving, and response to *E. coli* J-5 vaccination.

Zinc and ruminant immune function

The link between dietary Zn and ruminant immune function was discovered during studies of an inherited, lethal trait in Black Pied Danish calves of Friesian descent (Brummerstedt *et al.*, 1971). Symptoms of this condition included hair loss, parakeratosis around the mouth, eyes, and jaw, diarrhea, conjunctivitis, rhinitis, and bronchopneumonia. Symptoms typically began at 4 to 6 weeks of age, and most affected calves died at approximately 4 months of age. At necropsy the calves were found to have atrophy of the thymus and Peyer's patches (Brummerstedt *et al.*, 1971). These symptoms, similar to those in humans possessing inborn errors of Zn metabolism, improved with oral administration of Zn oxide (Brummerstedt *et al.*, 1971). Zn supplementation has been used to reduce effects of infectious pododermatitis in Friesian cattle (Suttle and Jones, 1989). However, studies to determine whether *in vivo*

Zn status affects cellular immune function in the ruminant are lacking. Droke and Spears (1993) found that severe Zn deficiency was required to increase numbers of blood neutrophils and decrease numbers of lymphocytes. Patterns of lymphocyte blastogenesis may have been altered in this study, but these results were equivocal. Serum IgG response to lysozyme or *in vivo* response to phytohemagglutinin injection was not altered.

Zinc and mastitis

Data addressing the specific mechanisms of Zn in resistance to mastitis are limited. Zn deficiency in ruminants has been postulated to weaken skin and other stratified epithelia (i.e. keratinocytes) as well as reducing the magnitude of increase of basal metabolic rate following infectious challenge (reviewed in Suttle and Jones, 1989). Because the mammary gland is essentially a skin gland and the importance of the keratin lining of the streak canal in prevention of infection is known (Capuco *et al.*, 1992), speculation that Zn supplementation may enhance resistance to mastitis is tempting.

Moderate to severe blood hypozincemia during mastitis, particularly acute coliform mastitis, has been documented. Experimental intramammary infection with *E. coli* resulted in depression of plasma Zn by 6 hours post infusion (Lohuis *et al.*, 1990). A significant positive correlation between plasma Zn and *E. coli* counts in secretion was observed. Similarly, experimental endotoxin mastitis in dairy cattle resulted in hypozincemia (Verheijden *et al.*, 1983). Zn is a cofactor for many proteins and enzymes involved in the acute phase response to infection and inflammation (Prasad, 1979; Vallee and Galdes, 1984). Mastitis-associated hypozincemia in cattle is likely linked to the magnitude of the acute phase response, but the possibility that normal blood or tissue levels of Zn could be a limiting factor in the bovine acute phase response has not been explored. Consequently the role of Zn supplementation during mastitis is not clear.

Most studies in this area have focused on reduction of SCC during supplementation of organic forms of Zn which are reported to be more bioavailable to the ruminant compared with inorganic Zn (Madsen, 1993). Kellogg (1990) summarized results of eight trials evaluating effects of Zn-methionine compared with equivalent amounts of Zn oxide and methionine. Overall, supplementation of Zn-methionine (180 or 360 mg Zn, 360 or 720 mg methionine) resulted in 22% decrease in SCC when feeding the lower level. Feeding the higher level of Zn-methionine lowered SCC by 50%. However, decreased SCC were not observed in another study of Zn-methionine supplementation (Moore *et al.*, 1988).

A limited number of studies have evaluated the effects of organic Zn on new mastitis cases. Galton (1990) saw no effect of Zn-methionine on rate of new infections resulting from experimental challenge with *Streptococcus agalactiae*, although SCC were significantly decreased in supplemented cows. In contrast, Spain (1993) reported beneficial effects of Bioplex Zinc (providing 50% of a total 800 mg Zn per cow per day as proteinate) on rate of new, naturally occurring intramammary infections. No effects of Bioplex Zn supplementation on SCC or milk yield were observed compared with Zn oxide. However, numbers of new infections were doubled in the Zn oxide group compared to the Bioplex Zn supplemented animals (11 vs five new infections). The majority of isolates were environmental mastitis pathogens. Spain (1993) suggested that organic Zn is beneficial in enhancing resistance to mastitis pathogens because of the postulated role of Zn in maintaining skin integrity and the keratin lining of the streak canal.

Bioplex minerals and somatic cell counts

Several studies have demonstrated a reduction in SCC in dairy cattle which were supplemented with mineral proteinates (Table 3). Harris (1995) reported results of a 90-day field trial in which one group of 70 cows received a TMR supplemented with 400 mg Zn per cow per day as Bioplex Zn and the control group was fed the normal TMR. The mean SCC in the Zn proteinate group decreased 24% and the SCC in the control group increased 36%; SCC was 57% lower in the group supplemented with Bioplex Zn at trial end. Adjustment for the lower SCC in the Bioplex group at initiation of the trial would still show an estimated 30 to 40% lower SCC in the Bioplex Zn cows. Boland *et al.* (1996) reported the results of three different trials in which a combination of mineral proteinates were supplemented to normal dairy cow diets; i.e., the control diet was the same as the proteinate-supplemented diet but without the mineral proteinates (Table 3). The mineral proteinates (Bioplex and Sel-Plex 50) provided the following supplemental minerals per cow per day in the diets during all three trials: Cu, 100 mg; Zn, 300 mg; Se, 2 mg). Blood mineral profiles were normal for both groups suggesting mineral status was adequate in both groups and unaffected by the supplements. In the groups receiving mineral proteinates in the three trials the SCC were reduced by 52%, 45%, and 35% over the duration of the trial. In the last trial SCC were reduced

Table 3. Influence of mineral proteinate (Bioplex) supplementation on somatic cell counts.

Form of proteinate supplemented	Mineral supplied daily as Bioplex (mg)	Reduction in SCC (%)	Reference
Zn	400	57 (~ 40%; adjusted)	Harris. 1995 (n = 70 per group)
Cu Zn Se	100 300 2	52	Boland *et al.*, 1996 (n = 7 per group)
Cu Zn Se	100 300 2	45	Boland *et al.*, 1996 (n = 28 per group)
Cu Zn Se	100 300 2	35; wk 0 to 12 52; wk 9 to 12	Boland *et al.*, 1996 (n = 23 per group)

52% during the final 4 weeks. Boland *et al.* (1996) indicated that these data showed a greater improvement in SCC the longer the treatment continued. Overall the results of these studies suggest a beneficial effect of Bioplex mineral supplementation on SCC in the herd and, thus, on udder health.

Economics of Bioplex copper and zinc supplementation

The economic returns of supplementing Bioplex minerals can be calculated using DHIA figures for gains in milk production associated with changes in SCCS (Table 2) and the cost of the mineral proteinate. Of course, the return realized will depend upon the farm price received for the additional milk production. Since a linear relationship exists between SCCS and milk production loss (or gain) at all SCC levels above 50,000, Table 4 was constructed to show the approximate milk production and financial gains realized per cow in a 305-day lactation, assuming various percentage reduction in SCC. The rule holds that a 50% reduction in SCC results in approximately 400 lb increase in milk yield. This translates to an additional $48.00 to $60.00 in income per cow per lactation, depending on the milk price to the farmer. Two examples will be given to demonstrate the economic implications of Bioplex minerals.

Example 1 assumes a 50% reduction in SCC when feeding 100 mg Cu and 300 mg Zn per head per day as proteinate. The daily cost of supplementation would be slightly less than $0.02 per head or $5.92 per head per lactation (Sel-Plex 50 is not included as it is not currently marketed in the US). The additional 400 lb of milk at $12 per 100 lb would be worth $48.00; at $15 per 100 lb the milk value would be $60.00. This represents between $8.11 and $10.14 return for every $1.00 invested in proteinate. At a 30% reduction in SCC, returns of $4.11 to $5.22 could be realized.

Example 2 also assumes 50% reduction in SCC when feeding 400 mg Zn per head per day as proteinate as reported by Harris (1995). The cost of supplementation would be approximately $5.19 per head per lactation. With a 400 lb gain in milk yield, the return per $1.00 spent on proteinate would be $9.25 (at milk value of $12 per 100 lb) to $11.56 (at $15 per 100 lb). If SCC were reduced 30%, the returns per $1.00 invested would be between $4.76 and $5.96.

These estimates demonstrate the potential economic returns which can be

Table 4. Relationship of lowering somatic cell count (SCC) and gain in 305–day milk production and income per cow.

Decrease in SCC (%)	Estimated increase in milk production (lb/305 days)	Increased income if farm milk price is:	
		$12 per 100 lb	$15 per 100 lb
50	400	$48.00	$60.00
40	300	$36.00	$45.00
30	206	$24.72	$30.90

realized from supplementation with Cu and Zn proteinates. Returns per dollar invested appear to range from $4.00 to $11.00, dependent upon milk prices and SCC responses. A conservative estimated return of $6.00 to $7.00 per dollar invested is comparable to the estimated return of $5.66 per dollar invested in conventional mastitis control measures.

Summary

There is increasing evidence that Cu and Zn play important roles in host defense and in limiting tissue damage during inflammation. This idea, combined with the fact that there is a massive influx of neutrophils into the mammary gland during inflammation, suggests that Cu or Zn inadequacy in the dairy cow could limit protection against intramammary infection and may increase the opportunity for tissue damage. A potential role for Zn in mastitis resistance may be maintenance of skin or keratin integrity to prevent colonization of the streak canal by mastitis pathogens. Alternatively Zn may directly influence function of cells involved in defense of the mammary gland. Cu may play a role in mammary gland resistance through altered antioxidant capacity and function of inflammatory cells or perhaps by influencing endogenous modulators of inflammation. The resulting impact could be altered clinical severity as well as reduced incidence of mastitis.

Supplementation of balanced dairy diets with mineral proteinates has been shown in field trials and controlled studies to have beneficial effects on SCC. When the potential economic impact of these udder health benefits are evaluated, estimated returns ranging from $4.00 to $11.00 per dollar invested in proteinate supplementation can be calculated. An average return of $6.00 to $7.00 per dollar invested is likely a realistic value which is comparable to that of traditional mastitis control measures. However, these nutritional approaches must be considered tools which enhance the resistance of the cow and aid in the control of mastitis; sound nutrition will not replace sound mastitis management.

References

Andrewartha, K.A. and I.W. Caple. 1980. Effects of changes in nutritional copper on erythrocyte superoxide dismutase activity in sheep. Res. Vet. Sci. 28:101.

Arthur, J.R. and R. Boyne. 1985. Superoxide dismutase and glutathione peroxidase activities in neutrophils from selenium-deficient and copper-deficient cattle. Life Sci. 36:1569.

Beisel, W.R. 1982. Single nutrients and immunity. Am. J. Clin. Nutr. 35:417.

Boland, M.P., G. O'Donnell and D. O'Callaghan. 1996. The contribution of mineral proteinates to production and reproduction in dairy cattle. Page 95 in Biotechnology in the Feed Industry. T.P. Lyons and K. Jacques, eds. Nottingham University Press, Nottingham, UK.

Boyne, R. and J.R. Arthur. 1981. Effects of selenium and copper deficiency on neutrophil function in cattle. J. Comp. Path. 91:271.

Bramley, A.J., J.S. Cullor, R.J. Erskine, L.K. Fox, R.J. Harmon, J.S. Hogan, Nickerson, S.C., S.P. Oliver, K.L. Smith and L.M. Sordillo. 1996. Current Concepts of Bovine Mastitis. National Mastitis Council, Madison, WI.

Broadley, C. and R.L. Hoover. 1989. Ceruloplasmin reduces the adhesion and scavenges superoxide during the interaction of activated polymorphonuclear leukocytes with endothelial cells. Amer. J. Path. 135:647.

Brummerstedt, E., T. Flagstad, A. Basse and E. Andresen. 1971. The effect of zinc on calves with hereditary thymus hypoplasia (lethal trait A 46). Acta Pathol. Microbiol. Scand. 79:686.

Capuco, A.V., S.A. Bright, J.W. Pankey, D.L. Wood, R.H. Miller and J. Bitman. 1992. Increased susceptibility to intramammary infection following removal of teat canal keratin. J. Dairy Sci. 75:2126.

Conner, J.G., P.D. Eckersall, M. Doherty and T.A. Douglas. 1986. Acute phase response and mastitis in the cow. Res. Vet. Sci. 41:126.

Craven, N. and M.R. Williams. 1985. Defenses of the bovine mammary gland against infection and prospects for their enhancement. Vet. Immunol. Immunopathol. 10:71.

Droke, E.A. and J.W. Spears. 1993. *In vitro* and *in vivo* immunological measurements in growing lambs fed diets deficient, marginal or adequate in zinc. J. Nutr. Immunol. 2:71.

Du, Z., R.W. Hemken and T.W. Clark. 1995. Copper proteinate may be absorbed in chelated form by lactating Holstein cows. Page 315 in Biotechnology in the Feed Industry. T.P. Lyons and K. Jacques, eds. Nottingham University Press, Nottingham, UK.

Eberhart, R.J., L.J. Hutchinson and S.B. Spencer. 1982. Relationships of bulk tank somatic cell counts to prevalence of intramammary infection and to indices of herd production. J. Food Prot. 45:1125.

Erskine, R.J. 1993. Nutrition and mastitis. Page 551 in Vet. Clinics North America: Food Anim. Practice. Vol. 9. K.L. Anderson, ed. Saunders Co., Philadelphia, PA.

Erskine, R.J., R.J. Eberhart, P.J. Grasso and R.W. Scholz. 1989. Induction of *Escherichia coli* mastitis in cows fed selenium-deficient or selenium-supplemented diets. Am. J. Vet. Res. 50:2093.

Galdston, M., V. Levytska and M.S. Schwartz. 1984. Ceruloplasmin: increased serum concentration and impaired antioxidant activity in cigarette smokers, and ability to prevent suppression of elastase inhibitory capacity of α-proteinase inhibitor. Am. Rev. Respir. Dis. 129:258.

Galton, D. 1990. Effect of feeding Zinpro® to lactating dairy cows on udder health. Zinpro Technical Bulletin # D-8911.

Grasso, P.J., R.W. Scholz, R.J. Erskine and R.J. Eberhart. 1990. Phagocytosis, bactericidal activity, and oxidative metabolism of milk neutrophils from dairy cows fed selenium-supplemented or selenium-deficient diets. Am. J. Vet. Res. 51:269.

Gutteridge, J.M.C. 1983. Antioxidant properties of ceruloplasmin towards iron and copper-dependent oxygen radical formation. FEBS Lett. 157:37.

Harmon, R.J., T.W. Clark, D.S. Trammell, B.A. Smith, P.M. Torre and R.W. Hemken. 1994. Influence of copper status in heifers on response to intramammary challenge with *Escherichia coli* endotoxin. J. Dairy Sci. 77(Suppl. 1): 198.

Harris, B. 1995. The effect of feeding zinc proteinate to lactating dairy cows. Page 299 in Biotechnology in the Feed Industry. T.P. Lyons and K. Jacques, ed. Nottingham University Press, Nottingham, UK.

Hart, E.B., H. Steenbock, J. Waddell and C.A. Elvehjem. 1928. Iron in nutrition. VII. Copper as a supplement to iron for hemoglobin building in the rat. J. Biol. Chem. 77:797.

Hemken, R.W., T.W. Clark and Z. Du. 1993. Page 35 in Biotechnology in the Feed Industry. T.P.Lyons, ed. Alltech Technical Publ., Nicholasville, KY.

Holmberg, C.G. and C.B. Laurell. 1948. Investigations in serum copper II: Isolation of copper containing protein and description of some of its properties. Acta. Chem. Scand. 2:550.

Jones, D.G. and N.F. Suttle. 1981. Some effects of copper deficiency on leukocyte function in sheep and cattle. Res. in Vet. Sci. 31:151.

Jones, D.G. and N.F. Suttle. 1983. The effects of copper deficiency on the resistance of mice to infection with *Pasteurella haemolytica*. J. Comp. Path. 93:143.

Jones, D.G. and N.F. Suttle. 1988. Copper and disease resistance. In: Trace Substances in Environmental Health. 21:514. University of Missouri, Columbia, MO.

Kellogg, D.W. 1990. Zinc methionine affects performance of lactating cows. Feedstuffs 62:15.

Libbey, C.W., D.W. Mamar, M. Gerlach and J.L. Johnson. 1985. Linking copper with lactexia with abomasal ulcers in beef calves. Vet. Med. 80:85.

Lohuis, J.A.C.M., Y.H. Schukken, J.H.M. Verheijden, A. Brand and A.S.J.P.A.M. Van Miert. 1990. Effect of severity of systemic signs during the acute phase of experimentally induced *Escherichia coli* mastitis on milk production losses. J. Dairy Sci. 73:333.

Madsen, F.C. 1993. Disease and stress: A reason to consider the use of organically complexed trace elements. Page 147 in Biotechnology in the Feed Industry. T.P. Lyons, ed. Alltech Technical Publications, Nicholasville, KY.

McCord, J.M. and I. Fridovich. 1969. Superoxide dismutase: an enzymic function for erythrocuprein (hemocuprein). J. Biol. Chem. 244:6049.

Moore, C.L., P.M. Walker, M.A. Jones and J.W. Webb. 1988. Zinc methionine supplementation for dairy cows. J. Dairy Sci. 71(Suppl. 1):152.

Mulhern, S.A. and L.D. Koller. 1988. Severe or marginal copper deficiency results in a graded reduction in immune status in mice. J. Nutr. 118:1047.

Paynter, D.I., R.J. Moir and E.J. Underwood. 1979. Changes in activity of the Cu-Zn superoxide dismutase enzyme in tissues of the rat with changes in dietary copper. J. Nutr. 109:1570.

Prasad, A.S. 1979. Clinical, biochemical, and pharmacological role of zinc. Ann. Rev. Pharmacol. Toxicol. 19:393.

Prohaska, J.R. and O.A. Lukasewycz. 1981. Copper deficiency suppresses the immune response of mice. Science 213:559.

Prohaska, J.R. and O.A. Lukasewycz. 1990. Effects of copper deficiency on the immune system. In: Antioxidant Nutrients and Immune Functions. A. Bendich, M. Phillips, and R.P. Tengerdy, eds. Adv. Exp. Med. Biol. 262:123. Plenum Press, New York, NY.

Puls, R. 1988. Mineral levels in animal health: Diagnostic data. Sherpa Int., Clearbrook, British Columbia, Canada.

Smith, K.L., J.H. Harrison, D.D. Hancock, D.A. Todhunter and H. R. Conrad. 1984. Effect of vitamin E and selenium supplementation on incidence of clinical mastitis and duration of clinical symptoms. J. Dairy Sci. 67:1293.

Smith, K.L., H.R. Conrad, B.A. Amiet and D.A. Todhunter. 1985. Incidence of environmental mastitis as influenced by dietary vitamin E and selenium. Kieler Milchwirtschaftliche Forschungsberichte 37:482.

Spain, J. 1993. Tissue integrity: A key defense against mastitis infection: The role of zinc proteinates and a theory for mode of action. Page 53 in Biotechnology in the Feed Industry. T.P. Lyons, ed. Alltech Technical Publications, Nicholasville, KY.

Suttle, N.F. and D.G. Jones. 1989. Recent developments in trace element metabolism and function: Trace elements, disease resistance and immune responsiveness in ruminants. J. Nutr. 119:1055.

Torre, P.M., R.J. Harmon, R.W. Hemken, T.W. Clark, D.S. Trammell and B.A. Smith. 1996. Mild dietary copper insufficiency depresses blood neutrophil function in dairy cattle. J. Nutr. Immunol. 4(3):3.

Vallee, B.L. and A. Galdes. 1984. The metallobiochemistry of zinc enzymes. Adv. Enzymol. 56:284.

Verheijden, J.H.M., A.S.J.P.A.M. Van Miert, A.J.H. Schotman and C.T.M. Van Duin. 1983. Pathophysiological aspects of *E. coli* mastitis in ruminants. Vet. Res. Commun. 7:229.

Williams, D.M., R.E. Lynch, C.R. Lee and G.E. Cartwright. 1975. Superoxide dismutase activity in copper-deficient swine. Proc. Soc. Exp. Biol. Med. 149:534.

Woolliams, C., N.F. Suttle, J.A. Woolliams, D.J. Jones and G. Wiener. 1986. Studies on lambs genetically selected for low and high copper status. 1. Differences in mortality. Anim. Prod. 43:293.

Xin, Z., R.W. Hemken, D.F. Waterman and R.J. Harmon. 1991. Effects of copper status on neutrophil function, superoxide dismutase, and copper distribution in steers. J. Dairy Sci. 74:3078.

INDEX

ALLTECH AROUND THE WORLD

A Listing of Alltech International Offices and Distributors

AUSTRALIA
ALLTECH AUSTRALIA
Mr. Kim Turnley
9/810 Princes Highway
Springvale, Victoria 3171
TEL: 61–3–9574–2333
FAX: 61–3–9574–2444

BOLIVIA
ALLTECH BOLIVA
Ing. Clodys A. Menacho R.
Av. Alemania calle Valencia #63
Casilla 1450
Santa Cruz
TEL: 591–342–1395
FAX: 591–342–1395

BRAZIL
ALLTECH DO BRASIL
Aidan Connolly
Caixa Postal 10808
Cep 81170–610
CIC-Curitiba -Parana
TEL: 55–41–246–6515
FAX: 55–41–246–5188

CANADA
ALLTECH CANADA
Dr. Ted Sefton
449 Laird Road
Guelph, Ontario
N1G 4W1
TEL: 519–763–3331
FAX: 519–763–5682

CHILE
ALLTECH CHILE
Dr. Mario Román
Padre Orellana 1315
Las Condes
Santiago
TEl/FAX: 56–2–5519041

CHINA
ALLTECH ASIA PACIFIC
BIOSCIENCE CENTER
Dr. Wim de Koning
No. 30 Baisbiqiao Road
Beijing 100081
TEL: 86–10–6218–7751
FAX: 86–10–6218–7750

COLOMBIA
INVERSIONES AMAYA
(INVERAMAYA)
Dr. Luis Londoño J.
Calle 85 No. 20–25
Oficina 401A
Bogota
TEL: 57–1–218–2829
FAX: 57–1–218–5317

COSTA RICA
NUTEC, S.A.
Carlos Lang, Ing.
Apartado 392
P.O. Box 392
Tibas
TEL/FAX: 506–2–33–31–10

CYPRUS
CHRONEL BIOTECHNOLOGY Ltd.
Christoforos Kyriacou
P.O. Box 2792
Larnaca
TEL: 357–4–638082
FAX: 357–4–638082

CZECH REPUBLIC
ALLTECH CZECH REPUBLIC
Dr. Valdimir Siske
Mezirka 13
60200 Brno
TEL: 42–5–41–21–57–40
FAX: 42–5–41–21–57–41

DENMARK
ALLTECH DENMARK
Søren Healy
2, Graabroedrestraede
DK-8900 Randers
TEL: 45–86–439700
FAX: 45–86–429300

DOMINICAN REPUBLIC
SANUT, S.A.
Miguel A Lajara P.
Km 10 1/2 Aut. Duarte
Apartado Postal 30–004
Santo Domingo
TEL: 809–560–5840
FAX: 809–564–4070

ECUADOR
ALLTECH ECUADOR
John Weir
Victor E. Estrada
117 y Bálsamos
Guayacquil
FAX: 59–3–4–887481

EGYPT
EGYTECH
Dr. Mohy Z. Sabry
13, El-Solouly St.
P.O. Box 442
Dokki, Cairo 12311
TEL: 202–361–0605
FAX: 202–361–5909

FINLAND
BERNER Ltd.
Antti Rinta Harri
Eteläranta 4B
SF 00130 Helsinki
TEL: 358–0–134–511
FAX: 358–0–134–51380

FRANCE
ALLTECH FRANCE
Marc Larousse
2–4, Avenue du 6 juin 1944
95190 Goussainville
TEL: 33–1–398–86351
FAX: 33–1–398–80778

GERMANY
ALLTECH DEUTSCHLAND GMBH
Dr. Des Cole
Esmarchstrasse 6
23795 Bad Segeberg
TEL: 49–4551–88700
FAX: 49–4551–887099

GREECE
LAPAPHARM, INC.
A. Mantis & George Lappas
73 Menandrou Str.
10437 Athens
TEL: 30–1–522–7208
FAX: 30–1–522–7152

HONDURAS
S.B.F. INTERNATIONAL
Ing. Sigfrido Burgos
Colonia Palermo
No. 1862
Tegucigalpa
TEL/FAX: 504–32–39–64

HONG KONG
PING SHAN ENTERPRISE Co. Ltd.
SINGAPORE TRADING Co., Ltd.
Oscar Lam
2/F Seaview Commercial Building,
21–24 Connaught Road West
TEL: 852–2 858–1188
FAX: 852–2 858–1452

HUNGARY
ALLTECH HUNGARY
Levente Gati
Kresz Geza utca 13–15
Budapest, II-1132
TEL/FAX: 36–1–269–5384

INDIA
VETCARE
Bharat Tandon
I.S. 40
K.H.B. Industrial Area
Yelahanka New Town
Bangalore 560–064
TEL : 91–80–8460060
FAX: 91–80–8461240

INDONESIA
P.T. ROMINDO PRIMAVETCOM
Dr. Lukas
Dr. Saharjo
No. 266
Jakarta 12870
TEL: 62–21–830–0300
FAX: 62–21–828–0678

IRELAND
ALLTECH IRELAND
Mr. Aidan Brophy
Unit 28, Cookstown Industrial Estate
Tallaght, Dublin 24
TEL: 353–14–510276
FAX: 353–14–510131

ITALY
ASCOR CHIMICI, S.R.L.
Sig. Arnaldo Valentini
Via Piana, 265
47032 Capocolle (FO)
TEL: 39–543–448070
FAX: 39–543–448644

JAMAICA, W.I.
WINCORP
Dr. Leon Headley
38–39 Caracas Avenue
Kingston Export Free Zone
Kingston 15
TEL: 809–923–6894
FAX: 809–923–6856

JAPAN
BUSSAN BIOTECH Co., Ltd.
Mr. Nick Koyama
3rd Floor, Shiba Daimon
Makita Building
5–8, 2-Chome Shiba Daimon
Minato-Ku
Tokyo
TEL: 81 3 5470 6601
FAX: 81 3 5470 6606

MALAYSIA
FARM CARE SDN. BHD.
Ong Seng Say
No. 48–3, Jalan Radin Tengah
Seri Petaling
57000 Kuala Lumpur
TEL : 60–3–957–3669
FAX: 60–3–957–3648

MEXICO
ALLTECH MEXICO, S.A. DE C.V.
Dra. Gladys Hoyos
Dr. Enrique Gonzalez MTZ No. 244
Col. Sta. Maria La Ribera
C.P. 06400
TEL: 525–547–5042
FAX: 525–547–5040

NEPAL
NEPA PHARMAVET PVT, Ltd.
Suvas Bishet
GA-1–481, Wotu Tole
Katmandu-3
NEPAL
TEL: 977–1–217–952
FAX: 977–1–224–627

NETHERLANDS
ALLTECH NETHERLANDS
Timm Neelsen
Hollandsch Diep 63
2904 EP Capelle aan den IJssel
TEL: 31–10–450–1038
FAX: 31–10–442–3798

NEW ZEALAND
CUNDY TECHNICAL SERVICES
Mike E. Cundy
P.O. Box 69–170
Glendene
Auckland 8
TEL: 64–9–837–3243
FAX: 64–9–837–3214

PERU
ALLTECH PERU
Dr. Jorge Arias
Av. Comandante Espinar 260
3er Piso, Oficina 301
Miraflores
Lima 18
TEL: 51–12–411–794
FAX: 51–14–476–982

PHILIPPINES
FERMENTATION INDUSTRIES CORP.
Rodney Vincent Choa Yu
Suite 1305, Far East Bank Bldg.
560 Quintin Parades St.
Binondo, Manila
PHILIPPINES
TEL: 632–241–0846
FAX: 632–241–0870

POLAND
POLMARCHÉ
Dr. Wojciech Zalewski
Koczargi Nowe
Ul. Warszawska 72
05–082 Babice Stare k/ Warszawy
TEL: 48–90–216–293
FAX: 48–90–219–939

PORTUGAL
ALLTECH PORTUGAL
Orlando de Sousa
Rúa Alvaro de Brée No. 6
Leceia P-2745 Queluz
PORTUGAL
TEL: 351–1–421–8029
FAX: 351–1–421–8100

RUSSIA
ALLTECH RUSSIA
Vincent Roze
Novinski boulvard, dom 16
Korpus 3, Appt 51
Moscow 121 069
TEL/FAX: 7–095–202–2120

SAUDI ARABIA
ARASCO
M. Abdulla S. Al-Rubaian
P.O. Box 53845
Riyadh 11593
New Akaria Center (Olaya)
Suite 625
FAX: 966–1–4645375

SOUTH AFRICA
ALLTECH SOUTH AFRICIA
Derek Foster
P.O. Box 2654
Somerset West 7129
Cape Province
TEL: 27–21–8517–052
FAX: 27–21–8517–000

SOUTH KOREA
YOONEE CHEMICAL Co., Ltd.
Mr. Yun Tae Lee, D.V.M.
ALLTECH Technical Manager-Dr. Han
C.P.O. Box 6161
Seoul
TEL: 82–2–585–1801
FAX: 82–2–584–2523

SPAIN
PROBASA
Juan Rosell Lizana
c/o Argenters, 9 Nave 3
Pol Ind Satiga, Sta
Perpetua de la Mogoda
Barcelona
TEL: 34–3718–2215
FAX: 34–3719–1307

SWEDEN
VETPHARMA, AB
Thord Bengtsson
Annedalsvägen 9
S-227 64 LUND
TEL: 46–46–12–81–00
FAX: 46–46–14–65–55

SWITZERLAND
INTERFERM AG
Dr. Fritz Näf
Postfach 112
Hardlurmstrasse 175
CH-8037 Zürich
SWITZERLAND
TEL: 41–1–272–8024
FAX: 41–1–273–1844

TAIWAN
JARSEN Co., Ltd.
Wen Ho Lin
12th floor, No. 1337
Chung Cheng Road
Tao-Yuan City
Tao Yuan Hsien
TEL: 886–3–356–6678
FAX: 886–3–356–5527

THAILAND
DIETHELM TRADING Co., Ltd.
Dr. Chai Wacharonke
ALLTECH Technical Manager – Mr. Steve Caskey
2533 Sukhumvit Road
Bangchack
Prakhanong
Bangkok, 10250
TEL: 66–2361–8276
FAX: 66–2361–8280

UNITED KINGDOM
ALLTECH UK
Mr. Jem Clay
16/17 Abenbury Way
Wrexham Industrial Estate
Wrexham Clwyd, LL 13 9UZ
TEL: 44–1978–660198
FAX: 44–1978–661475

UNITED STATES
CORPORATE HEADQUARTERS
ALLTECH BIOTECHNOLOGY CENTER
3031 Catnip Hill Pike
Nicholasville, Kentucky, 40356
TEL: 606–885–9613
FAX: 606–885–6736
WEB: http//www.alltech-bio.com

VENEZUELA
SIDELAC
Dr. Rafael Alonso
P.O. Box 1813
Maracaibo
TEL: 58–62–47079
FAX: 58–61–918889

VIETNAM
BAYER AGRITECH SAIGON
Nicholas Brand
1/3 Xom Moi Hamlet
Phuoc Long Village
Thu Duc District
Ho Chi Minh City
TEL: 84–8–960127
FAX: 84–8–961–523

YUGOSLAVIA
ALLTECH YUGOSLAVIA
Dr. M. Vukic Vranjes
Somborska 57
Yu-21000 Novi Sad
TEL/FAX: 381–21–301–083